T0199600

ANALYSIS OF
MIXED DATA

METHODS & APPLICATIONS

ANALYSIS OF
MIXED DATA
METHODS & APPLICATIONS

EDITED BY
ALEXANDER R. de LEON
KEUMHEE CARRIÈRE CHOUGH

CRC Press
Taylor & Francis Group
Boca Raton London New York

CRC Press is an imprint of the
Taylor & Francis Group, an **informa** business
A CHAPMAN & HALL BOOK

CRC Press
Taylor & Francis Group
6000 Broken Sound Parkway NW, Suite 300
Boca Raton, FL 33487-2742

First issued in paperback 2019

© 2013 by Taylor & Francis Group, LLC
CRC Press is an imprint of Taylor & Francis Group, an Informa business

No claim to original U.S. Government works

ISBN-13: 978-1-4398-8471-3 (hbk)
ISBN-13: 978-0-367-38041-0 (pbk)

This book contains information obtained from authentic and highly regarded sources. Reasonable efforts have been made to publish reliable data and information, but the author and publisher cannot assume responsibility for the validity of all materials or the consequences of their use. The authors and publishers have attempted to trace the copyright holders of all material reproduced in this publication and apologize to copyright holders if permission to publish in this form has not been obtained. If any copyright material has not been acknowledged please write and let us know so we may rectify in any future reprint.

Except as permitted under U.S. Copyright Law, no part of this book may be reprinted, reproduced, transmitted, or utilized in any form by any electronic, mechanical, or other means, now known or hereafter invented, including photocopying, microfilming, and recording, or in any information storage or retrieval system, without written permission from the publishers.

For permission to photocopy or use material electronically from this work, please access www.copyright.com (http://www.copyright.com/) or contact the Copyright Clearance Center, Inc. (CCC), 222 Rosewood Drive, Danvers, MA 01923, 978-750-8400. CCC is a not-for-profit organization that provides licenses and registration for a variety of users. For organizations that have been granted a photocopy license by the CCC, a separate system of payment has been arranged.

Trademark Notice: Product or corporate names may be trademarks or registered trademarks, and are used only for identification and explanation without intent to infringe.

**Visit the Taylor & Francis Web site at
http://www.taylorandfrancis.com**

**and the CRC Press Web site at
http://www.crcpress.com**

To my family
— A. R. L.

To Nari and Hanna
— K. C. C.

Contents

10 Joint analysis of mixed discrete and continuous outcomes via copula models   **139**

by Beilei Wu, Alexander R. de Leon, and Niroshan Withanage

11 Analysis of mixed outcomes in econometrics: Applications in health economics   **157**

by David M. Zimmer

Preface

Multivariate data of mixed types occur frequently in many fields of science and social science. The analysis of such data has created new challenges that have made it necessary to develop new statistical techniques and methodologies. Statisticians, working in collaboration with biologists, economists, epidemiologists, social scientists, and many others, have met these challenges with many remarkable advances over the past two decades. These include, among others, applications of mixed models to mixed outcomes in clustered and longitudinal studies, advances in dependence modeling of mixed data via copula and graphical models, extensions of Bayesian methods to mixed data settings, and adaptations of entropy- and divergence-based association measures to mixed outcomes.

Despite the attention researchers have given to mixed data analysis in recent years, there is yet no single book that focuses purely on this important topic. A close scrutiny of the literature reveals the following textbooks and monographs that contain discussions, albeit mostly brief, of mixed data analysis:

- H. Goldstein (2011). *Multilevel Statistical Models*. 4th ed. Wiley & Sons, Inc.

- D. M. Berridge and R. Crouchley (2011). *Multivariate Generalized Linear Mixed Models in R*. Chapman & Hall/CRC Press.

- L. Wu (2010). *Mixed Effects Models for Complex Data*. Chapman & Hall/CRC Press.

- C. E. McCulloch, S. R. Searle, and J. M. Neuhaus (2008). *Generalized, Linear, and Mixed Models*. 2nd ed. Wiley & Sons, Inc.

- M. J. Daniels and J. W. Hogan (2008). *Missing Data in Longitudinal Studies: Strategies for Bayesian Modeling and Sensitivity Analysis*. Chapman & Hall/CRC Press.

- P. X.–K. Song (2007). *Correlated Data Analysis: Modeling, Analytics, and Applications*. Springer.

- G. Molenberghs and G. Verbeke (2005). *Models for Discrete Longitudinal Data*. Springer.

- A. Skrondal and S. Rabe–Hesketh (2004). *Generalized Latent Variable Modeling: Multilevel, Longitudinal, and Structural Equation Models*. Chapman & Hall/CRC Press.

- R. J. Little and D. B. Rubin (2002). *Statistical Analysis with Missing Data*. 2nd ed. Wiley & Sons, Inc.

- J. L. Schafer (1997). *Analysis of Incomplete Multivariate Data*. Chapman & Hall/CRC Press.

- D. R. Cox and N. Wermuth (1996). *Multivariate Dependencies: Models, Analysis and Interpretations*. Chapman & Hall/CRC Press.

- G. A. F. Seber (1984). *Multivariate Observations*. Wiley & Sons, Inc.

In addition, the following edited volumes devote separate chapters to mixed data:

- G. Fitzmaurice, M. Davidian, G. Verbeke, and G. Molenberghs (Eds.) (2009). *Longitudinal Data Analysis*. Chapman & Hall/CRC Press.

- M. Aerts, H. Geys, G. Molenberghs, and L. M. Ryan (Eds.) (2002). *Topics in Modelling of Clustered Data*. Chapman & Hall/CRC Press.

However, since these books provide only snapshots of contemporary developments in mixed data analysis, there is thus a need for an authoritative book on mixed data analysis that traces important developments, systematizes terminology and methodologies, and gives an overview of applications. Our intention in producing this book is to show the depth and diversity of current research in the field by bringing together the work of as many researchers as possible, thus providing synthesis as well as development of directions for future research.

The thirteen chapters in this book were written by leading researchers who have made important and sustained contributions to mixed data analysis, and who were selected with a view to covering as much ground as possible in this broad area. While each chapter can be read independently of the others, we have informally organized them into groups to facilitate smooth thematic transitions (see Section 1.4). With the technical nature of the subject and this being an edited volume with numerous chapter authors, we have endeavored to maintain a certain degree of clarity and harmony in the presentation style. We believe that we have achieved this goal for the following reasons:

- The book was carefully and thoroughly edited for smooth readability and seamless transitions between chapters. All the chapters follow a common structure, with an introduction and a concluding summary, and include illustrative examples, many drawn from real-life case studies in developmental toxicology, economics, medicine and health, marketing, and genetics.

- We have included ample cross-references between chapters to enable readers to connect the book's various topics and research strands and to facilitate self-study.

- We have, as much as possible, unified notations, table formats, and terminologies across chapters. In particular, we have adopted, whenever possible, a common set of notations for mathematical and statistical quantities, such as vectors and matrices (both random and fixed) as well as distributions.

- To facilitate easy referencing by readers, we have come up with a combined index as well as a single up-to-date bibliography for the entire book. The references collected at the end of the book provide the most comprehensive, most current, and most complete list of published material on mixed data analysis.

- As a unique feature of the book, we have included an introductory chapter that provides a "wide angle" overview and comprehensive survey of mixed data analysis. The chapter contains useful background material that should prepare readers for the rest of the book.

To enable more focused readers to skip those chapters that are not of particular interest to them without disrupting the continuity of the material, we have made the chapters self-contained by allowing some degree of duplication in the literature reviews included in each chapter's introduction. As with any similar collection, there are bound to be omissions of some topics. While we strived for a good balance between theory and applications, it is impossible to include all topics in a single volume. For example, mixed-data graphical models are not discussed, apart from a brief mention in Chapter 1. Other mixed-data settings such as those with continuous (and/or binary) and time-to-event outcomes (Rizopoulos, 2012) or those with mixed continuous and semi-continuous responses are similarly not considered. Nevertheless, the book's technical level along with the many examples and case studies should make the book appeal to a broad audience. We believe that it would be a valuable resource to methodologically oriented as well as subject matter-motivated researchers. These include graduate students, applied statisticians, biostatisticians, and researchers in subject matter areas like medicine, health, genetics, and epidemiology, among many others. The book should be an excellent supplement to the textbooks and monographs enumerated above.

This book was completed with considerable help from several people. Our deepest thanks go to all the contributors, for their patience, enthusiasm, and expertise. Likewise, we are indebted to the following reviewers for their input and comments: Dipankar Bandyopadhyay (Division of Biostatistics, University of Minnesota, Minneapolis, MN), Kenneth A. Bollen (Department of Sociology, University of North Carolina, Chapel Hill, NC), Claudia Czado (Zentrum Mathematik, Technische Universität München, Munich, Germany), Ruzong Fan (Biostatistics and Bioinformatics Branch, National Institute of Child Health and Human Development, Bethesda, MD), Garrett Fitzmaurice (Department of Biostatistics, Harvard School of Public Health, Boston, MA), Helena

Geys (Center for Statistics, Hasselt University, Diepenbeek, Belgium), Harry Joe (Department of Statistics, University of British Columbia, Vancouver, BC), Stefan Lang (Department of Statistics, Universität Innsbruck, Innsbruck, Austria), Huilin Li (Department of Environmental Medicine, New York University, New York, NY), Mingliang Li (Department of Economics, SUNY at Buffalo, Buffalo, NY), Irini Moustaki (Department of Statistics, London School of Economics, London, UK), Raffaella Piccarreta (Department of Decision Sciences, Università Bocconi, Milan, Italy), David S. Siroky (School of Politics & Global Studies, Arizona State University, Tempe, AZ), and Qiaohao Zhu (Cross Cancer Institute, Edmonton, AB). We thank as well Michael J. Daniels, Christel Faes, Jian Kang, Helga Wagner, and David M. Zimmer, for doubling as reviewers.

Special thanks go to Beilei Wu, for her invaluable help at crucial stages, Yin Li, for computational assistance, and Hanna Carrière, for the lovely cover art. We are also grateful to Wiley & Sons, Inc., for permission to reproduce Figure 5.1 in Chapter 5.

This book was made possible by funding from the Natural Science and Engineering Research Council (NSERC) of Canada. Part of the work was done in Australia while Keumhee C. Chough was visiting Prof. Richard Huggins at the University of Melbourne. The contributors also acknowledge financial support from various research organizations and agencies: NSERC and Le Fonds québécois de la recherche sur la nature et les technologies, for grants to Denis Larocque and François Bellavance; National Institute on Drug Abuse, for Grants 5K05DA000017 and 5P01DA001070 to Peter M. Bentler, who also acknowledges a financial interest in EQS and its distributor Multivariate Software; Japan Science and Technology Agency, for PRESTO 3709 award to Takahiro Hoshino; Tsinghua–Yue–Yuen Medical Science Fund and Natural Science Foundation of China, for Grant 11071137 to Ying Yang; Portuguese National Foundation for Science and Technology, for Grant PTDC/SAU-ESA/100841/2008 to Armando Teixeira–Pinto; National Institutes of Health, for Grant NIH CA85295 to Michael J. Daniels.

It was a pleasure to work with John Kimmel and his colleagues in the production department at CRC/Chapman & Hall. John's commitment to and encouragement of this project from first to last have been remarkable. We thank him for giving us the opportunity to work on this book.

Finally, we express our gratitude to all our family and friends. Keumhee, in particular, is especially grateful to Jean Chough and James Osadczuk, for their loving support throughout the book production.

Alexander R. de Leon
Calgary, Alberta
Keumhee C. Chough
Edmonton, Alberta

Editors

Alexander R. de Leon is an associate professor in the Department of Mathematics and Statistics at the University of Calgary. Originally from the Philippines, he obtained his BSc and MSc from the School of Statistics, University of the Philippines. After a research studentship at Tokyo University of Science, he completed his PhD in statistics in 2002 at the University of Alberta. His research interests include methods for analyzing correlated data, multivariate models and distances for mixed discrete and continuous outcomes, pseudo- and composite likelihood methods, copula modeling, assessment of diagnostic tests, statistical quality control, and statistical problems in medicine.

Keumhee Carrière Chough is a professor of statistics in the Department of Mathematical and Statistical Sciences at the University of Alberta. After completing her BSc in agriculture from Seoul National University, in Seoul, Korea, she earned her MSc from the University of Manitoba, and her PhD in statistics from the University of Wisconsin–Madison in 1989. Since 1996, she has been with the Department of Mathematical and Statistical Sciences, University of Alberta, after stints as an assistant professor at the University of Iowa (1990–1992) and University of Manitoba (1992–1996). She was also the director of the Statistics Consulting Center at the University of Iowa (1990–1992). Her research interests include design and analysis for repeated measures data, missing data methods, high dimensional data analysis methods, multivariate methods, designs for clinical trials, item response data, variable selection methods, and survival analysis. As well, she specializes in such biostatistical methods as small area variation analysis techniques with applications to health care utilization. She has been a Health Scientist funded through the Alberta Heritage Foundation for Medical Research (1996–2011). She is a Fellow of the American Statistical Association, the Institute of Health Economics, and the Manitoba Centre for Health Policy.

Contributors

Qi An   Department of Mathematical Sciences
                                  University of Memphis
                                  Memphis, TN, USA

François Bellavance   Department of Management Sciences
                                  HEC Montréal
                                  Montréal, QC, Canada

Peter M. Bentler   Departments of Psychology & Statistics
                                  University of California–Los Angeles
                                  Los Angeles, CA, USA

Dale Bowman   Department of Mathematical Sciences
                                  University of Memphis
                                  Memphis, TN, USA

Keumhee Carrière Chough   Department of Mathematical & Statistical Sciences
                                  University of Alberta
                                  Edmonton, AB, Canada

Michael J. Daniels   Department of Statistics
                                  University of Florida
                                  Gainesville, FL, USA

Alexander R. de Leon   Department of Mathematics & Statistics
                                  University of Calgary
                                  Calgary, AB, Canada

Abdessamad Dine   Department of Management Sciences
                                  HEC Montréal
                                  Montréal, QC, Canada

Christel Faes   Interuniversity Institute for Biostatistics & Statistical Bioinformatics
                                  Hasselt University
                                  Diepenbeek, Belgium

Jeremy T. Gaskins   Department of Statistics
                                  University of Florida
                                  Gainesville, FL, USA

E. Olusegun George   Department of Mathematical Sciences
                                  University of Memphis
                                  Memphis, TN, USA

Ralitza Gueorguieva   School of Public Health
                             Yale University
                             New Haven, CT, USA

Jaroslaw Harezlak   Department of Biostatistics
                             Indiana University School of Medicine
                             Indianapolis, IN, USA

Takahiro Hoshino   Department of Economics & Business Administration
                             Nagoya University
                             Chikusa–ku, Nagoya, Japan

Jian Kang   Department of Biostatistics & Bioinformatics
                             Emory University
                             Atlanta, GA, USA

Minjung Kwak   Office of Biostatistics Research
                             National Heart, Lung, and Blood Institute
                             Bethesda, MD, USA

Denis Larocque   Department of Management Sciences
                             HEC Montréal
                             Montréal, QC, Canada

Armando Teixeira–Pinto   Department of Health Information & Decision Sciences
                             University of Porto
                             Porto, Portugal

Regina Tüchler   Department of Statistics
                             Austrian Federal Economic Chamber
                             Wien, Austria

Helga Wagner   Department of Applied Statistics
                             Johannes Kepler University
                             Linz, Austria

Niroshan Withanage   Department of Mathematics & Statistics
                             University of Calgary
                             Calgary, AB, Canada

Beilei Wu   Department of Mathematics & Statistics
                             University of Calgary
                             Calgary, AB, Canada

Colin O. Wu   Office of Biostatistics Research
                             National Heart, Lung, and Blood Institute
                             Bethesda, MD, USA

Ying Yang   Department of Mathematical Sciences
                             Tsinghua University
                             Beijing, P. R. China

Gang Zheng

Office of Biostatistics Research
National Heart, Lung, and Blood Institute
Bethesda, MD, USA

David M. Zimmer

Department of Economics
Western Kentucky University
Bowling Green, KY, USA

List of Figures

List of Tables

Chapter 1

Analysis of mixed data: An overview

Alexander R. de Leon and Keumhee Carriére Chough

1.1 Introduction

The advent of sophisticated tools of measurement has given rise to new modes of data collection. As a result, data often come with complex dependence structures. These complex structures typically require non-standard statistical approaches that usually entail computationally intensive methodologies. Conventional tools generally rely on the assumption that the data, or some suitable transformations of them, follow a normal distribution. This assumption no longer directly applies in these contexts. Over the past 20 years, there have been remarkable developments in statistical methodology for the analysis of such data. The development of statistical software and packages has unfortunately not kept pace with these methodological advances, but practitioners nonetheless now have a host of increasingly sophisticated tools available to them for handling the complex data. This has made possible their adoption and application in solving important substantive problems across a number of disciplines, particularly in engineering and finance, and in medicine and health.

Multivariate data comprising mixtures of discrete (i.e., categorical, binary, count) and continuous measurements (also referred to as "non-commensurate" outcomes) are a particularly common example of non-standard correlated data in practice. Aerts *et al.* (2002) gave several examples from developmental toxicology, where fetal data from laboratory animals include binary, categorical, and continuous outcomes. More recently, Daniels and Normand (2006) analyzed mixed patient data to profile the performance of regional networks of health care providers in the United States. Mixed data of this sort present unique challenges to analysts in that often, one of the goals in such analysis is to characterize the nature of relationships between measurements, of different and/or the same subjects, either over time or cross-sectionally in one or more spatial dimensions. Besides the ad hoc approach of carrying out separate analyses for the discrete and continuous variables in the data, which are clearly deficient in many applications, this is not a straightforward undertaking.

In this chapter, we review many of the major advances that have been made concerning the analysis of mixed data. We take a brief historical tour of the literature and highlight significant methodological developments of the last 20 years. Our review is an attempt to synthesize the various research strands in the literature; as such, it provides a useful prism through which to view previous and current advances in mixed data analysis. Our focus is on important and lasting contributions to mixed data methodology. While omissions are inevitable, the chapter is the perfect starting point for the book and should prepare the reader for the broad range of topics discussed in the remaining chapters.

1.2 Early developments in mixed data analysis

A number of simple, albeit ad hoc, approaches to the analysis of mixed data have been used in applications. If, on the one hand, the discrete variables can be subjected to some numerical scoring scheme, then all the variables can be treated as continuous. On the other hand, all the variables can be treated as discrete if the continuous variables can be discretized through some grouping criteria.

Another approach would be to analyze the discrete and continuous variables separately, and then to synthesize the two sets of results. However, as Krzanowski (1983) noted, "all these options involve some element of subjectivity, with possible loss of information, and do not appear very satisfactory in general." The first approach introduces considerable subjectivity in the numerical scoring scheme adopted, the second results in information loss due to categorization of the continuous variables, while the third ignores any associations that exist between the mixed variables.

The ideal general approach is to first specify a model for the joint distribution of the mixed variables, then to fit the model to the data at hand, and finally to use the parameter estimates to draw inferences. One way to specify the joint distribution of a number of variables is to express it as the product of the conditional distribution of a subset of the variables multiplied by the marginal distribution of the remaining variables. This suggests two routes that can be taken to formulate the joint distribution in the mixed case: (1) specify the marginal distribution of the discrete variables and the conditional distribution of the continuous variables, given the discrete variables; or (2) specify the marginal distribution of the continuous variables and the conditional distribution of the discrete variables, given the continuous variables.

The second approach was first mentioned by Cox (1972), who suggested that the joint distribution of a mixture of binary and continuous variables could be written as a logistic conditional distribution for the binary variables given the continuous variables multiplied by a marginal multivariate normal distribution for the latter. Cox and Wermuth (1992) pursued this idea further and pointed out its connection to probit-style and latent variable models. Such models are now known as conditional Gaussian regression models (see, e.g., Edwards, 1995).

The first approach has received much attention in the literature in the context of the analysis of data with mixtures of categorical and continuous variables. Here it is assumed that the continuous variables have a different multivariate normal distribution for each possible setting of the categorical variable values, while the categorical variables have an arbitrary marginal multinomial distribution. This model, which has been termed the conditional Gaussian distribution (CGD), forms the central plank of graphical association models for the analysis of mixed categorical and continuous variables (Whittaker, 1990; Lauritzen and Wermuth, 1989). We elaborate on this in the following sections.

1.2.1   Early models

The models for mixed data mentioned in the previous section were originally developed as a device for testing hypotheses of independence or conditional independence. Lauritzen and Wermuth (1989) provided an all-encompassing treatment of multivariate dependencies in mixed data with the introduction of conditional Gaussian families. Denote D categorical and C continuous variables as $\mathbf{U} = (U_1, \cdots, U_D)^\top$ and $\mathbf{Y} = (Y_1, \cdots, Y_C)^\top$, respectively. Suppose that the dth categorical variable U_d has s_d categories, so that there are a total of $S = \prod_{d=1}^{D} s_d$ possible patterns of discrete response, or states, for $\mathbf{U}$. A full CGD for $\mathbf{U}$ and $\mathbf{Y}$ assumes that the joint probability density of observing state s of $\mathbf{U}$ with $\mathbf{Y} = \mathbf{y}$ is

$$\pi_s (2\pi)^{-C/2} |\mathbf{\Sigma}_s|^{-1/2} \exp \left\{ -\frac{1}{2} (\mathbf{y} - \boldsymbol{\mu}_s)^\top \mathbf{\Sigma}_s^{-1} (\mathbf{y} - \boldsymbol{\mu}_s) \right\}. \tag{1.1}$$

That is, it assumes that if $\mathbf{U}$ falls in the sth state (or sth discrete response pattern), then $\mathbf{Y}$ has the multivariate normal distribution $N_C(\boldsymbol{\mu}_s, \mathbf{\Sigma}_s)$ with mean vector $\boldsymbol{\mu}_s$ and covariance matrix $\mathbf{\Sigma}_s$, while the probability that $\mathbf{U}$ falls in state s is π_s ($\sum_{s=1}^{S} \pi_s = 1$). The density in (1.1) can be rewritten in the form

$$\exp \left(\phi_s + \boldsymbol{\psi}^\top \mathbf{y} - \frac{1}{2} \mathbf{y}^\top \mathbf{\Sigma}_s \mathbf{y} \right). \tag{1.2}$$

The triple $(\pi_s, \boldsymbol{\mu}_s, \mathbf{\Sigma}_s)$ comprising, respectively, the sth state probability, the sth state mean and the sth state dispersion matrix, in (1.1) are called the moment parameters of the CGD, while the parameters in (1.2) are its canonical parameters. Here the ϕ_s are discrete canonical parameters and the $\boldsymbol{\psi}_s$ are $C \times 1$ vectors of linear canonical parameters. The technical aspects of fitting these models, including likelihood-based estimation and hypothesis-testing, are covered in the references cited earlier.

Moustafa (1957) was the first to consider the full CGD model in the analysis of multi-way tables. Another CGD model introduced by Olkin and Tate (1961) for mixed binary and continuous data, is known as the general location model (GLOM) (Little and Rubin, 2002; Schafer, 1997; Little and Schluchter, 1985). This particular model assumes a uniform dispersion matrix $\boldsymbol{\Sigma}$ across the states and is called a homogeneous CGD in the graphical modeling literature. Olkin and Tate (1961), while considering canonical correlations between the binary and continuous variables, established results connecting these canonical correlations and the state means.

Another approach to handling mixed data assumes that the discrete variables are coarsely measured versions of unobservable continuous variables called latent variables, and are obtained by partitioning or thresholding the space of the latent variables into non-overlapping intervals. Models for discrete data specified this way were first suggested by Pearson (1904), and they have been further developed over the years (Skrondal and Rabe–Hesketh, 2004). One such model, called the grouped continuous model (GCM) (de Leon, 2005; Anderson and Pemberton, 1985), considers the multivariate normal distribution as the distribution for the latent variables. In it, a discrete vector $\mathbf{Z} = (Z_1, \cdots, Z_Q)^\top$ is observed, where Z_q has discrete values $1 < \cdots < L_q$, $q = 1, \cdots, Q$, and corresponding to $\mathbf{Z}$ is a vector of unobservable continuous latent variables $\mathbf{Y}^* = (Y_1^*, \cdots, Y_Q^*)^\top$, distributed according to $N_Q(\mathbf{0}, \mathbf{R}^*)$ with mean vector $\mathbf{0}$ and correlation matrix $\mathbf{R}^*$, such that $Z_q = \ell_q$ if and only if $\alpha_q^{\ell_q - 1} < Y_q^* \leq \alpha_q^{\ell_q}$, $\ell_q = 1, \cdots, L_q$, with $\{\alpha_q^0 = -\infty < \alpha_q^1 < \cdots < \alpha_q^{L_q} < \alpha_q^{L_q+1} = +\infty\}$ the unknown cutpoints or thresholds for Z_q, $q = 1, \cdots, Q$; the correlations in $\mathbf{R}^*$ are usually called polychoric correlations. This model is a generalization of the univariate GCM discussed earlier by Anderson and Philips (1981) and McCullagh (1980), and is closely linked to probit models in latent variable theory. The extension of GCM to the case of mixed discrete and continuous data has been studied by Anderson and Pemberton (1985) and by Poon and Lee (1987, 1986), and is referred to as the conditional GCM (CGCM) in the literature. This approach involves the assumption that the continuous variables share a joint multivariate normal distribution with the latent variables, and the thresholds and polychoric correlations are defined in terms of the conditional distribution of the latent variables (or the discrete data) given the continuous data. In addition to these parameters, the model introduces additional parameters representing the polyserial correlations, the correlations between the discrete and continuous variables. Early references on CGCM include Poon and Lee (1992), Poon et al. (1990), and Lee et al. (1989). A recent application to sample size determination in clinical trials is discussed in Sozu et al. (2012).

The most general case of mixed data encountered in practice are those which include mixtures of nominal, ordinal and continuous variables. One way to go about the analysis of such data is to use CGCM to model their joint distribution, thus implicitly assuming an underlying latent variable structure for the nominal data. Although this approach has been previously used for dichotomous nominal variables (see, e.g., Bock, 1972), it is generally inappropriate in the polychotomous nominal case. In principle, GLOM may also be used in this case, but it may be inadequate, and hence inappropriate, for two reasons. Firstly, there is no clear-cut manner of accounting for ordinal information, and secondly, there is no explicit way of incorporating correlations between the nominal and ordinal variables, and between the ordinal and continuous data. de Leon and Carrière (2007) recently introduced a general model for mixed nominal, ordinal and continuous data called the general mixed-data model (GMDM). The model is made up of two components: (1) a GLOM for the joint distribution of the nominal and continuous variables, and (2) a CGCM for the joint distribution of the ordinal and continuous variables, given the nominal data. The hybrid nature of GMDM not only accounts for the ordinal information in the data but also incorporates associations between nominal and ordinal, nominal and continuous, and ordinal and continuous variables. It is flexible enough to be applicable to various types of mixed data and includes the GLOM and CGCM as special cases.

In this respect, GMDM provides a unified treatment of these two conventional mixed data models. The model has served as a platform for extending conventional multivariate methods to the case of mixed data with discrete and continuous variables (e.g., de Leon *et al.*, 2011).

To illustrate the model, let the mixed data consist of a vector $\mathbf{X} = (X_1, \cdots, X_S)^\top$ of binary variables with $\sum_s X_s = 1$, a vector $\mathbf{Y} = (Y_1, \cdots, Y_C)^\top$ of continuous variables, and a vector $\mathbf{Z} = (Z_1, \cdots, Z_Q)^\top$ of ordinal variables with Z_q having $L_q + 1 \geq 1$ ordinal levels. The binary vector $\mathbf{X}$ represents nominal categorical data from a contingency table with $S = \prod_d s_d$ nominal states (or cells) defined by each possible value of vector $\mathbf{U} = (U_1, \cdots, U_D)^\top$ of nominal categorical variables, each with s_d possible categories. In this case, we have $\mathbf{X} = \mathbf{x}_{(s)}$ (i.e., $X_s = 1$ and $X_{s'} = 0$, for all $s' \neq s$) if $\mathbf{U}$ falls in state $s = 1, \cdots, S$.

For the vector $\mathbf{Z}$, an underlying continuous latent vector $\mathbf{Y}^* = (Y_1^*, \cdots, Y_Q^*)^\top$ is assumed linked to $\mathbf{Z}$ by the threshold relationship $Z_q = \ell_q$ if and only if $\alpha_q^{\ell_q - 1} < Y_q^* \leq \alpha_q^{\ell_q}$, where $\{\alpha_q^0 = -\infty, \alpha_q^1, \cdots, \alpha_q^{L_q}, \alpha_q^{L_q + 1} = +\infty\}$ are unknown cutpoints, $\ell_q = 1, \cdots, L_q + 1$, are ordinal scores for Z_q, and $var(Y_1^*) = \cdots = var(Y_Q^*) = 1$.

A GMDM models the marginal density $f_{\mathbf{X}}(\cdot)$ of $\mathbf{X}$ as multinomial and the joint density $f_{\mathbf{Y},\mathbf{Y}^*|\mathbf{X}}(\cdot)$ of $\mathbf{Y}$ and $\mathbf{Y}^*$ given $\mathbf{X}$, as multivariate normal whose mean depends on $\mathbf{X}$ but whose covariance matrix is constant across states, so that the joint density $f_{\mathbf{X},\mathbf{Y},\mathbf{Y}^*}(\cdot)$ of $\mathbf{X}$, $\mathbf{Y}$, and $\mathbf{Y}^*$, is GLOM; de Leon and Carrière (2007) showed that the joint density $f_{\mathbf{X},\mathbf{Y},\mathbf{Z}}(\cdot)$ of $\mathbf{X}$, $\mathbf{Y}$, and $\mathbf{Z}$, can then be written as

$$f_{\mathbf{X},\mathbf{Y},\mathbf{Z}}(\mathbf{x}_{(s)}, \mathbf{y}, \boldsymbol{\ell}) = \pi_s \phi_C(\mathbf{y} - \boldsymbol{\mu}_s; \boldsymbol{\Sigma}) \int_{\mathscr{S}(s,\mathbf{y},\boldsymbol{\ell})} \phi_Q(\mathbf{v}; \mathbf{R}) d\mathbf{v}, \qquad (1.3)$$

for $s = 1, \cdots, S$, with $\mathscr{S}(s, \mathbf{y}, \boldsymbol{\ell}) = (v_{s1}^{\ell_1 - 1}, v_{s1}^{\ell_1}] \times \cdots \times (v_{sQ}^{\ell_Q - 1}, v_{sQ}^{\ell_Q}]$, where $v_{sq}^{\ell_q} = \gamma_q^{\ell_q} - \tau_{sq} - \boldsymbol{\beta}_q^\top \mathbf{y}$, and $\phi_K(\cdot; \mathbf{H})$ is a K-dimensional normal density with mean $\mathbf{0}$ and covariance matrix $\mathbf{H}$. We say $\mathbf{X}$, $\mathbf{Y}$ and $\mathbf{Z}$ are jointly distributed according to GMDM if their joint density $f_{\mathbf{X},\mathbf{Y},\mathbf{Z}}(\cdot)$ is given by (1.3). The symmetric matrix $\mathbf{R}$ contains (conditional) polychoric correlations $r_{qq'}$ of $\mathbf{Z}$, $\boldsymbol{\pi} = (\pi_1, \cdots, \pi_S)^\top$ contains multinomial state probabilities ($\sum_s \pi_s = 1$), $\boldsymbol{\mu}_s$ is the mean vector of $\mathbf{Y}$ for state s, $\boldsymbol{\Sigma}$ is the covariance matrix of $\mathbf{Y}$, and $\boldsymbol{\ell} = (\ell_1, \cdots, \ell_Q)^\top$ is the vector of ordinal scores of $\mathbf{Z}$. Note that *i*) the standardized cutpoints $\gamma_q^{\ell_q}$ account for ordinal information in $\mathbf{Z}$; *ii*) the state effects τ_{sq} induce associations between $\mathbf{X}$ and $\mathbf{Z}$; and *iii*) the regression effects $\boldsymbol{\beta}_q$ represent polyserial correlations between $\mathbf{Y}$ and $\mathbf{Z}$. For model identifiability, state S is arbitrarily designated as the reference state. Parameters can then be represented by $\boldsymbol{\Theta}$, the stacked vector of $\boldsymbol{\Theta}_1$ and $\boldsymbol{\Theta}_2$, where $\boldsymbol{\Theta}_1$ is the vector containing "location" parameters $\boldsymbol{\pi}$, $\{\boldsymbol{\mu}_s : s = 1, \cdots, S\}$, $\{\gamma_q^{\ell_q} : q = 1, \cdots, Q; \ell_q = 1, \cdots, L_q\}$, and $\{\tau_{sq} : s = 1, \cdots, S - 1; q = 1, \cdots, Q\}$; and $\boldsymbol{\Theta}_2$ is the vector containing the rest of the parameters. A restricted GMDM may be defined by imposing restrictions on $\boldsymbol{\Theta}$ to reduce its dimension and streamline its structure. See de Leon and Carrière (2007) for details. Note that we put "location" in quotes, as $\boldsymbol{\Theta}_1$ contains the state effects τ_{sq}, which measure associations between ordinal and nominal data.

Several models are obtained as special cases of GMDM. If $Q = 0$ (i.e., no ordinal variables), then GMDM specializes to GLOM. Similarly, GMDM reduces to CGCM when $S = 1$ (i.e., no nominal variables), in which case $\mathbf{Y}$ is multivariate normal and $\mathbf{Z}$ depends on $\mathbf{Y}$ via a multivariate probit model; GCMs for ordinal data are obtained by taking $C = 0$ and $S = 1$. The choice of $f_{\mathbf{Y},\mathbf{Y}^*|\mathbf{X}}(\cdot)$ is completely arbitrary; however, modeling it by the multivariate normal distribution with constant covariance matrix across the states, as in GMDM, is convenient because of the normal distribution's nice marginal and conditional distributions. While normality may not hold in many cases, it can be easily checked in practice; in addition, transformations are readily available for normalizing non-normal data. Non-normal latent distributions may also be considered; however, estimates tend to be robust with respect to the latent distribution even if the latter is skewed (Tan *et al.*, 1999).

GLOMs are used in Chapter 2 for tree-based predictions of mixed outcomes. They are also generalized in Chapters 5, 6 and 7 to longitudinal and clustered mixed data regression settings. The CGCM is adopted in Chapter 3 to develop joint tests for mixed data in genetics.

1.2.2 Multivariate methods for mixed data

Methods for the analysis of multivariate continuous data are well documented (e.g., Seber, 1984; Mardia *et al.*, 1979), as are those for multivariate discrete data (Bishop *et al.*, 1975), but methods for multivariate mixed data are not well developed.

One aspect of mixed data inference that has received little attention so far is the so-called location hypothesis, for which the construction of reasonable statistical tests remains an important problem in such applications as quality control (de Leon and Carrière, 2000) and clinical studies (Afifi and Elashoff, 1969). Consider the GLOM with location parameter $\boldsymbol{\Theta}^\top = (\boldsymbol{\pi}^\top, \boldsymbol{\mu}^\top)$, with $\boldsymbol{\mu}^\top = (\boldsymbol{\mu}_1^\top, \cdots, \boldsymbol{\mu}_S^\top)$ as the $CS \times 1$ vector of state means. The problem of interest is to test

$$H : \boldsymbol{\Theta} = \boldsymbol{\Theta}_0 \qquad \text{against} \qquad K : \boldsymbol{\Theta} \neq \boldsymbol{\Theta}_0, \tag{1.4}$$

for some specified $\boldsymbol{\Theta}_0$. Hypothesis H in (1.4) is referred to in the literature as the one-sample location hypothesis, and much work has been done on the case of continuous data. Afifi and Elashoff (1969) tackled the two-sample mixed data problem and obtained two global tests, one based on the Kullback–Leibler divergence (Kullback, 1968) and another on the likelihood ratio approach. Despite the ubiquity of mixed data in practice, only a few authors have considered similar problems. Moustafa (1957) studied a multi-factor experiment where the response variables consist of continuous and discrete variables jointly distributed according to the full CGD. He considered hypotheses concerning the independence and conditional independence of the responses, and proceeded to construct asymptotic likelihood ratio tests, the theory for which was previously studied in Ogawa *et al.* (1957). A related problem was addressed by Olkin and Tate (1961), who derived tests of independence for GLOMs via canonical correlation theory.

Recent work by Morales *et al.* (1998) introduced a general class of dissimilarity or entropy-type measures to obtain test statistics for various hypotheses, including (1.4), involving mixed continuous and categorical data, and used the asymptotic theory of these statistics to construct the tests. Exact likelihood ratio tests in the bivariate two-sample case have likewise been obtained by de Leon and Carrière (2000), and were later generalized to general multivariate multi-sample settings in de Leon and Zhu (2008) and de Leon (2007). These tests can be viewed as extensions of classical univariate and multivariate analysis of variance tests (ANOVA) for continuous data to mixed data settings (see, e.g., Mardia *et al.*, 1979). Closely related tests in genetic association studies are discussed in Chapter 3.

GLOMs have also received much attention in the literature in the context of classification and discrimination (Krzanowski, 1993). Krzanowski (1976, 1975) and Chang and Afifi (1974) studied GLOM-based location linear discriminant functions as generalization of classical linear discrimination rules for continuous data. These were further extended by de Leon *et al.* (2011) to mixed data modeled by GMDM. Chapter 2 concerns mixed data classification and prediction via GLOM-based random forests.

The development of mixed data distances for use in classification and discrimination has also been the focus of a number of researchers (e.g., Cuadras *et al.*, 1997; Cuadras, 1992). Distance-based methods have the advantage of being able to handle disparate types of data including those with mixtures of discrete and continuous variables. Methods based on this approach rely on various distance functions between individual observations, and thus the choice of distance function is a crucial consideration. Among the first to develop mixed data distances was Krzanowski (1984, 1983), who adopted the GLOM in the calculation of Matusita's distance (Matusita, 1956) between two mixed data populations. Bar–Hen and Daudin (1995), in contrast, applied the Kullback–Leibler divergence to the GLOM and obtained a distance that specializes to the Mahalanobis distance in the absence of discrete variables. Nakanishi (1996) proposed another mixed data distance that includes Bar–Hen and Daudin's (1995) and Krzanowski's (1984) distances. More recently, Bedrick *et al.* (2000) used the CGCM to obtain another generalization of the Mahalanobis distance. Finally, de Leon and Carrière (2005) derived the Kullback–Leibler divergence between two GMDMs, resulting in a distance that specializes to Bar–Hen and Daudin's (1995) and Bedrick *et al.*'s (2000) generalized Mahalanobis distances. More recent closely related work can also be found in Nuñez *et al.* (2003).

1.3   Joint analysis of mixed outcomes

Research on mixed data analysis took a different turn in the early 1990s with major advances in statistical methodologies for longitudinal and clustered data (Fitzmaurice *et al.*, 2009). Prior to this period, much work in the area was confined mainly to generalizing and extending conventional multivariate methods for continuous data (for example, discriminant analysis and other distance-based methodologies) to mixed data. While this work continues, a new research strand started with the need to analyze data from a series of developmental toxicity studies undertaken by the U. S. National Toxicology Program in the late 1980s (see Aerts *et al.*, 2002, for detailed statistical description of the program). The purpose of such studies is to determine toxicity levels of chemical and pharmaceutical substances with the goal of properly regulating their use. They are typically done on laboratory animals, mostly pregnant mice, which are exposed to increasing dose levels of toxicants whose adverse effects (e.g., low birth weight, congenital malformations) on offspring are then observed. Such studies often involve correlated mixed discrete and continuous outcomes measured on clustered fetuses, and the main interest is in the dose-response relationships. Traditionally, risk assessment has been undertaken for each outcome separately; however, simultaneously analyzing the outcomes and then carrying out a joint risk assessment may be more appropriate in practice, given that the outcomes are correlated. Such an approach has many potential advantages, both in statistical and practical terms. For example, it enables analysts to account for relationships between outcomes and assess the joint influence of dose and other covariates on them. Joint analysis also avoids multiple testing and naturally leads to global tests, thus resulting in increased power and better control of Type I error rates (de Leon and Zhu, 2008). Significant efficiency gains over separate marginal analyses have also been reported, especially in settings with missing data (McCulloch, 2008; Gueorguieva and Sanacora, 2006). There was thus a need for flexible joint models that can meaningfully capture the dose-response relationships and the associations between outcomes on different and/or the same fetuses, including the outcomes' marginal and conditional characteristics. As pointed out by Ryan (2002), such joint models should ideally have marginally interpretable dose-response models for the outcomes; in addition, a desirable joint model should account for intra-litter (i.e., between fetuses from same litter) effects, and directly incorporate the intra-fetus association between mixed outcomes.

Model specification for mixed outcomes, in general, and in developmental toxicology, in particular, is therefore not straightforward, and conventional approaches do not directly apply. A number of joint models have nonetheless been proposed and studied in the literature (see de Leon and Carrière Chough, 2010, for a recent review; see also Teixeira–Pinto and Normand, 2009). Among the first to be developed were those based on the factorization models GLOM and CGCM in Section 1.2. We review these models, among others, and highlight connections between various modeling strategies in the literature, providing background material on the assorted challenges by paying particular attention to their advantages and disadvantages.

1.3.1   Joint models for mixed outcomes

When analyzing data comprising mixtures of discrete and continuous outcomes in clustered and longitudinal settings, analysts usually emphasize determining the mixed outcomes' joint distribution, from which they obtain specific aspects of their relationships, such as marginal and conditional distributions, and associations. Let the data be represented by vectors $\mathbf{X}$ and $\mathbf{Y}$, where $\mathbf{X}$ comprises discrete (e.g., binary, count, ordinal categorical) outcomes and $\mathbf{Y}$ consists of continuous responses. These outcomes may be measurements taken from subjects belonging to the same cluster (i.e., clustered data), or from the same subject over time (i.e., longitudinal data). In either case, joint analysis of $\mathbf{X}$ and $\mathbf{Y}$ requires either direct or indirect specification of the joint density $f_{\mathbf{X},\mathbf{Y}}(\cdot)$.

1.3.1.1 Direct approaches

Catalano and Ryan (1992) and Catalano (1997) adopted the CGCM by describing discrete outcome $\mathbf{X}$ as a thresholded normally distributed latent vector $\mathbf{Y}^*$ and using a conditional probit regression model for $\mathbf{Y}$, given $\mathbf{X}$. That is, $f_{\mathbf{X},\mathbf{Y}}(\mathbf{x},\mathbf{y}) = f_{\mathbf{Y}}(\mathbf{y})f_{\mathbf{X}|\mathbf{Y}}(\mathbf{x}|\mathbf{y})$ is specified through $f_{\mathbf{Y}^*,\mathbf{Y}}(\mathbf{y}^*,\mathbf{y}) = f_{\mathbf{Y}}(\mathbf{y})f_{\mathbf{Y}^*|\mathbf{Y}}(\mathbf{y}^*|\mathbf{y})$, where $f_{\mathbf{Y}^*,\mathbf{Y}}(\cdot)$ (hence, $f_{\mathbf{Y}}(\cdot)$ and $f_{\mathbf{Y}^*|\mathbf{Y}}(\cdot)$) is conveniently modeled as a multivariate normal distribution, ensuring reproducibility of their joint model in the process. The mean models $\boldsymbol{\mu}_{\mathbf{Y}} = E(\mathbf{Y})$ and $\boldsymbol{\mu}_{\mathbf{Y}^*|\mathbf{Y}} = E(\mathbf{Y}^*|\mathbf{Y})$ are linked to covariates and then imbedded in $f_{\mathbf{Y}^*,\mathbf{Y}}(\cdot)$; note that $\boldsymbol{\mu}_{\mathbf{Y}^*|\mathbf{Y}}$ is conditional on $\mathbf{Y}$, resulting in a conditional probit regression model for $\mathbf{X}$, given $\mathbf{Y}$ and the covariates. However, the model, while able to account for cluster effects, lacks marginal interpretability and does not permit direct estimation of the between-outcome correlation. Fedorov *et al.* (2012) recently applied this model to study optimality in dose-finding clinical studies with mixed categorical and continuous responses.

By reversing the direction of factorization, Fitzmaurice and Laird (1997, 1995) extended the GLOM to incorporate regression models for $\mathbf{X}$ and $\mathbf{Y}$. While dependence between $\mathbf{X}$ and $\mathbf{Y}$ is accounted for in the model, this is done indirectly. In addition, the model is not reproducible and computation becomes an issue. George *et al.* (2007) recently adopted the same approach and exploited the exchangeability of outcomes within clusters, a reasonable assumption in many cases, to facilitate estimation. A recent application in oncology trials is discussed in Hirakawa (2012). Chapter 7 summarizes recent work in this area.

Further extensions of the same approach have been studied by Yang and Kang (2010) and Yang *et al.* (2007), who applied the model to jointly analyze incomplete mixed Poisson and continuous longitudinal data. They update their contribution in Chapter 5.

Although factorization models are straightforward to construct, deciding the direction of conditioning is an important consideration, as they are not invariant to the factorization adopted. While outcomes are typically ordered temporally in practice, factorization models use a structural approach to classify them into continuous or discrete. A possibly artificial hierarchy in the outcomes is thus established, with the conditioning outcomes treated as an intermediate response, and the conditioned outcomes as the ultimate response (Cox and Wermuth, 1996, p. 3). For example, Fitzmaurice and Laird's (1995) model suggests a predictive model where $\mathbf{X}$ is the response of true interest with $\mathbf{Y}$ serving as an "explanatory" variable. While ideally dictated by subject-matter considerations, the choice of the conditioning outcomes is made mainly for statistical convenience. Because the resulting models are not comparable, it is also possible for different models to yield very different inferences, especially of the associations; a more symmetrical treatment of outcomes is thus needed in practice. Teixeira–Pinto and Normand (2009) recently provided a detailed discussion of these issues in the context of clustered mixed binary and continuous outcomes; see Chapter 6 for a summary and additional results.

General forms of the multivariate exponential family that incorporate both discrete and continuous variables have also been studied by a number of people. Zhao and Prentice (1990) introduced the quadratic exponential model for the joint distribution of correlated binary outcomes; see also Betensky and Whittemore (1996) for its extension to the multivariate setting, and Zhao *et al.* (1992), for the partly exponential model. It was adapted to the mixed data context by Prentice and Zhao (1991). However, due to the computational demands of calculating the model's normalizing constant, so that a fully likelihood-based approach becomes infeasible in high-dimensional cases, Prentice and Zhao (1991) instead adopted the GEE approach for estimation.

1.3.1.2 Indirect approaches

Generalized linear mixed models (GLMMs) are a natural framework within which to analyze mixed-outcome data. Unlike factorization models which generally focus on marginal modeling, GLMMs incorporate subject-specific effects in the analysis (McCulloch, 2008). The inclusion of random effects is used to indirectly build joint models that embed an association structure between clustered (or longitudinal) measurements of same/different outcomes. This is done as follows:

$$f_{\mathbf{X},\mathbf{Y}}(\mathbf{x},\mathbf{y}) \quad = \quad \int f_{\mathbf{X},\mathbf{Y}|\mathbf{B}}(\mathbf{x},\mathbf{y}|\mathbf{b})f_{\mathbf{B}}(\mathbf{b})d\mathbf{b}, \tag{1.5}$$

where $\mathbf{B}$ is the vector of random effects, and $f_{\mathbf{X},\mathbf{Y}|\mathbf{B}}(\cdot|\mathbf{b})$ is the conditional density of $\mathbf{X}$ and $\mathbf{Y}$ given $\mathbf{B}$. Typically, it is assumed that $f_{\mathbf{X},\mathbf{Y}|\mathbf{B}}(\mathbf{x},\mathbf{y}|\mathbf{b}) = f_{\mathbf{X}|\mathbf{B}}(\mathbf{x}|\mathbf{b})f_{\mathbf{Y}|\mathbf{B}}(\mathbf{y}|\mathbf{b})$, i.e., $\mathbf{X}$ and $\mathbf{Y}$ are conditionally independent.

Gueorguieva and Agresti (2001) introduced a correlated probit model for $\mathbf{X}$ and $\mathbf{Y}$, which is a multivariate GLMM with a probit model for $\mathbf{X}$ and a Gaussian linear model for $\mathbf{Y}$; see also Gueorguieva and Sanacora (2006). The joint model is based on a CGCM for $f_{\mathbf{X},\mathbf{Y}|\mathbf{B}}(\cdot)$ — similar to Catalano and Ryan's (1992) — with correlated normally distributed residual errors. A recent application of the model is discussed in Najita et al. (2009); a Bayesian treatment of the model is given in Bello et al. (2012). Their joint model is closely related to that of Regan and Catalano (1999a), who relied on complete exchangeability of $\mathbf{X}$ and $\mathbf{Y}$; a less restrictive semiparametric joint model was also studied in Regan and Catalano (2002, 1999b), where GEE (generalized estimating equations) ideas are used to incorporate cluster effects. A general discussion of GLMMs in mixed-outcome longitudinal settings is given in Faes et al. (2009); see also Faes et al. (2008) for applications in high dimensional cases. These and more are discussed in Chapters 8 and 9.

Note that these models can be essentially viewed within Rabe–Hesketh et al.'s (2001; see also Skrondal and Rabe–Hesketh, 2004) generalized linear latent and mixed model (GLLAMM) framework. GLLAMMs are a general class that unifies multilevel, structural equation, latent class, and longitudinal models. Sammel et al.'s (1997) and Moustaki and Knott's (2000) latent trait models with continuous, dichotomous, and ordinal responses may also be considered as special cases of GLLAMM; see also Shi and Lee (2000) for a Bayesian perspective. An application of GLLAMM in factor score regression is presented in Chapter 4.

A related general Bayesian framework for modeling mixed outcomes via GLMMs was introduced by Dunson (2003, 2000); see also Miglioretti (2003). The approach similarly relies on latent formulations of discrete data, and can be considered as a Bayesian reformulation of GLLAMM. Important related recent work in this area includes Wagner and Tüchler (2010), Liu et al. (2009), Daniels and Normand (2006), and Quinn (2004), among others. Timely summaries are given in Chapters 12 and 13.

GLMMs developed for mixed outcomes have all assumed normally distributed random effects and residual errors, allowing for correlations in the latter to account for conditional dependence between outcomes. One recent exception is the model studied in Lin et al. (2010), where correlated random effects are incorporated in Fitzmaurice and Laird's (1995) joint model to build in intra-cluster correlations between $\mathbf{X}$ and $\mathbf{Y}$, and where, for interpretational ease, the random effect for $\mathbf{X}$ is assumed to have a marginal bridge distribution. A Gaussian copula (Song, 2007) was then used to construct its joint distribution with a normally distributed random effect for $\mathbf{Y}$. However, a normal distribution for residual errors was still assumed.

Copulas are particularly well-suited for constructing joint models for $\mathbf{X}$ and $\mathbf{Y}$, where the relevant joint distribution is either not available or difficult to specify but marginal distributions for $\mathbf{X}$ and for $\mathbf{Y}$ can be specified with confidence. While not new, applications of copulas to discrete data (e.g., Nikoloulopoulos and Karlis, 2010, 2009, 2008; Zimmer and Trivedi, 2006; Trégouët et al., 2004; Meester and MacKay, 1994) have only recently been elucidated and clarified (Genest and Nešlehovà, 2007). Recent important references include de Leon and Wu (2011), Dobra and Lenkoski (2011), Song et al. (2009), and Song (2007), among others. An earlier application is given in Geys et al. (2001), which adopted the Plackett copula (Plackett, 1965) to construct a joint model for mixed outcomes in a developmental toxicity study; see also Molenberghs et al. (2001) for a related application to the evaluation of surrogate endpoints in randomized experiments. Following de Leon and Wu (2011), Wu and de Leon (2012) recently developed flexible GLMMs for clustered mixed outcomes that permit non-normally distributed residual errors and random effects. Specifically, they assumed a threshold model that links $\mathbf{X} = (X_1,\cdots,X_D)^\top$ to a latent vector

$\mathbf{Y}^* = (Y_1^*, \cdots, Y_C^*)^\top$, and then proceeded to model $f_{\mathbf{Y}^*, \mathbf{Y}|\mathbf{B}}(\cdot|\mathbf{b})$ via a copula $C(\cdot)$ as

$$F_{\mathbf{Y}^*, \mathbf{Y}|\mathbf{B}}(\mathbf{y}^*, \mathbf{y}|\mathbf{b}) = C\left(\begin{array}{c} F_{Y_1^*|\mathbf{B}}(y_1^*|\mathbf{b}), \cdots, F_{Y_D^*|\mathbf{B}}(y_D^*|\mathbf{b}), \\ F_{Y_1|\mathbf{B}}(y_1|\mathbf{b}), \cdots, F_{Y_C|\mathbf{B}}(y_C|\mathbf{b}) \end{array} \right), \qquad (1.6)$$

where $F_{Y_1^*|\mathbf{B}}(\cdot|\mathbf{b}), \cdots, F_{Y_D^*|\mathbf{B}}(\cdot|\mathbf{b})$, and $F_{Y_1|\mathbf{B}}(\cdot|\mathbf{b}), \cdots, F_{Y_C|\mathbf{B}}(\cdot|\mathbf{b})$, are the marginal (conditional) distributions for $\mathbf{Y}^*$ and $\mathbf{Y} = (Y_1, \cdots, Y_C)^\top$, respectively, given random effect $\mathbf{B}$; the conditional density $f_{\mathbf{X}, \mathbf{Y}|\mathbf{B}}(\cdot|\mathbf{b})$ is then obtained from (1.6) via the corresponding density $f_{\mathbf{Y}^*, \mathbf{Y}|\mathbf{B}}(\mathbf{y}^*, \mathbf{y}|\mathbf{b}) = \partial^{C+D} F_{\mathbf{Y}^*, \mathbf{Y}|\mathbf{B}}(\mathbf{y}^*, \mathbf{y}|\mathbf{b})/\partial \mathbf{y}^* \partial \mathbf{y}$. Note that $F_{Y_1^*|\mathbf{B}}(\cdot|\mathbf{b}), \cdots, F_{Y_D^*|\mathbf{B}}(\cdot|\mathbf{b})$, and $F_{Y_1|\mathbf{B}}(\cdot|\mathbf{b}), \cdots, F_{Y_C|\mathbf{B}}(\cdot|\mathbf{b})$, can be any continuous distributions and $C(\cdot)$ can be any copula, thus allowing researchers great flexibility in modeling the conditional density $f_{\mathbf{X}, \mathbf{Y}|\mathbf{B}}(\cdot|\mathbf{b})$ of $\mathbf{X}$ and $\mathbf{Y}$ given the random effect $\mathbf{B}$. In fact, using a Gaussian copula with normal (i.e., Gaussian) margins in (1.6) results in a correlated probit model, thus extending Gueorguieva and Agresti's (2001) model to non-Gaussian settings. Note that instead of specifying the residual error distribution (typically Gaussian) to construct the conditional joint distribution of $\mathbf{X}$ and $\mathbf{Y}$ (given random effect $\mathbf{B}$), as is usually done in GLMMs, Wu and de Leon's (2012) approach specifies the latter directly by specifying the (conditional) marginal response distributions and coupling them together using a copula. The corresponding residual error distributions can then be obtained from the response distribution (given random effects) by transformation methods. Hence, their joint model does not require Gaussian residual errors, and is therefore more flexible than conventional GLMMs. A detailed discussion of this approach is given in Chapter 10.

Note that it is also possible to similarly use a copula to build the density $f_{\mathbf{B}}(\cdot)$ of the random effect $\mathbf{B}$. This was the approach taken by Lin et al. (2010) in lieu of the usual normality assumption for the random effects distribution. This affords flexibility in specifying the $f_{\mathbf{B}}(\cdot)$ in practice; in the case of a bivariate binary-continuous outcome, for example, a typical normally distributed random effect B_1 may be assumed for the continuous outcome Y, along with a bridge distributed random effect B_2 for the binary outcome X, to facilitate interpretability of marginal effects with a logistic regression model for X (i.e., a logistic latent distribution for Y^*). Chapters 10 and 11 outline and illustrate this approach via real data examples.

1.3.2 Estimation

Implementing various models has been hindered in practical applications by computational issues related to estimation, due in large part to problems associated with the analysis of discrete data. The normal distribution can often be relied upon to capture important characteristics of continuous data. Generally, there is no analogous distribution for discrete data, especially in complex clustered and longitudinal settings. These difficulties are necessarily inherited by models like the GLOM and GMDM. In many applications, sparseness of data becomes a vexing concern for such models, especially in high-dimensional situations. In addition, models constructed from latent variables (e.g., CGCM, GMDM) typically require high-dimensional integration, as in (1.3), which may prohibit their adoption in practice.

Nevertheless, the literature offers a variety of approaches for model estimation. Given a full specification of the model, a likelihood-based approach is an obvious choice; however, evaluation and direct maximization of the full likelihood frequently lack an analytical solution. Approximations, mostly numerical, have thus been proposed in the literature. These include approaches based on numerical integration (e.g., Gauss–Hermite quadrature methods) and Markov Chain Monte Carlo algorithms; see Liu et al. (2009) and Sammel et al. (1997), among others. Chapter 8 provides details on these methods.

Bayesian methods have also been considered in a number of applications; important references include Daniels and Normand (2006) and Shi and Lee (2000). See also Craiu and Sabeti (2012) and Dunson (2003, 2000). Surveys of these methods are found in Chapters 12 and 13.

There has also been much research on alternative approaches based on some simplified formulation of the full likelihood to address concerns about the robustness of the likelihood specification, an important issue quite apart from any computational difficulties. To circumvent these issues, the full likelihood may be replaced by a more computationally tractable pseudo-likelihood. For example, one could obtain composite pairwise likelihoods over the data (see Varin *et al.*, 2011 and Varin, 2008, for recent surveys). Faes *et al.* (2008) adopted this strategy in a high-dimensional mixed-outcome analysis via GLMMs; de Leon and Carrière (2007) and de Leon (2005) demonstrated the potential of composite pairwise likelihood methods for mixed data in the case of the GMDM and CGCM. Hoff (2007) introduced an extended rank likelihood for semiparametric inference in copula models. Working with such modifications of the likelihood function generally yields consistent estimates of model parameters, including correlations under a range of possible models for higher-order dependency as captured by the full joint distribution. The computational and statistical performance of these methods has been shown to range from acceptably good to excellent. See Chapter 9 for an illustration.

Non-likelihood based approaches have also been studied by several researchers. The development of GEE methods by Liang and Zeger (1986) has enabled the analysis of correlated data, including those with mixed outcomes, without the specification of the full model. This convenient simplification of the modeling process, surprisingly achieved with minimal loss in estimation efficiency, has allowed researchers to circumvent difficulties in analyzing mixed outcomes. It has proved quite useful and computationally more feasible in practice than likelihood-based methods. Regan and Catalano (2002), Fitzmaurice and Laird (1997; 1995), and Catalano and Ryan (1992), among others, have successfully adopted GEE methods in their contributions. Chapter 5 shows an application of GEE ideas to mixed count-continuous longitudinal data analysis.

1.4   Highlights of book

The following 12 chapters describe some of the many recent developments in the analysis of mixed data. They show the diversity of areas in which mixed data methodologies have been applied and expanded. With the tremendous growth in the mixed data literature in the past decade, it is clear that the many contributors to this growth, and the breadth and depth of methodology and practice are now far too large to cover with any degree of completeness in a single book like this one. However, what the succeeding chapters do show is the variety and ongoing vitality of this important area.

Chapters 2 to 4 concern multivariate analysis of mixed data and form the first part of the book. In Chapter 2, **A. Dine**, **D. Larocque**, and **F. Bellavance** discuss the application of regression trees to the prediction of mixed outcomes. They illustrate how ensembles of univariate and multivariate trees can be combined to improve predictions in mixed data settings. With appropriate weighting, these random forests are shown to generally exhibit very good predictive performance. In Chapter 3, **M. Kwak**, **G. Zheng**, and **C. O. Wu** take us into the realm of genetics, where mixed binary and quantitative traits are becoming more and more ubiquitous. They review various approaches for testing these mixed traits jointly, and demonstrate empirically the benefit of doing so in genetic association studies. In Chapter 4, **T. Hoshino** and **P. M. Bentler** flesh out a number of issues connected to estimation in structural equation models involving mixed discrete and continuous variables. In particular, they provide important insights into the sources of bias in factor score regression. Asymptotic theory is exploited to provide an alternative stepwise approach to conventional maximum likelihood and least-squares estimation methods. They show the validity of their approach empirically and illustrate it with data on web browsing behavior and purchasing intent.

Chapters 5 to 9 comprise the second part of the book on joint analysis of mixed outcomes in longitudinal and clustered settings. In Chapter 5, **J. Kang** and **Y. Yang** review recent developments in the analysis of mixed Poisson and continuous outcomes in longitudinal settings. They present and illustrate adaptations and extensions of Fitzmaurice and Laird's (1995) factorization approach. Moving through the complete to the incomplete data case, through fully parametric and semiparametric model formulations, they describe various ways of jointly analyzing count and continuous longitudinal data. In Chapter 6, **A. Teixeira–Pinto** and **J. Harezlak** update an earlier *Statistics in Medicine* paper of Teixeira–Pinto and Normand (2009) on joint models for correlated binary and continuous outcome data. They focus on factorization models, drawing comparisons and contrasts between various formulations, including those based on latent variables. In Chapter 7, **E. O. George**, **D. Bowman**, and **Q. An** exploit the exchangeability at the cluster level of clustered mixed binary and continuous observations, to formulate fully parametric regression models. The surprising tractability of the resulting model likelihoods allows the calculation of maximum likelihood estimates. With data from a developmental toxicity study, they show how quantitative risk assessment can be carried out using their methodology. In Chapter 8, **R. Gueorguieva** lays out the development of the GLMM approach to mixed-outcome data analysis that was catalyzed in the early work of Gueorguieva and Agresti (2001). She reviews the rich developments in estimation that have paralleled those in correlated data analysis. Data from another developmental toxicity study and a clinical trial are used to illustrate the methodologies. In Chapter 9, **C. Faes** provides a companion to Chapter 8 with a comprehensive review of multi-level modeling of mixed data using GLMMs. She illustrates how greater modeling flexibility can be achieved with increasingly complex specifications of the random effect and residual error distributions. These are used to justify consideration of pseudo-likelihood methods as a remedy for the computational difficulties associated with maximum likelihood estimation. To illustrate, the developmental toxicity data analyzed in Chapter 8 is revisited.

The third part of the book concerns recent applications of copula models to analysis of mixed outcomes. In Chapter 10, **B. Wu**, **A. R. de Leon**, and **N. Withanage** outline a class of flexible copula models for joint analysis of outcomes of mixed types. Their models extend and generalize conventional GLMMs discussed in Chapters 8 and 9, by allowing the adoption of non-normally distributed residual errors and random effects. Data previously considered in Chapters 8 and 9 are re-analyzed to show the practical utility of their methodology. In Chapter 11, **D. M. Zimmer** studies the application of random effects and copula models to econometric data with mixed discrete and continuous variables. Using data on health and drug spending/usage, he highlights the drawbacks of either approach, and in the process provides important insights into the relationship between drug spending and health problems as well as that between nondrug spending and drug usage.

The last two chapters concern Bayesian analysis of mixed data and form the fourth and final part of the book. In Chapter 12, **H. Wagner** and **R. Tüchler** consider data augmentation techniques for estimation of regression and random effects models for cross-sectional and longitudinal data, respectively; additionally, they employ variable selection as a flexible tool for modeling complex data. Application of their method is illustrated by two econometric data sets, one on household income and living conditions and the other on consumer behavior. Finally, **M. J. Daniels** and **J. T. Gaskins** give a comprehensive account of Bayesian methods for mixed data in Chapter 13, placing particular emphasis on longitudinal settings. They employ diverse and interesting examples to demonstrate the various methodological issues arising from a Bayesian analysis, and showcase the wide array of computational remedies available.

Chapter 2

Combining univariate and multivariate random forests for enhancing predictions of mixed outcomes

Abdessamad Dine, Denis Larocque, and François Bellavance

2.1 Introduction

Drawing inferences about the parameters of a model and obtaining predictions for future observations are two common goals of a statistical analysis.

When multiple outcomes are present, drawing inference by using a model for each of them independently of the others is one possibility. However, multiple testing (or multiple confidence intervals) is a concern that must be appropriately dealt with to obtain valid inferences. This is why using a single multivariate model can be attractive. Moreover, a single multivariate model can, in some circumstances, provide a more powerful analysis since it can effectively use the outcomes' covariance structure. These facts were recognized a long time ago and this is why models for multiple outcomes are well developed, at least when all outcomes are of the same type, e.g., all continuous or all binary. Classic monographs like Anderson (1984) and Bishop *et al.* (1975) are devoted to such models. The mixed outcomes situation, where some outcomes are continuous and some are categorical or discrete, is more difficult to handle even though it frequently occurs in practice. However, the interest in this setting has grown in recent years as indicated by the diverse work in this volume.

Drawing inference for parameters is not the goal of interest in this chapter. We are mainly concerned with predictions of future observations. In the multivariate linear regression context, methods with better predictive performance than the usual multivariate ordinary least squares (OLS) regression, which amounts to fitting separately an OLS regression model to each outcome, have been proposed (see for example, Breiman and Friedman, 1997). Recently, attention has been given to the case where the number of predictors is greater than the number of observations, the so-called $p > n$ case, (see Rothman *et al.*, 2010, and Koch and Naito, 2010). Hence, multivariate models can be useful for the sole purpose of predictions, and not only to handle the multiple testing problem.

For classification and regression tasks, tree-based methods like CART (Breiman *et al.*, 1984) are valuable alternatives to parametric methods. Their ability to automatically detect certain types of interactions and their ease of interpretation and visualization make them tools of choice for practitioners.

The classical tree-based methods can accommodate one continuous or one categorical outcome at a time. When multiple outcomes are available, one possibility is to grow as many univariate trees as there are outcomes. But understanding the overall structure by inspecting many trees might be difficult. Hence, it may be preferable to build a single multivariate tree for all outcomes, since it may provide a parsimonious and simple-to-interpret model. Few generalizations of tree-based methods to multiple outcomes have been proposed. Ciampi *et al.* (1991) propose to treat multiple continuous outcomes using a maximum likelihood criterion, under normality assumption, within their recursive partition and amalgamation process. Segal (1992) proposes a tree that can analyze continuous longitudinal outcomes using the Mahalanobis distance as a within-node homogeneity measure. Zhang (1998) develops a tree for multiple binary outcomes using a generalized entropy criterion that is proportional to the maximum likelihood of the joint distribution of multiple binary outcomes. Siciliano and Mola (2000) develop a general two-stage approach to build trees with either multivariate continuous outcomes or multivariate categorical outcomes. Lee (2005) proposes a general method based on GEE for multiple outcomes of the same type (continuous, binary, ordinal, polytomous).

However, the predictive performance of a single tree can often be improved by using an ensemble of trees such as random forests (Breiman, 2001). Both theoretical results (Biau *et al.*, 2008) and empirical studies (e.g., Breiman, 2001; Hamza and Larocque, 2005) have demonstrated the benefits of random forests, which mainly operate by reducing the variability associated with split selections and by expanding the model space. Moreover, random forests are very fast to compute since usually only a small subset of predictors is used to find the best split at each node of each tree. The main disadvantage is that an ensemble of trees is usually more difficult to interpret than a single tree. Recent surveys about random forests and ensemble methods can be found in Rokach (2009), Siroky (2009) and Verikas *et al.* (2011). With multiple outcomes, Segal and Xiao (2011) study multivariate random forests using the approach of Segal (1992).

All the above multivariate extensions of trees and forests can only accommodate one variable type at a time, i.e., the outcomes must all be continuous or all be binary, for example. Recently, Dine *et al.* (2009) proposed a tree-based method with a maximum likelihood splitting criterion derived from the general location model (GLOM) (Olkin and Tate, 1961). This new approach can accommodate multiple outcomes of mixed types. In a related work, Barcena and Tusell (2004) propose a method that combines univariate trees, one for each outcome, and imputation to compute multivariate predictions. This approach does not build a single multivariate tree but can be used to compute predictions with mixed outcomes. Finally, Piccarreta (2010) proposes a method to build a tree to relate a set of predictors to an outcome consisting of a matrix of dissimilarity measures (as commonly used in cluster analysis). Piccarreta (2010) also discusses the case where the dissimilarity measures are computed from multiple outcomes of mixed types.

Since the obvious advantage of a single multivariate tree over many univariate trees is interpretability, the literature on multivariate trees has focused mainly on the descriptive, rather than the predictive, aspect. Zhang (1998) and Siciliano and Mola (2000) are rare instances where limited predictive comparisons between univariate and multivariate trees are performed. Hence, it is still an open question whether or not multivariate trees are preferable to many univariate trees (one per outcome) for prediction tasks. However, since the predictive performance is the focus of this chapter, we use random forests instead of single trees.

In this chapter, we investigate the use of univariate and multivariate forests of trees to predict mixed outcomes. More precisely, we compare the predictive performances of univariate random forests, multivariate random forests, and of combined predictions from multivariate and univariate forests, with three different weighting schemes.

This chapter is organized as follows. In Section 2.2, we present the data setting for multiple outcomes of mixed types, the multivariate tree and forest methods for mixed outcomes, and the different prediction methods that are compared. Section 2.3 details the design and gives the results of a simulation study. Concluding remarks are presented in Section 2.4.

2.2   Predictions from univariate and multivariate random forests

2.2.1   *Data setting*

We consider the case in which we have a mixture of p continuous responses $\mathbf{Y}_i = (Y_{1i}, \cdots, Y_{pi})^\top$ and u categorical responses $\mathbf{F}_i = (F_{1i}, \cdots, F_{ui})^\top$, with $i = 1, \cdots, N$. Let s_j be the number of different values of the categorical outcome F_j, for $j = 1, \cdots, u$. Note that $\min(p, u) \geq 0$ and $\max(p, u) > 0$, i.e., we may have only continuous or categorical responses. The data settings in Ciampi *et al.* (1991) and in Segal (1992) are thus special cases with $p > 1$ and $u = 0$. Also, the data setting in Zhang (1998) is a special case with $p = 0$, $u > 1$, and $s_j = 2$ for $j = 1, \cdots, u$. Finally, the data settings in Siciliano and Mola (2000) and in Lee (2005) are also special cases with either $p > 1$ and $u = 0$, or $p = 0$ and $u > 1$.

2.2.2 Basic tree building

The classical way to build a tree is to recursively split the sample in order to partition it into more and more homogeneous nodes. Starting from a root node containing all the data, a best split, defined with the covariates, is found. Even though more complicated splits involving linear combinations are possible, only simple splits involving one predictor are usually considered. For a continuous (or at least ordinal) covariate x, the possible splits take the form $x \le c$, where c is a specified cutpoint. For a categorical covariate x, the possible splits take the form $x \in \{c_1, \cdots, c_l\}$, where $\{c_1, \cdots, c_l\}$ is a subset of the possible values of x. Once the best split is found by scanning through all possible splits, the data is partitioned into two children nodes, a left and a right node. The process is then repeated recursively for each resulting nodes. Typically, a large tree is built and then a right-size subtree is found by a pruning algorithm in order to avoid overfitting. Since we are using forests of trees which do not require pruning, we do not discuss this aspect further but the interested reader can refer to Breiman *et al.* (1984) for more details. One crucial aspect of tree building is finding an appropriate splitting criterion. The least-squares criterion is usually used with a single continuous outcome. With this criterion, the best split is the one minimizing $\sum_{i \in t^L} (Y_i - \overline{Y}^L)^2 + \sum_{i \in t^R} (Y_i - \overline{Y}^R)^2$, where $\overline{Y}^L$ ($\overline{Y}^R$) is the average of the response in the left (right) node t^L (t^R). We next describe the splitting criterion that we use for multiple mixed outcomes.

2.2.3 Multivariate tree for mixed outcomes

The multivariate tree for multiple outcomes of mixed types proposed in Dine *et al.* (2009) is based on the GLOM of Olkin and Tate (1961); see Chapter 1 for a brief survey of mixed data models. This method is described next. The vector $(F_1, \cdots, F_u)^\top$ is cast into one variable D, taking $S = \prod_{j=1}^{u} s_j$ possible states. A state from the variable D corresponds to one unique combination of the original categorical responses. To simplify the presentation, we describe the splitting of the root node with all N observations. The same process is repeated recursively to split children nodes using only the observations in them. The splitting criterion is the observed log-likelihood of a two-nodes GLOM. This type of approach is advocated in Su *et al.* (2004) in the case of univariate regression trees. This model assumes node- and state-specific mean vectors for the continuous outcomes, node-specific probabilities of being in each state, and a common covariance matrix across the two nodes and the states, for the continuous outcomes. Specifically, to partition the observations of the root node into left and right children nodes t^L and t^R, the splitting criterion is given by

$$\ell\ell(2 \text{ nodes}) \quad = \quad \sum_{k=1}^{S} (n_k^L \log \widehat{\pi}_k^L + n_k^R \log \widehat{\pi}_k^R) - \frac{N}{2} \log |\widehat{\Sigma}_2| - \frac{Np}{2} \log(2\pi) - \frac{Np}{2} \quad (2.1)$$

where $\widehat{\pi}_k^L = n_k^L / N^L$, $\widehat{\pi}_k^R = n_k^R / N^R$, $\widehat{\mu}_k^L = \sum_{i \in J_k^L} Y_i / n_k^L$, $\widehat{\mu}_k^R = \sum_{i \in J_k^R} Y_i / n_k^R$, for $k = 1, \cdots, S$, and

$$\widehat{\Sigma}_2 \quad = \quad \frac{1}{N} \sum_{k=1}^{S} \left\{ \sum_{i \in J_k^L} (Y_i - \widehat{\mu}_k^L)(Y_i - \widehat{\mu}_k^L)^\top + \sum_{i \in J_k^R} (Y_i - \widehat{\mu}_k^R)(Y_i - \widehat{\mu}_k^R)^\top \right\},$$

J_k^L denotes the subset of indices from $\{1, \cdots, N\}$ that are in the left node t^L and that are falling into state k, N^L is the number of observations in t^L, and n_k^L is the number of these observations falling into state k. Hence, $\widehat{\pi}_k^L$ is the proportion of observations from state k in node t^L, and $\widehat{\mu}_k^L$ is the mean vector of the continuous outcomes of these observations. The quantities are defined similarly for the right node t^R. Finally, $\widehat{\Sigma}_2$ is an estimate of the covariance matrix of the continuous outcomes, assumed common across the two nodes and the states.

The best split is the one that maximizes (2.1). This splitting criterion is equivalent to the likelihood ratio test introduced in de Leon and Zhu (2008) for differences among several groups with mixed data. This likelihood ratio test, derived from the GLOM for the joint distribution of mixed data, can be viewed as a generalization to the mixed data setting of the classical analysis of variance (ANOVA) for continuous data. In our case, since we have two groups defined by the two nodes, the best split would be the one making the two nodes look the most different according to this test statistic. It is easy to see that this splitting criterion is equivalent to the entropy criterion of CART if used with one categorical outcome and is equivalent to the least-squares criterion of CART if used with one continuous outcome. However, when the outcome vector contains only continuous outcomes, this criterion is different from the one developed by Segal (1992), in which the covariance matrix, obtained from the parent node, can be modeled to impose a specific structure. It is also different from the criterion developed by Ciampi et al. (1991), in which the covariance matrix is not the same in the two children nodes. Otherwise, when the outcome vector contains only categorical outcomes, this criterion amounts to casting all outcomes into a single categorical outcome ending up with the classical entropy criterion.

Focusing on single trees, Dine et al. (2009) adopt the classical strategy, which grows a large tree and then prunes it in order to avoid overfitting, to select the final tree. The single multivariate tree can then be used to gain insight about the relationship between the predictors and the multiple outcomes in a parsimonious way, and it can also be used to predict new observations. To predict the multiple outcomes of a new observation, it is sent down the multivariate tree and the terminal node where it ends up, is noted. For a continuous response, the mean of the response in the terminal node is used as prediction. For a categorical response, the class with the largest probability in the terminal node is used as prediction. More details and examples are given in Dine et al. (2009).

2.2.4   Random forests

A random forest (Breiman, 2001) is an ensemble of trees. Each tree is fitted to a bootstrap sample from the training data set and only a subset of predictors is chosen randomly to find the best split at each node. The trees are not pruned. Many trees are built this way and their predictions are averaged. Let T be the number of trees in a random forest. For a continuous response, the prediction from a random forest is $\widehat{Y} = \sum_{t=1}^{T} \widehat{Y}_t / T$, where $\widehat{Y}_t$ is the prediction from the tth tree. For a categorical response, we compute the estimated probability for each class c, over all trees in the forest, as

$$P\widehat{(Y=c)} \quad = \quad \frac{1}{T} \sum_{t=1}^{T} P_t\widehat{(Y=c)}, \tag{2.2}$$

where $P_t\widehat{(Y=c)}$ is the estimated probability for class c from the tth tree in the forest.

In this chapter, we construct univariate and multivariate forests, where the univariate forests consist of forests built separately for each outcome, and the multivariate forests are built using the multivariate tree method for mixed outcomes from Dine et al. (2009).

2.2.5   Description of predictive models

We next compare the predictive performances of the multivariate random forests for mixed outcomes, separate univariate random forests for each outcome, and the combination of predictions from the multivariate and univariate random forests with three different weighting schemes. We define w_{qU} and w_{qM} as, respectively, the weights assigned to the univariate and multivariate random forest predictions in the final prediction of the qth outcome, where $w_{qU} + w_{qM} = 1$. A different set of weights is allowed for each outcome; however, for simplicity, we drop the subscript q from now on. For a continuous response, the weighted prediction is

$$\widehat{Y} = w_U \widehat{Y}_U + w_M \widehat{Y}_M,$$

where $\widehat{Y}_U$ is the prediction made from the univariate random forest and $\widehat{Y}_M$ is the one made from the multivariate random forest. For a categorical response, the weighted probability estimate for class c is

$$\widehat{P(Y = c)} = w_U \widehat{P_U(Y = c)} + w_M \widehat{P_M(Y = c)},$$

where $\widehat{P_U(Y = c)}$ and $\widehat{P_M(Y = c)}$ are defined as in (2.2). We predict the class as that with the highest weighted probability

$$\widehat{Y} = \underset{c}{\mathrm{argmax}} \; \widehat{P(Y = c)}. \tag{2.3}$$

We consider the following weighting schemes for the predictive performance comparisons.

Benchmark method 1: This method relies solely on the univariate trees by putting $w_U = 1$ and $w_M = 0$. This implies that only the prediction from the univariate random forest is retained in the final prediction.

Benchmark method 2: This method relies solely on the multivariate tree by putting $w_U = 0$ and $w_M = 1$. This implies that only the prediction from the multivariate random forest is retained in the final prediction.

The other three strategies use estimated weights as described next. When building a tree in a forest, observations not selected in the bootstrap sample are called the out-of-bag (OOB) sample. The basic idea is to estimate the performance of the trees with these OOB samples and design the weights around them. The same bootstrap sample is used to grow a multivariate tree and separate univariate trees.

Scheme 1: Proportional tree-by-tree.
For each tree in the forest, we compute on the OOB sample the predictive mean squared error (PMSE) for each continuous outcome and the predictive misclassification rate (PMR) for each categorical outcome. The weight w_M is defined as the proportion of OOB samples where the multivariate trees did better than the univariate trees for a particular outcome and $w_U = 1 - w_M$.

We define the following intermediate quantities for the two remaining weighting schemes. For a continuous response, we compute the average mean squared errors $\overline{\mathrm{PMSE}}_U$ and $\overline{\mathrm{PMSE}}_M$ over all OOB samples for the univariate and multivariate random forests, respectively. More weight will be given to the prediction from the multivariate random forest if $\overline{\mathrm{PMSE}}_U$ is greater than $\overline{\mathrm{PMSE}}_M$ and vice versa. Similarly for a categorical response, we compute the average misclassification rates $\overline{\mathrm{PMR}}_U$ and $\overline{\mathrm{PMR}}_M$ over all OOB samples for the univariate and multivariate random forests, respectively.

Scheme 2: Proportional-overall.
For a continuous response, the weight for the prediction from the multivariate forest is

$$w_M = \frac{\overline{\mathrm{PMSE}}_U}{\overline{\mathrm{PMSE}}_M + \overline{\mathrm{PMSE}}_U}.$$

Consequently, the weight for the prediction from the univariate forest is

$$w_U = \frac{\overline{\mathrm{PMSE}}_M}{\overline{\mathrm{PMSE}}_M + \overline{\mathrm{PMSE}}_U}.$$

Similarly for a categorical response, the weight for the prediction from the multivariate forest is

$$w_M \;=\; \frac{\overline{\text{PMR}}_U}{\overline{\text{PMR}}_M + \overline{\text{PMR}}_U}.$$

Consequently, the weight for the prediction from the univariate forest is

$$w_U \;=\; \frac{\overline{\text{PMR}}_M}{\overline{\text{PMR}}_M + \overline{\text{PMR}}_U}.$$

Scheme 3: All-or-nothing.
This is an all or nothing weighting scheme. For any response (continuous or categorical), the prediction from the random forest, multivariate or univariate, with the lowest $\overline{\text{PMSE}}$ ($\overline{\text{PMR}}$) is given a weight of 1 and the other, a weight of 0 in the final prediction.

2.3   Simulation study

2.3.1   Simulation design

We performed a simulation study to evaluate the performance of the aforementioned prediction methods. Nine different scenarios were used to generate six responses using one multivariate tree or separate independent univariate trees (Table 2.1). Seven predictors, $X_1, \cdot, X_7$, were first generated independently from the uniform distribution $U(0, 10)$. Only some predictors were informative for the responses and the others were used as noise.

Table 2.1 *Generation of responses for simulation study.*

Scenario	Outcomes	Generation
1	6 continuous	1 multivariate tree
2	6 binary	1 multivariate tree
3	3 continuous 3 binary	1 multivariate tree
4	6 continuous	6 univariate trees
5	6 binary	6 univariate trees
6	3 continuous 3 binary	3 univariate trees 3 univariate trees
7	6 continuous	1 multivariate tree 3 univariate trees
8	6 binary	1 multivariate tree 3 univariate trees
9	2 continuous 1 binary 1 continuous 2 binary	1 multivariate tree 3 univariate trees

To generate the responses in Scenarios 1 to 3, each observation is sent down the multivariate tree with five terminal nodes (Figure 2.1), according to the generated values of the predictors. Then, six equicorrelated (i. e., $\rho = 0.3$) node-specific outcomes are generated. In Scenarios 1 and 3, the six multivariate responses are generated from a multivariate normal distribution with node-specific mean vectors given in Table 2.2 and unit variances. In Scenario 3, the last three variables are further dichotomized according to whether they are greater or less than 0 to obtain the three binary responses Y_4, Y_5, Y_6. In Scenario 2, the six multivariate binary equicorrelated (i.e., $\rho = 0.3$) responses in each of the five terminal nodes are generated with the algorithm provided by Oman (2009).

In Scenarios 4 to 6, the six outcomes are generated independently using the six different univariate trees in Figure 2.2. In Scenarios 4 and 6, each continuous outcome is generated using a normal distribution with the node-specific mean given in Table 2.3 and unit variance. In Scenarios 5 and

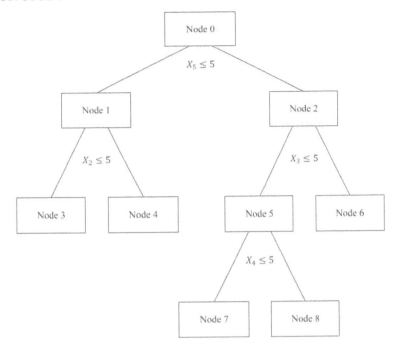

Figure 2.1 *Multivariate tree used in simulation study.*

6, each binary outcome is independently generated according to a Bernoulli distribution with the node-specific probability given in Table 2.3.

In Scenarios 7 to 9, for each set of predictors generated, three of the six outcomes are generated using an equicorrelated (i.e., $\rho = 0.3$) multivariate normal distribution with unit variances (Scenarios 7 and 9) or a multivariate binary distribution (Scenario 8) based on the tree in Figure 2.1 and the node-specific mean or probability vectors given in Table 2.4. The three other outcomes are generated independently with either a normal distribution with unit variance or a Bernoulli distribution with a node-specific mean or probability given in Table 2.4. In Scenario 9, the generated variable Y_4 is further dichotomized according to whether it is greater or less than 0 to obtain a binary response.

For each scenario, 500 training data sets were generated, each containing 500 observations, and one test data set with 10,000 observations. The number of trees in each random forest is set to 100 and the number of predictors chosen randomly in each node is set to 3. For the stopping criteria, the maximum depth allowed is set to 7, the smallest number of observations that a node must have to consider splitting it is set to 20, and the smallest number of observations that a node must have is set to 10. The multivariate tree algorithm with splitting criterion given in (2.1) was implemented using the Ox language (Doornik, 2007). All univariate and multivariate trees were built with it. Hence, for univariate trees with a single categorical (continuous) outcome, this is equivalent to using the entropy (i.e., least-squares) splitting criterion.

2.3.2 Results

We assessed the predictive performance of the five methods by comparing their PMSEs for the continuous outcomes and their PMRs for the categorical outcomes, over the 500 replications. The trees were built with the training data sets and the PMSE and PMR were calculated using the test data sets. We present and discuss the results along two dimensions of interest: *i*) compare the performance of the multivariate random forests to one of the separate univariate forests, and *ii*) investigate whether or not the methods which combine the univariate and multivariate predictions with the proposed

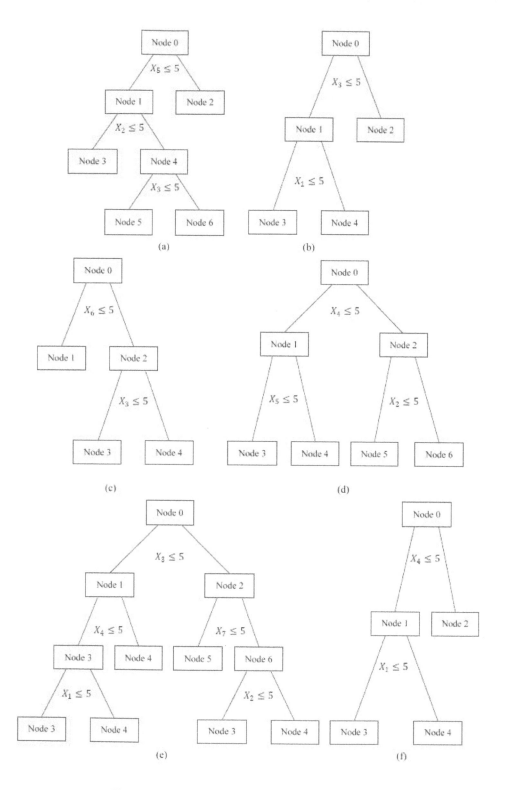

Figure 2.2 *Univariate trees used in the simulation study.*

Table 2.2 *Means of the six outcomes in each terminal node of the multivariate tree for Scenarios 1 to 3.*

Scenario	Outcome	Type	Terminal node of tree in Fig. 2.1				
			3	4	6	7	8
1	Y_1	Continuous	1	1.5	1	0.5	2.5
	Y_2	Continuous	2	2.5	2.4	3.5	3
	Y_3	Continuous	3	4	3	2.5	3.5
	Y_4	Continuous	4	5.5	4.2	4.5	5
	Y_5	Continuous	5.5	6	6	4.5	4
	Y_6	Continuous	7.5	2	11	2.5	9
2	Y_1	Binary	0.3	0.75	0.55	0.75	0.3
	Y_2	Binary	0.25	0.65	0.65	0.65	0.25
	Y_3	Binary	0.3	0.7	0.7	0.7	0.3
	Y_4	Binary	0.15	0.65	0.7	0.65	0.15
	Y_5	Binary	0.25	0.6	0.75	0.6	0.25
	Y_6	Binary	0.25	0.65	0.8	0.65	0.3
3	Y_1	Continuous	1	1.5	1	0.5	2.5
	Y_2	Continuous	2	2.5	2.4	3.5	3
	Y_3	Continuous	3	4	3	2.5	3.5
	Y_4^*	Binary	−0.25	0.67	−0.52	−0.52	0.39
	Y_5^*	Binary	−0.25	0.39	−0.25	−0.25	0.13
	Y_6^*	Binary	−0.52	0.52	−0.67	−0.67	−0.67

*Variables Y_4, Y_5, Y_6, in Scenario 3 were first generated with a multivariate normal distribution with means given in the table and then dichotomized using the cutpoint 0.

estimated weights, can be a good compromise knowing that according to the simulation design, multivariate random forests are expected to have the best performance for Scenarios 1 to 3, and the separate univariate random forests should perform better than the other methods for Scenarios 4 to 6. We expect that the weighted combinations would perform better in Scenarios 7 to 9, where we have a mix of three multivariate and three independent univariate outcomes.

Figures 2.3 to 2.11 display boxplots of the distributions of the predictive performance measures, over the 500 replications, for the six responses (continuous and binary), for each of the nine scenarios. As expected, it can be seen in Figures 2.3 to 2.5 (i.e., Scenarios 1 to 3 with multivariate trees) that the multivariate random forests produced better predictions with smaller variations than the univariate random forests, regardless of the type of the outcome (continuous or binary). It can also be seen that, in general, the three weighting schemes outperformed the separate univariate random forests and they performed almost as well as the multivariate random forests. On the other hand, except for Scenario 5, the separate univariate random forests produced, on average, better predictions (see Figures 2.6 to 2.8) when the six outcomes are generated through separate independent univariate trees (i.e., Scenarios 4 to 6). Once again, the predictive performance of the three weighting combination schemes is relatively close to that of the separate univariate trees. Figures 2.9 to 2.11 present the results of the simulation for Scenarios 7 to 9, respectively, where we generated a mixture of multivariate and separate independent univariate responses. For the outcomes generated with the multivariate tree, the predictions from the multivariate random forests provided generally a better performance compared to the other methods, followed relatively closely by at least one of the three weighting schemes. For the outcomes generated with univariate independent trees, the separate univariate tree predictions often gave a better performance (see Figure 2.9 for outcomes Y_5 and Y_6, Figure 2.10 for outcome Y_6, and Figure 2.11 for outcomes Y_3 and Y_6), but the multivariate tree predictions performed as well (see Figure 2.9 for outcome Y_4, and Figure 2.11 for outcome Y_5) or even surprisingly better in some cases (see Figure 2.10 for outcomes Y_4 and Y_5).

Table 2.3 Means of the six outcomes in each terminal node of the separate univariate trees for Scenarios 4 to 6.

Scenario	Tree	Outcome	Type	Terminal node of trees in Fig. 2.2									
				1	2	3	4	5	6	7	8	9	10
4	Fig. 2.2 a	Y_1	Continuous		10	4							
	Fig. 2.2 b	Y_2	Continuous		15	4	10	2	7				
	Fig. 2.2 c	Y_3	Continuous	5		8	13						
	Fig. 2.2 d	Y_4	Continuous			8	11	16	21				
	Fig. 2.2 e	Y_5	Continuous				12	13		5	7	16	
	Fig. 2.2 f	Y_6	Continuous		13	5	9						20
5	Fig. 2.2 a	Y_1	Binary		0.4	0.2		0.3	0.7				
	Fig. 2.2 b	Y_2	Binary		0.6	0.7	0.25						
	Fig. 2.2 c	Y_3	Binary	0.7		0.3	0.8						
	Fig. 2.2 d	Y_4	Binary			0.35	0.8	0.35	0.65	0.3	0.6	0.3	0.65
	Fig. 2.2 e	Y_5	Binary				0.7	0.7					
	Fig. 2.2 f	Y_6	Binary		0.7	0.65	0.3						
6	Fig. 2.2 a	Y_1	Continuous		10	4							
	Fig. 2.2 b	Y_2	Continuous		15	4	10	2	7				
	Fig. 2.2 c	Y_3	Continuous	5		8	13						
	Fig. 2.2 d	Y_4	Binary			0.35	0.8	0.35	0.65	0.3	0.6	0.3	0.65
	Fig. 2.2 e	Y_5	Binary				0.7	0.7					
	Fig. 2.2 f	Y_6	Binary		0.7	0.65	0.3						

Table 2.4 *Means of the outcomes in each terminal node of the multivariate tree for three of the six outcomes and of the separate independent univariate trees for the other three outcomes in Scenarios 7 to 9.*

Scenario	Tree	Outcome	Type	Terminal node of trees in Figures 2.1 and 2.2									
				1	2	3	4	5	6	7	8	9	10
7	Fig. 2.1	Y_1	Continuous			1	1.5		1	0.5	2.5		
	Fig. 2.1	Y_2	Continuous			2	2.5		2.4	3.5	3		
	Fig. 2.1	Y_3	Continuous			7.5	2		11	2.5	9		
	Fig. 2.2 d	Y_4	Continuous			8	11	16	21				
	Fig. 2.2 e	Y_5	Continuous				12	13		5	7	16	20
	Fig. 2.2 f	Y_6	Continuous		13	5	9						
8	Fig. 2.1	Y_1	Binary			0.25	0.8		0.6	0.75	0.3		
	Fig. 2.1	Y_2	Binary			0.25	0.7		0.7	0.65	0.25		
	Fig. 2.1	Y_3	Binary			0.3	0.7		0.7	0.7	0.3		
	Fig. 2.2 d	Y_4	Binary			0.35	0.8	0.35	0.65				
	Fig. 2.2 e	Y_5	Binary				0.7	0.7		0.3	0.6	0.3	0.65
	Fig. 2.2 f	Y_6	Binary		0.7	0.65	0.3						
9	Fig. 2.1	Y_1	Continuous			1	1.5		1	0.5	2.5		
	Fig. 2.1	Y_2	Continuous			2	2.5		2.4	3.5	3		
	Fig. 2.1	Y_4^*	Binary			−0.25	0.67		−0.39	0.52	−0.52		
	Fig. 2.2 c	Y_3	Continuous	5		8	13						
	Fig. 2.2 e	Y_5	Binary				0.7	0.7		0.3	0.6	0.3	0.65
	Fig. 2.2 f	Y_6	Binary		0.7	0.65	0.3						

*Variable Y_4 in Scenario 9 was first generated with a multivariate normal distribution with Y_1 and Y_2 with the means given in the table and then dichotomized using the cutting value of 0.

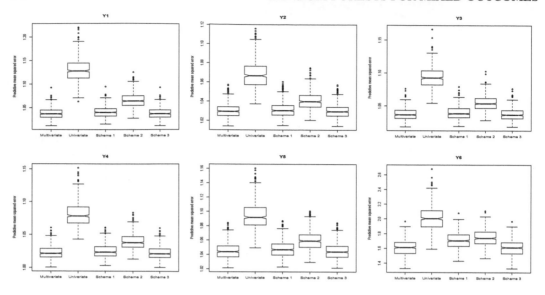

Figure 2.3 *Distributions of PMSE in Scenario 1 for outcomes* $Y_1, \cdots, Y_6$, *generated with 1 multivariate tree.*

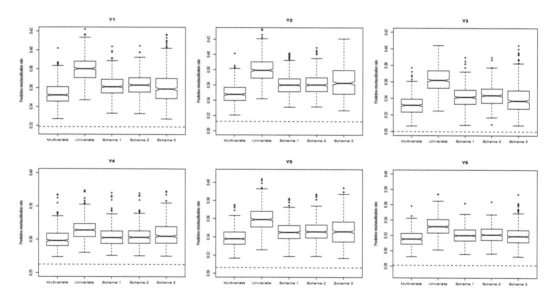

Figure 2.4 *Distributions of PMR in Scenario 2 for outcomes* $Y_1, \cdots, Y_6$, *generated with 1 multivariate tree.*

In order to get a more global view, we also considered the relative performance of each method compared to the best performer for each individual simulation run and outcome. We have 500 runs multiplied by 9 scenarios multiplied by 6 outcomes, giving 27,000 estimations of the PMSE (if the outcome is continuous) or the PMR (if the outcome is binary) for each of the five methods. For a given scenario, run-by-outcome (continuous for example) combination, let PMSE_i be the PMSE of method i with $i \in \{\text{Multivariate,Univariate,Scheme 1,Scheme 2,Scheme 3}\}$. Then, the % increase in PMSE of method i' with respect to the best performer is

$$\frac{\text{PMSE}_{i'} - \min_i\{\text{PMSE}_i\}}{\min_i\{\text{PMSE}_i\}} \times 100.$$

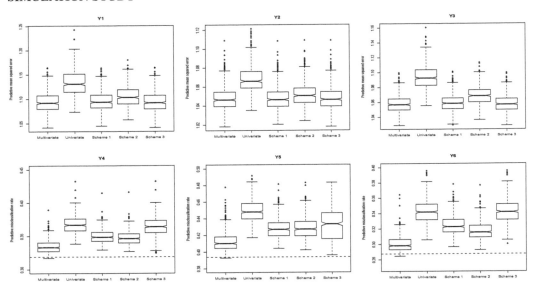

Figure 2.5 *Distributions in Scenario 3 of PMSE for outcomes $Y_1, \cdots, Y_3$, and of PMR for outcomes $Y_4, \cdots, Y_6$, generated with 1 multivariate tree.*

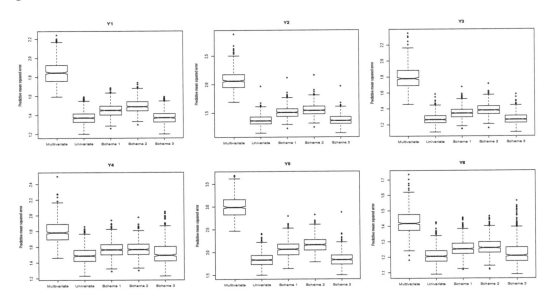

Figure 2.6 *Distributions of PMSE in Scenario 4 for outcomes $Y_1, \cdots, Y_6$, generated with 6 univariate trees.*

For a binary outcome, the same formula is used using the PMR instead. Table 2.5 presents a summary of the results. The first block on the top of the table contains the summary statistics for all $27,000$ runs. The next three blocks contain a breakdown of the results according to the data generating process. More precisely, the second block contains the results for the $9,000$ runs when the data are generated with multivariate trees (Scenarios 1–3). The next block contains the results for the $9,000$ runs when the data are generated with univariate trees (Scenarios 4–6). Finally, the last block contains the results for the $9,000$ runs when the data are generated with a combination of multivariate and univariate trees (Scenarios 7–9).

Looking at the first block, we see that scheme 3 has the best performance in terms of mean and

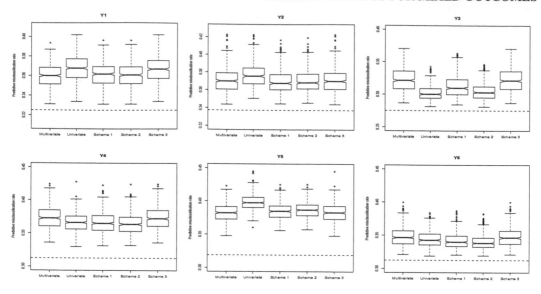

Figure 2.7 *Distributions of PMR in Scenario 5 for outcomes $Y_1, \cdots, Y_6$, generated with 6 univariate trees.*

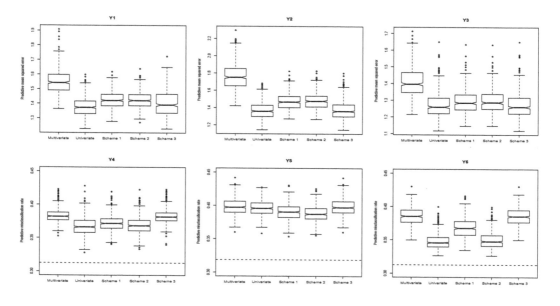

Figure 2.8 *Distributions in Scenario 6 of PMSE for outcomes $Y_1, \cdots, Y_3$, and of PMR for outcomes $Y_4, \cdots, Y_6$, generated with 6 univariate trees.*

median % increase with respect to the best performer, with values of 2.32% and 0.00%. In fact, the median value of 0.00% indicates that scheme 3 has the lowest estimated error for more than half of the 27,000 runs (all scenarios and outcomes combined). Moreover, all three schemes have a better mean % increase compared to the multivariate (i.e., 7.57%) and univariate (i.e., 3.92%) random forests. However, interestingly, the multivariate random forests has the second best median % increase (i.e., 0.35%), but its performance is more variable (i.e., highest standard deviation (SD), interquartile range $Q_3 - Q_1$, and maximum) than other methods. From a worst-case-scenario point of view, scheme 1 is interesting. Indeed, it has the smallest maximum % increase (i.e., 24.18%) and has quite a good mean and median % increase (i.e., 2.67% and 1.25%, respectively).

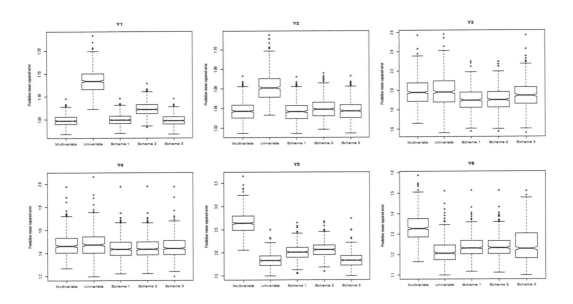

Figure 2.9 *Distributions of PMSE in Scenario 7 for outcomes $Y_1, \cdots, Y_3$, generated with 1 multivariate tree, and outcomes $Y_4, \cdots, Y_6$, generated with 3 univariate trees.*

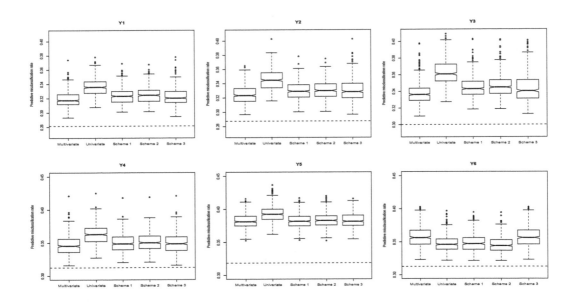

Figure 2.10 *Distributions of PMR in Scenario 8 for outcomes $Y_1, \cdots, Y_3$, generated with 1 multivariate tree, and outcomes $Y_4, \cdots, Y_6$, generated with 3 univariate trees.*

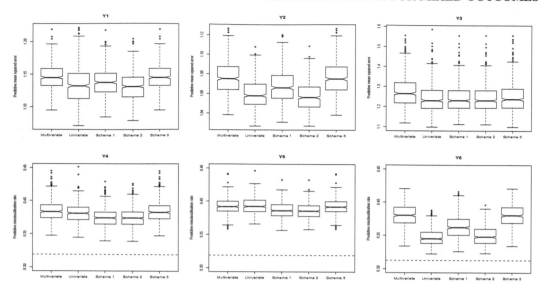

Figure 2.11 *Distributions in Scenario 9 of (1) PMSE for outcomes Y_1, Y_2, and of PMR for outcome Y_4, all generated with 1 multivariate tree; and of (2) PMSE for outcome Y_3 and of PMR for outcomes Y_5, Y_6, all generated with 3 univariate trees.*

Table 2.5 *Relative performance of the five methods.*

| | Method | % increase in PMSE (or PMR) with respect to best performer | | | | | | |
		Mean	Median	Q1	Q3	SD	Minimum	Maximum
Scenarios 1–9 All scenarios	Multivariate	7.57	0.35	0.00	6.98	15.06	0.00	115.49
	Univariate	3.92	2.39	0.00	5.84	5.13	0.00	59.56
	Scheme 1	2.67	1.25	0.16	4.05	3.41	0.00	24.18
	Scheme 2	3.00	1.56	0.35	3.79	4.03	0.00	30.22
	Scheme 3	2.32	0.00	0.00	3.01	4.08	0.00	48.18
Scenarios 1–3 Multivariate trees	Multivariate	0.03	0.00	0.00	0.00	0.20	0.00	6.50
	Univariate	7.68	5.95	3.84	9.53	6.22	0.00	59.56
	Scheme 1	2.10	0.65	0.15	3.42	2.66	0.00	23.89
	Scheme 2	2.69	1.94	1.08	3.58	2.38	0.00	16.91
	Scheme 3	2.42	0.00	0.00	1.52	4.79	0.00	33.75
Scenarios 4–6 Univariate trees	Multivariate	17.57	9.57	2.23	27.42	19.93	0.00	115.49
	Univariate	0.96	0.00	0.00	1.30	1.78	0.00	14.47
	Scheme 1	4.12	3.05	0.62	6.50	4.06	0.00	24.18
	Scheme 2	4.48	2.08	0.26	7.32	5.41	0.00	30.22
	Scheme 3	2.40	0.00	0.00	3.40	3.95	0.00	48.18
Scenarios 7–9 Univariate+ multivariate trees	Multivariate	5.11	1.22	0.00	4.61	10.99	0.00	87.26
	Univariate	3.13	1.87	0.00	5.02	3.67	0.00	33.13
	Scheme 1	1.80	0.71	0.07	2.22	2.84	0.00	19.99
	Scheme 2	1.83	0.66	0.01	2.26	3.19	0.00	24.07
	Scheme 3	2.15	0.55	0.00	2.90	3.39	0.00	33.14

Looking at the second block (i.e., only the runs involving data generated with multivariate trees), it is not surprising to see that the multivariate forests have the best performance; however, scheme 3 also has a median % increase of 0.00%. But recall that this method is the all-or-nothing scheme, where the prediction comes either from the multivariate or the univariate forest, as chosen by the data themselves. Hence, it seems that the scheme selects the right forest most of the time, but when it does not, the error can be high (i.e., it has the largest maximum % increase among the three schemes with 33.75%). In any case, we see that the three schemes are preferable to using univariate forests for these scenarios.

Looking at the third block (i.e., only the runs involving data generated with univariate trees), it is again not surprising to see that the univariate forests have the best performance. Once again, scheme 3 has a median % increase of 0.00%, but has the largest maximum % increase among the three schemes. The three schemes are preferable to using multivariate forests for these scenarios.

Looking at the fourth block, we see that scheme 1 has the lowest mean % increase (i.e., 1.80%), followed closely by scheme 2 (i.e., 1.83%). However, scheme 3 has the lowest median % increase (i.e., 0.55%). Scheme 1 also has the least variable performance with the smallest SD, interquartile range, and maximum. This time, the three schemes are preferable to using either multivariate or univariate forests for these scenarios.

These results show that combining both multivariate and univariate random forests provides satisfactory predictions across different data generating processes and for different types of outcomes. Moreover, schemes 1 and 3 seem to have a slight edge over scheme 2 in terms of global performance.

2.4 Discussion

While the parsimony of using a single multivariate tree for exploration is clear, it is still only one model. This is in contrast with, for example, multivariate regression models, where each outcome has its own equation. Similarly, a forest of multivariate trees is still built with single multivariate trees. Thus, it was not clear if predicting multiple outcomes from a single forest of multivariate trees would be worthwhile. One of the main conclusions of this chapter is that multivariate forests can indeed be useful for prediction.

We used a simulation study to investigate the predictive performances of using only the prediction from the multivariate forest, only the one from the univariate forest, and of three different weighting schemes to combine both predictions. Results have shown that a random forest of multivariate trees is generally preferable to separate random forests of univariate trees when the outcomes are generated from a multivariate tree. Similarly, random forests of separate univariate trees are generally preferable to a random forest of multivariate trees when the outcomes are generated from separate independent univariate trees. However, there were some instances where the predictions from the multivariate random forests were preferable even when the outcomes are generated from univariate trees.

But it would be hard to recommend using either the multivariate or the univariate approach in practice because we ignored if the data generating processes would be favorable to them. We would rather recommend using one of the proposed combination methods, more particularly scheme 1, where the weights are defined as the proportion of OOB samples, where the multivariate (univariate) trees did better than the univariate (multivariate) trees for a particular outcome; or scheme 3, where the prediction from the random forest, multivariate or univariate, with the lowest mean squared error or misclassification rate, is given a weight of 1 and the other is given a weight of 0.

The advantage of these combination methods is that they let the data determine automatically which weight to give to the multivariate and univariate forests' predictions. Moreover, the weights may be different across outcomes; thus, the prediction for one outcome may give more weight to the multivariate forest while the one for another outcome may give more weight to the univariate forest. The combination methods are thus very adaptive. The simulation results have shown that regardless of the type of outcome and data generating process, these automatic weighting strategies generally come close to the best performing approach.

The goal of this chapter was to introduce and show the potential of combining predictions from univariate and multivariate forests. Even though they are rather ad hoc, the proposed weighting strategies performed reasonably well. But it could be worthwhile to perform a more complete investigation of this aspect. Simulations with data generated under more complex data generating processes (in particular non-normal data) and the use of many real data sets would help clarify the benefits of the weighting strategies, and perhaps provide more precise guidelines for practitioners (e.g., in which precise circumstances is one strategy better than the others?).

Another aspect that could be worthy of future work concerns the situation where outcomes are missing. Indeed, the case of missing predictors with trees has been investigated before, and many methods to handle them are available (surrogate splits and imputation for instance). But this is not the case when the outcomes are missing, and this is where multivariate trees could be useful because the dependence structure between the outcomes can be exploited.

In this chapter, we used the same number of randomly selected predictor at each node, which is 3, to build all trees. However, the best number could be different for univariate forests and for the multivariate forest. It could depend on the type of outcome, for instance. Investigating these aspects, in general, and especially in the large p (i.e., number of predictors) small n case might be worthwhile.

Finally, this chapter dealt with mixed outcomes without censoring. Our general idea of combining multivariate and univariate forests could also be applied to survival trees with censored outcomes, and this might be interesting to pursue. The literature on single univariate survival trees and forests is vast, and a few proposals have been made to build trees with multiple survival outcomes. A recent survey about survival trees is given in Bou−Hamad *et al.* (2011) and could serve as the starting point for such an investigation.

Chapter 3

Joint tests for mixed traits in genetic association studies

Minjung Kwak, Gang Zheng, and Colin O. Wu

3.1 Introduction

In population-based genetic association studies, one tests association between a genetic marker and a phenotype (trait). A common genetic marker is a single-nucleotide polymorphism (SNP), which consists of two alleles. Each person has one of the three genotypes formed by the two alleles. The two most common types of phenotype in genetic association studies are binary (discrete) and quantitative (continuous) traits. Case-control is a typical binary trait, and blood pressure is an example of a quantitative trait. The true disease loci, or functional loci, contributing to the variations of a trait are often unknown. A large number of SNPs, about half to one million, are genotyped in genome-wide association studies (GWASs) for many common and complex diseases. When a SNP is in linkage disequilibrium with a disease locus, the association between the SNP and the trait can be tested. When a SNP has strong association with the trait, it may indicate that the disease locus is nearby. Follow-up studies, including regional haplotype analysis, meta-analysis, imputation and resequencing, may provide more evidence about the underlying disease loci.

In genetic association studies, a single trait (binary or quantitative) is often tested with a single SNP or multiple SNPs (haplotype analysis or multi-marker analysis) with or without covariates, gene-environment, and gene-gene interactions. In this chapter, we only consider single-marker analysis.

Denote the alleles of a SNP to be tested as A and B and the three genotypes of the SNP as $G_0 = AA$, $G_1 = AB$ and $G_2 = BB$. Without loss of generality, let B be the risk allele when the association is present, that is, the more copies of allele B in the genotype, the greater probability of developing a disease or having higher trait value. The risk allele, however, may not be known in practice. The population frequency of allele B is denoted as $p = P(B)$, and the frequency of allele A is $q = P(A) = 1 - p$. The frequency of an allele is called the minor allele frequency (MAF) if it is no more than 0.5. An allele with MAF may also be the risk one for some traits. The genotype frequencies are denoted as $P(G_i)$, $i = 0, 1, 2$. Let W be Wright's coefficient of inbreeding (Weir, 1996). Then, from population genetics, $P(G_0) = q^2 + Wpq$, $P(G_1) = 2(1 - W)pq$, and $P(G_2) = p^2 + Wpq$. When $W = 0$, $P(G_0) = q^2$, $P(G_1) = 2pq$, and $P(G_2) = p^2$ are often referred to as Hardy–Weinberg proportions or Hardy–Weinberg equilibrium (HWE).

Penetrance is often used for a binary trait, which is defined as $f_i = P(\text{disease}|G_i)$, $i = 0, 1, 2$. Under the alternative hypothesis H_1 of association and when allele B is the risk one, we often assume $f_2 \geq f_1 \geq f_0$ and $f_2 > f_0$, while under the null hypothesis H_0, we have $f_0 = f_1 = f_2 = P(\text{disease})$, which is also the disease prevalence. A genetic model for the binary trait is called recessive, additive, or dominant if $f_0 = f_1$, $f_1 = (f_0 + f_2)/2$ or $f_1 = f_2$, respectively. The true genetic model (also known as the mode of inheritance) is rarely known. In that case, an additive model is often used. Denote the genotype relative risks (GRRs) as $\lambda_j = f_j/f_0$, for $j = 1, 2$. Therefore, the recessive, additive, and dominant models correspond to $\lambda_1 = 1$, $\lambda_1 = (1 + \lambda_2)/2$, and $\lambda_1 = \lambda_2$, respectively. To integrate the three genetic models into one general model, we write $\lambda_1 = 1 - x + x\lambda_2$ for $x \in [0, 1]$, with $x = 0, 1/2, 1$, for the three common genetic models.

For a binary trait, the case-control data are often summarized in genotype counts in a 2×3 contingency table, if there is no other covariate to adjust for. Suppose r cases and s controls are collected from the case and control populations respectively, and each individual is genotyped. Denote their genotype counts for (G_0, G_1, G_2) as (r_0, r_1, r_2) among r cases, and (s_0, s_1, s_2) among s controls. Let $n_i = r_i + s_i$, $i = 0, 1, 2$, and $n = r + s$.

Notation for a quantitative trait is different. With the same SNP as discussed for the binary trait, denote the trait as Y. Conditional on G, a typical model for Y is given by

$$Y = \mu + c_1(G) + e, \tag{3.1}$$

where μ is the overall mean of the trait without the genetic effect, $c_1(G)$ is the coding of the genetic effect for a continuous trait, and e is a random environmental factor or a non-genetic factor. The value of $c_1(G)$ is given by $-a$, d, or a, when $G = G_0$, G_1, or G_2, respectively, where $a > 0$. We assume G and e are independent, and $E(e) = 0$ and $V(e) = \sigma_e^2$. It can be shown that $V(G) = \sigma_G^2 = 2pq\{a - (p - q)d\}^2 + 4p^2q^2d^2$. It can be seen that $\sigma_G^2 = 0$ if and only if $a = d = 0$. Note that the total variance of the trait Y is given by $V(Y) = \sigma_G^2 + \sigma_e^2$, part of which, explained by the genetic marker G is called heritability, given by $H^2 = \sigma_G^2/(\sigma_G^2 + \sigma_e^2)$. The null hypothesis of no association can be stated as $H_0 : H^2 = 0$ or $a = d = 0$. Under H_1, the recessive, additive, and dominant models are summarized as $d = -(1 - x)a + xa$ with $x = 0, 1/2, 1$, respectively. Here, we assume the true genetic model is known. Mostly, we only focus on the additive model. Discussion of the situations where the underlying genetic model is unknown is given at the end.

The data for a quantitative trait are denoted as (Y, G). Thus, for the jth individual, $j = 1, \cdots, n$, the data are given by (y_j, g_j), where g_j takes one of the three genotypes. If data consist of both binary and quantitative traits, we may also denote them as (y_{ijk}, g_{ijk}) for $i = 0, 1, 2$, for the three genotypes, $j = 1, \cdots, r$, when $k = 1$ (cases), and $j = 1, \cdots, s$, when $k = 0$ (controls).

The rest of this chapter is organized as follows. In Section 3.2, we first review common test statistics employed for single-marker association studies for a binary or quantitative trait. Then, in Section 3.3, we discuss how to analyze association with mixed types of traits (joint association with both binary and quantitative traits). In that section, for testing the genetic effect, we focus on two approaches. One is the simple Fisher's combination of the p-values of associations with the binary and quantitative traits, respectively, which was considered by Xing and Xing (2009). We investigate how the two tests are correlated and how it then affects the size of Fisher's combination test. The second approach is focused on outcome-dependent sampling, that is, quantitative traits are only measured on a subset of individuals while binary traits are observed among all individuals. We demonstrate how to do a joint analysis of mixed types of data in this case. Zheng $et\ al.$ (2012a) considered a similar testing problem under the outcome-dependent sampling. But we focus on a likelihood approach and study both testing and estimation problems. We report simulation results to provide insight into the properties of several test statistics and estimations that we discuss in this chapter. An application to real data is reported in Section 3.4. Discussion is given in Section 3.5. This chapter can serve as background knowledge and provide motivation for further study of joint associations of mixed types of traits to help identify novel markers associated with complex diseases.

3.2   Analysis of binary or quantitative traits

If we only test an association between a SNP and a binary trait (case-control data), logistic regression is a common approach although data are collected retrospectively (Prentice and Pyke, 1979). Many commonly used test statistics are derived from the logistic regression model, in which the SNP is entered as a covariate, coded based on the genetic model. The logistic regression model for case-control genetic association studies is given below. Given genotype G,

$$P(\text{case}|G) = \frac{\exp\{\alpha + \beta c_2(G)\}}{1 + \exp\{\alpha + \beta c_2(G)\}},$$

$$P(\text{control}|G) = 1 - P(\text{case}|G),$$

where $c_2(G)$ is the coding of the genetic effect of a binary trait given by $c_2(G) = 0$, x, and 1, for $G = G_0$, G_1, and G_2, respectively, where x is the same as before and depends on the underlying genetic model. In the above model, α is an intercept and β is the log odds ratio for the genetic effect. Then the likelihood function for r cases with genotype counts (r_0, r_1, r_2) and s controls with genotype counts (s_0, s_1, s_2) can be written as

$$\begin{aligned}
L_1 &= \frac{\exp\{\alpha r + \beta \sum_{i=1}^{r} c_2(G_i)\}}{\prod_{j=1}^{n} [1 + \exp\{\alpha + \beta c_2(G_j)\}]} \\
&= \frac{\exp\{\alpha r + \beta(x r_1 + r_2)\}}{\{1 + \exp(\alpha)\}^{n_0} \{1 + \exp(\alpha + \beta x)\}^{n_1} \{1 + \exp(\alpha + \beta)\}^{n_2}}.
\end{aligned} \tag{3.2}$$

Given x, the score test for testing $H_0 : \beta = 0$ is also known as the trend test, which can be written as (Freidlin et al., 2002; Sasieni, 1997)

$$Z(x) = \frac{\sum_{j=0}^{2} x_j \{(1-\phi) r_j - \phi s_j\}}{\sqrt{n\phi(1-\phi) \left\{ \frac{1}{n} \sum_{j=0}^{2} x_j^2 n_j - \left(\frac{1}{n} \sum_{j=0}^{2} x_j n_j \right)^2 \right\}}},$$

where $(x_0, x_1, x_2) = (0, x, 1)$ and ϕ is the proportion of cases. Under H_0, for a given x, $Z^2(x) \overset{a}{\sim} \chi_1^2$, where χ_m^2 is the chi-squared distribution with m degree of freedom and "$\overset{a}{\sim}$" is used for an asymptotic distribution.

The likelihood L_1 and the score test $Z(x)$ depend on the underlying genetic model. Alternatively, one can consider the following likelihood L_2, which is independent of the genetic model,

$$L_2 = \frac{\exp\{\alpha r + \beta_2 r_1 + (\beta_1 + \beta_2) r_2\}}{\{1 + \exp(\alpha)\}^{n_0} \{1 + \exp(\alpha + \beta_2 x)\}^{n_1} \{1 + \exp(\alpha + \beta_1 + \beta_2)\}^{n_2}}, \tag{3.3}$$

in which the SNP is coded with two dummy variables. The score test T for testing $H_0 : \beta_1 = \beta_2 = 0$ is equivalent to Pearson's chi-squared test for association for the 2×3 contingency table, which can be written as (Zheng et al., 2006)

$$T = \frac{1}{1 - \rho^2} \{Z^2(0) + Z^2(1) - 2\rho Z(0) Z(1)\}.$$

where $\rho = [n_0 n_2 / \{(n_1 + n_2)(n_0 + n_1)\}]^{1/2}$. Under H_0, $T \overset{a}{\sim} \chi_2^2$.

Note that both $Z(x)$ and T are trait-based analyses in that the analyses compare genotype distributions between cases and controls (conditional on the traits). To adjust for other covariates, they can be added to the models (3.2) and (3.3). The trend test and Pearson's test are the two most commonly used statistics for association of a binary trait. Other tests, especially those robust trait-based tests, are also studied (e.g., Zheng et al., 2009; Freidlin et al., 2002). They are less dependent on the genetic model such as Pearson's test T but are more powerful. In addition to the testing, the maximum likelihood estimates (MLEs) and their confidence intervals for log odds ratios β or (β_1, β_2) can be obtained routinely. We focus on $Z(x)$ with $x = 1/2$ in the following.

Typical methods for the analysis of quantitative trait association are based on a linear regression model, where the SNP is also entered as a covariate coded similarly to the analysis of case-control association. The linear regression model is given in (3.1). Hence, the likelihood for quantitative trait data analogous to (3.2) is given by

$$
\begin{aligned}
\widetilde{L}_1 &= \prod_{j=1}^{n} h\left(\frac{y_j - \mu - c_1(g_j)}{\sigma_e}\right) \\
&= \prod_{i=0}^{2}\left\{\prod_{j=1}^{r_i} h\left(\frac{y_{ij1} - \mu - c_1(g_{ij1})}{\sigma_e}\right)\prod_{j=1}^{s_i} h\left(\frac{y_{ij0} - \mu - c_1(g_{ij0})}{\sigma_e}\right)\right\},
\end{aligned}
\tag{3.4}
$$

where μ and σ_e are given in (3.1), y_i and g_i are the trait and genotype of the ith individual, $c_1(g_{ijk})$ can be written as $c_1(G_0) = -a$, $c_1(G_1) = a(2x - 1)$, and $c_1(G_2) = a$, for $x = 0, 1/2, 1$, and $h(\cdot)$ is the density for the trait, which is often a normal density (or after some transformations). The null hypothesis is $H_0 : a = 0$, and μ and σ_e are the nuisance parameters. Applying linear model theory, an F-test, denoted as F_1, given below can be obtained from (3.4) to test $H_0 : a = 0$,

$$
F_1(x) = \frac{(n-2)\left\{\sum_j (x_j - \bar{x})(Y_j - \bar{Y})\right\}^2}{\sum_j (x_j - \bar{x})^2 \sum_j (Y_j - \bar{Y})^2 - \left\{\sum_j (x_j - \bar{x})(Y_j - \bar{Y})\right\}^2},
$$

which has a χ_1^2-distribution under H_0, when the sample size n is large enough.

In deriving the F-test $F_1(x)$, we assume the quantitative trait Y is a primary outcome. In some situations, however, Y is regarded as a secondary trait if the primary outcome is case-control data. In this case, the distribution for the observed Y may be different from the distribution for Y in the general population due to the different sampling rates of cases and controls. This might result in using an incorrect asymptotic distribution for $F_1(x)$, unless the quantitative and binary traits are independent or the marker of interest is not associated with the binary trait. For the analysis of secondary outcomes, refer to Lin and Zeng (2009) and Monsees et al. (2009).

The likelihood for a quantitative trait, analogous to (3.3), is similar to $\widetilde{L}_1$, except that the genotypes are coded by two parameters a and d. That is, the null hypothesis is that the mean traits corresponding to the three genotypes are the same. Hence, an F-test can be obtained, which is equivalent to applying analysis of variance (ANOVA) by testing equal means of three genotype groups, which follows χ_2^2 under H_0, when n is large enough. Note that the above test statistics for quantitative trait association are marker-based analyses. In the following, we focus on $F_1(x)$ for $x = 1/2$.

3.3   Joint analysis of mixed traits

We have discussed association of a SNP with a single trait, either binary or quantitative. When a SNP is associated with both binary and quantitative traits, testing association of a single trait may not be powerful. However, modeling association jointly with both traits is not trivial due to the mixed types of data. Methods for analyzing mixed types of data have been developed since the 1950s. Tate (1955, 1954) study the correlation between discrete and continuous data, which is further investigated in Bedrick et al. (2000), Cox (1974), and Hannan and Tate (1965). We focus on genetic association in this chapter, but the analysis of mixed types of data has other applications (Poon and Lee, 1987; Olsson et al., 1982). The model that we describe here is a special case of the so-called conditional grouped continuous models (de Leon, 2005; Hannan and Tate, 1965); see Chapter 1 for a review.

In this chapter, we describe a simple model for the mixed types of traits, which is used in our simulation studies. Similar models have been discussed in Zheng et al. (2012a), Cox and Wermuth (1992), and Hannan and Tate (1965).

We assume there is another quantitative trait, denoted as Y_0, which determines the binary outcome. An individual is a case if $Y_0 > c$ and a control otherwise, where both Y_0 and the threshold c are latent. Also denote the observed quantitative trait as Y_1 (Y was used before). Moreover, we assume Y_0 and Y_1 jointly follow a bivariate normal distribution with respective means $\mu_0 + c_1(g_0)$

and $\mu_1 + c_1(g_1)$, where μ_i is the overall mean without genetic effect and $c_1(g_i)$ is the genetic effect, $i = 0, 1$, and $c_1(g_i) = -a_i, d_i$ and a_i, depending on the three genotypes given before. The correlation of the traits Y_0 and Y_1 is denoted as ρ_{01}. This model also includes the single marker association model. For example, if there is only association with the binary trait, we can set $a_0 > 0$ and $a_1 = 0$. Our null hypothesis is $H_0 : a_0 = a_1 = 0$, i.e., there is no association with either of the traits. Although the trend test, Pearson's test, and the F-test are derived to test a single trait, they are still valid to test $H_0 : a_0 = a_1 = 0$ (Zheng et al., 2012a).

Fisher's combination of p-values is a popular and simple approach to combine results from independent studies. It can be applied to combine p-values of the trend test $Z(x)$ and the F-test $F_1(x)$ as an analysis of association with both traits. Denote the p-values of the trend test and the F-test as p_Z and p_F, respectively. Then the Fisher's combination test, denoted as J, can be written as

$$J = -2\log p_Z - 2\log p_F \overset{a}{\sim} \chi_4^2, \text{ under } H_0.$$

However, to use the asymptotic distribution χ_4^2 for J, we need to examine whether or not the two p-values are independent under the null hypothesis H_0 that there is no association between the SNP and either of the traits. When the two p-values are correlated, J would not have the correct size.

Intuitively, when the two traits are independent (i.e., $\rho_{01} = 0$), the two tests are asymptotically independent under H_0. When $\rho_{01} > 0$, however, the two tests may be correlated. We conducted a simulation study to examine how ρ_{01} affects the Type I error of J. Genotypes were generated using genotype frequencies given MAF and W, and bivariate normal data were generated under H_0 with $a_0 = a_1 = 0$ (so $d_0 = d_1 = 0$). We also assumed $\mu_0 = \mu_1 = 0$ in the simulation. Then case-control data were obtained using the 80th percentile of $N(0, 1)$ as the threshold. The results presented here were based on equal number of cases and controls ($r = s = 500$) with 10,000 replicates. For comparison, we report Type I error rates of $Z(1/2)$, T, $F_1(1/2)$, and J, for the nominal levels 0.001, 0.01, and 0.05. The results reported in Table 3.1, with MAF $= 0.30$, show that the Fisher's combination test J has inflated Type I error when the two traits are correlated. The size of J does not depend on whether or not HWE holds. Although not reported, the size of J does not depend on MAF either.

The results presented in Table 3.1 assume asymptotic theory can be applied, that is, the total sample size n is large enough and the proportion of cases r/n is not too small or too large. Under extreme situations, the size of test J may deviate from the nominal level even though $\rho_{01} = 0$. For example, our simulations also show that, when $r = 500$ and $s = 50$, Type I error rate for J was 0.062 when $\rho_{01} = 0$. Overall, our empirical study shows that when $\rho_{01} = 0$, J has the correct size asymptotically to test joint association with mixed types of data. Hence, we also compare empirical power of the above tests with $a_0 = a_1 = 0.1$ and 1,000 replicates when $\rho_{01} = 0$ and HWE holds in the population. The results are reported in Table 3.2, which confirms that the joint analysis J is more powerful when the SNP is associated with both traits.

In practice, since we do not know if two traits are correlated, appropriate corrections for Type I error is needed in order to apply J. A simple parametric bootstrap correction is described as follows. The idea is to break down the correlation between the two traits and to simulate genotypes under H_0. For each individual with an observed trait, we discard his observed genotype and generate a new genotype for him under H_0 from the multinomial distribution with sample size 1, $Mul(1; \widehat{P}_0, \widehat{P}_1, \widehat{P}_2)$, where the probabilities of the three outcomes are given by $\widehat{P}_i = n_i/n$, $i = 0, 1, 2$. When genotypes are generated for all individuals, a new data set is formed, from which J can be calculated. After repeating M times, we obtain M J statistics, which can be regarded as random samples drawn from the distribution of J under H_0, from which we can determine critical values and p-values of J.

As an example, we simulated a data set of 500 cases and 500 controls, all having quantitative traits, with $a_0 = a_1 = 0.1$, MAF $= 0.3$ and $\rho_{01} = 0.95$, under the additive model (see "http://www.statisticalsource.com/source/statsource.htm"). The genotype counts are $(r_0, r_1, r_2) = (221, 235, 44)$ and $(s_0, s_1, s_2) = (255, 211, 34)$. The trend test is $Z^2(1/2) = 4.893$, with p-value 0.027, and the F-test is $F_1(1/2) = 9.763$, with p-value 0.002. Both p-values were obtained from

Table 3.1 *Type I error rates of trend test $Z(x)$, Pearson's test T, F-test $F_1(x)$ and Fisher's combination test J, with MAF $= 0.3$ and $r = s = 500$.*

Level	W	ρ_{01}	$Z(1/2)$	T	$F_1(1/2)$	J
0.001	0	0	0.0010	0.0011	0.0011	0.0008
0.010	0	0	0.0106	0.0110	0.0071	0.0088
0.050	0	0	0.0491	0.0533	0.0429	0.0474
0.001	0.05	0	0.0012	0.0014	0.0007	0.0012
0.010	0.05	0	0.0105	0.0107	0.0090	0.0092
0.050	0.05	0	0.0501	0.0480	0.0432	0.0464
0.001	0	0.5	0.0004	0.0007	0.0008	0.0022
0.010	0	0.5	0.0102	0.0107	0.0078	0.0163
0.050	0	0.5	0.0496	0.0496	0.0499	0.0572
0.001	0.05	0.5	0.0012	0.0009	0.0010	0.0023
0.010	0.05	0.5	0.0097	0.0102	0.0093	0.0156
0.050	0.05	0.5	0.0499	0.0505	0.0483	0.0608
0.001	0	0.95	0.0010	0.0006	0.0010	0.0069
0.010	0	0.95	0.0110	0.0100	0.0099	0.0269
0.050	0	0.95	0.0522	0.0497	0.0467	0.0784
0.001	0.05	0.95	0.0012	0.0015	0.0008	0.0067
0.010	0.05	0.95	0.0096	0.0116	0.0119	0.0294
0.050	0.05	0.95	0.0515	0.0529	0.0530	0.0812

Table 3.2 *Power of trend test $Z(1/2)$, Pearson's test, F-test $F_1(1/2)$ and Fisher's combination test J, with $\rho_{01} = 0$ and $r = s = 500$, under the recessive (REC), additive (ADD), and dominant (DOM) models. Significance level is 0.05.*

Model	MAF	$Z(1/2)$	T	$F_1(1/2)$	J
	0.15	0.064	0.099	0.085	0.088
REC	0.30	0.209	0.300	0.282	0.340
	0.45	0.434	0.517	0.569	0.707
	0.15	0.311	0.223	0.364	0.486
ADD	0.30	0.444	0.365	0.535	0.701
	0.45	0.495	0.387	0.604	0.767
	0.15	0.690	0.608	0.782	0.925
DOM	0.30	0.716	0.696	0.809	0.935
	0.45	0.519	0.599	0.658	0.814

χ_1^2. The Fisher's combination test is $J = 19.89$, which has p-value 0.000525, based on χ_4^2. To adjust for its size, we simulated genotype from the multinomial distribution for each individual with a given quantitative trait. The simulation was repeated 10,000 times to estimate the adjusted p-value, which is 0.0048, a p-value that is more reasonable after combining strength of p-values 0.027 and 0.002.

In some situations, however, not all quantitative traits are observed. One example considered in Zheng *et al.* (2012a) is data sharing or outcome-dependent sampling, in which quantitative traits are only obtained for some subgroup. For example, Genetic Analysis Workshop 16 (GAW16) contains data of a GWAS from North American Rheumatoid Arthritis Consortium (Amos *et al.*, 2009). The data consist of cases of rheumatoid arthritis (RA) and controls (RA free), where controls were sampled from the New York Cancer Project. A quantitative trait, anti-cyclic citrullinated peptide

Table 3.3 *Type I error rates of F-test $\widetilde{F}_1(1/2)$ using cases only and Fisher's combination test $\widetilde{J}$ with $W = 0$, $r = s = 500$, MAF = 0.3, and 10,000 replicates.*

Level	ρ_{01}	$\widetilde{F}_1(1/2)$	$\widetilde{J}$
0.001	0	0.0006	0.0003
0.01	0	0.0084	0.0100
0.05	0	0.0477	0.0503
0.001	0.5	0.0008	0.0010
0.01	0.5	0.0108	0.0092
0.05	0.5	0.0500	0.0497
0.001	0.95	0.0010	0.0016
0.01	0.95	0.0095	0.0104
0.05	0.95	0.0481	0.0488

(anti-CCP), was only measured among cases, but not among controls. It is true that higher values of anti-CCP are positively associated with RA.

In the case of GAW16, when testing association with the binary trait (RA vs. RA free), the trend test $Z(x)$ can be applied as usual. When testing association with anti-CCP, only cases can be used when applying the F-test, which is denoted as $\widetilde{F}_1(x)$. When the number of cases is large enough, $\widetilde{F}_1(x) \overset{a}{\sim} \chi_1^2$, for a given x under H_0. Using $\widetilde{F}_1(x)$ is less powerful than using $F_1(x)$ due to using partial data in the analysis. A similar version of joint analysis by Fisher's combination of the p-values of $Z(1/2)$ and $\widetilde{F}_1(1/2)$ was considered by Xing and Xing (2009). Denote this Fisher's combination test as $\widetilde{J}$. Unlike J, $\widetilde{J}$ has the correct size because $Z(1/2)$ and $\widetilde{F}_1(1/2)$ are asymptotically independent. Table 3.3 reports some results of a simulation study of Type I error rates of $\widetilde{J}$ and $\widetilde{F}_1(1/2)$. Overall, $\widetilde{J}$ has the correct size for a given nominal level.

Note that the analyses of a binary trait and a quantitative trait are related in the sense that if all observed quantitative trait values of the cases are replaced by a single value and of the controls by another single value, then the F-test becomes the squared trend test. An alternative joint analysis of mixed types of data proposed by Zheng *et al.* (2012a) is to replace unobserved quantitative traits of controls with a single value $Y_{ij0} = y^*$, and apply the usual F-test and its asymptotic distribution as if this single value were observed for controls. Their F-test is referred to as a modified F-test here. Their simulation results show that Type I error rates of their modified F-test are not affected by varying the value of y^*. But the power of the modified F-test may change as y^* varies. In the following, we give another view of their approach using likelihoods and demonstrate that while testing null hypothesis of association is fine, the estimation of the genetic effect may be biased by treating imputed single trait value y^* as if it were observed.

If all trait values of controls were observed, the likelihood of the data would be $\widetilde{L}_1$, given in (3.4). Denote $\widetilde{L}_1 = \widetilde{L}_{11}\widetilde{L}_{10}$, where $\widetilde{L}_{11}$ and $\widetilde{L}_{10}$ are the partial likelihoods for cases and controls, respectively. Since we only observe the trait values of cases, the likelihood is

$$\widetilde{L}_{11} = \prod_{i=0}^{2}\prod_{j=1}^{r_i} h\left(\frac{y_{ij1} - \mu - c_1(g_{ij1})}{\sigma_e}\right),$$

from which the F-test $\widetilde{F}_1(x)$ is derived for testing no genetic effect for a given genetic model. Using $\widetilde{L}_{11}$ for inference, the resulting test statistics are less powerful due to using the partial data for the analysis. There are multiple ways to add contribution of controls. The approach considered by Zheng *et al.* (2012a) is described as follows. In contrast to $\widetilde{L}_{10}$, which is for continuous trait values of controls, the likelihood for controls is the same as $\widetilde{L}_{10}$, except that all latent values Y_{ij0} are replaced by a single value y^*. That is, the partial likelihood for controls becomes

$$\widetilde{L}_{10}(y^*) = \prod_{i=0}^{2} \prod_{j=1}^{s_i} h\left(\frac{y^* - \mu - c_1(g_{ij0})}{\sigma_e}\right). \tag{3.5}$$

Then, the "full" likelihood becomes

$$\widetilde{L}_1(y^*) = \prod_{i=0}^{2} \left\{ \prod_{j=1}^{r_i} h\left(\frac{y_{ij1} - \mu - c_1(g_{ij1})}{\sigma_e}\right) \prod_{j=1}^{s_i} h\left(\frac{y^* - \mu - c_1(g_{ij0})}{\sigma_e}\right) \right\}. \tag{3.6}$$

It is interesting to compare $\widetilde{L}_1$, $\widetilde{L}_{11}$ and $\widetilde{L}_1(y^*)$. $\widetilde{L}_{11}$ is just the partial likelihood of $\widetilde{L}_1$. However, $\widetilde{L}_1(y^*)$ borrows information of genotype distributions of controls through $\widetilde{L}_{10}(y^*)$ as compared to $\widetilde{L}_{11}$. But it is also different from $\widetilde{L}_1$ in that the variability among latent controls' traits vanishes. If the variability among cases' traits also vanishes by replacing the observed values Y_{ij1} by another single value, the resulting statistic is equivalent to the trend test. Therefore, we argue $\widetilde{L}_1(y^*)$ is a likelihood mixing both binary and quantitative traits. In this sense, it is expected that when there is no case-control association, tests based on $\widetilde{L}_{11}$ would be more powerful than those based on $\widetilde{L}_1(y^*)$ regardless of the choice of y^*. On the other hand, if there is a strong case-control association, we expect tests based on $\widetilde{L}_1(y^*)$ would be more powerful than tests using $\widetilde{L}_{11}$. Under some situations, we may even see using $\widetilde{L}_1(y^*)$ as more powerful than using $\widetilde{L}_1$ because $\widetilde{L}_1(y^*)$ is a joint analysis of both traits, while $\widetilde{L}_1$ is an analysis of quantitative trait only. We regard analyses based on $\widetilde{L}_1(y^*)$ as inference for a mixed type of quantitative-control data. It takes into account the variability among Y_{ij1} (quantitative part) due to the genetic effect as well as the genotype distribution between y^* and the mean of Y_{ij1} (binary part in which the mean of case trait values is treated as *another single value* for cases). This insight also suggests what y^* one should choose. In order to mimic case-control data, we need to choose y^* away from the mean of the observed trait values of cases. In Zheng *et al.* (2012a), they chose y^* as the minimum observed trait value of cases. Although the performance of their F-test depends on the choice of y^*, their simulation results show that using the minimum observed trait value of cases is nearly optimum. In GAW16, one can choose $y^* = 20$, which is a typical cutoff point for high/low anti-CCP. In fact, all observed anti-CCP of cases in GAW16 are greater than 20.

If we assume the distribution for the trait is normal, subject to a constant, (3.6) can be written as

$$\widetilde{L}_1(y^*) = -n\log \sigma_e - \sum_{i=0}^{2} \left\{ \sum_{j=1}^{r_i} \frac{\{y_{ij1} - \mu - c_1(g_{ij1})\}^2}{2\sigma_e^2} + \sum_{j=1}^{s_i} \frac{\{y^* - \mu - c_1(g_{ij0})\}^2}{2\sigma_e^2} \right\}.$$

The null hypothesis is $H_0 : a = 0$. The usual F-test can still be applied, denoted as $\widetilde{F}_1(x|y^*)$ and Zheng *et al.* (2012a) show χ_1^2 can still be used to find critical values and p-values for $\widetilde{F}_1(x|y^*)$, regardless of the choice of y^*. Under the normality assumption, the usual least squares estimate of a has a closed form and is also the MLE.

Through simulation studies, we compare the performance of $Z(x)$ and the two F-tests $\widetilde{F}_1(x)$ (using observed trait values of cases only) and $\widetilde{F}_1(x|y^*)$ as well as the Fisher's combination test $\widetilde{J}$. The total sample size of cases was fixed at $r = 500$, but the number of controls s was varied. The MAF was fixed at 0.40 and HWE held. We chose y^* as the smallest observed trait value of cases, as in Zheng *et al.* (2012a). Three correlations were considered $\rho_{01} = 0$, 0.5, and 0.9; $\widetilde{F}_1(x|y^*)$ is expected to be more powerful when ρ_{01} is low than when it is high because the higher ρ_{01}, the more overlap there is of the information of the two traits, so the less that the partial likelihood $\widetilde{L}_1(y^*)$ can borrow. The results presented in Table 3.4 confirm this. However, the results also show that the power of $\widetilde{F}_1(1/2)$ decreases when ρ_{01} increases. When $\rho_{01} = 0$, the quantitative trait values of cases are random samples from the trait distribution. However, when ρ_{01} becomes larger, the observed quantitative trait values are random samples from a truncated trait distribution. Therefore, the loss

Table 3.4 *Power of trend test $Z(1/2)$, F-tests $\widetilde{F}_1(1/2)$ and $\widetilde{F}_1(1/2|y^*)$, and Fisher's combination test $\widetilde{J}$ of $Z(1/2)$ and $\widetilde{F}_1(1/2)$, with $r = 500$ cases, MAF $= 0.3$, and 1,000 replicates.*

s	ρ_{01}	$Z(1/2)$	$\widetilde{F}_1(1/2)$	$\widetilde{J}$	$\widetilde{F}_1(1/2|y^*)$
	0	0.090	0.335	0.278	0.408
20	0.5	0.076	0.103	0.083	0.165
	0.95	0.081	0.074	0.075	0.058
	0	0.354	0.333	0.419	0.597
100	0.5	0.352	0.087	0.234	0.396
	0.95	0.319	0.052	0.187	0.122
	0	0.529	0.330	0.551	0.722
200	0.5	0.536	0.111	0.384	0.552
	0.95	0.549	0.045	0.340	0.287

Table 3.5 *MLEs from $\widetilde{L}_{11}$ (using only observed cases' traits) and from $\widetilde{L}_1(y^*)$ (mixing likelihoods), $r = 500$, $a_0 = a_1 = a$, $d_0 = d_1 = 0$ (additive model), and MAF $= 0.4$.*

ρ_{01}	s	a	$\widetilde{L}_{11}$ Bias	MSE	$\widetilde{L}_1(y^*)$ Bias	MSE
	20	0	−0.0028	0.0037	0.0068	0.0046
0.5	20	0.1	−0.0231	0.0043	−0.0002	0.0049
	20	0.2	−0.0428	0.0055	−0.0021	0.0045
	100	0	0.0003	0.0035	0.0398	0.0080
0.5	100	0.1	−0.0178	0.0038	0.0806	0.0124
	100	0.2	−0.0393	0.0053	0.1129	0.0177
	200	0	0.0018	0.0036	0.0770	0.0125
0.5	200	0.1	−0.0195	0.0040	0.0129	0.0214
	200	0.2	−0.0437	0.0057	0.1701	0.0313
	20	0	0.0020	0.0028	0.0118	0.0035
0.9	20	0.1	−0.0414	0.0044	−0.0169	0.0036
	20	0.2	−0.0726	0.0080	−0.0382	0.0047
	100	0	0.0005	0.0027	0.0525	0.0071
0.9	100	0.1	−0.0398	0.0044	0.0533	0.0070
	100	0.2	−0.0762	0.0086	0.0636	0.0082
	200	0	0.0015	0.0029	0.0947	0.0133
0.9	200	0.1	−0.0344	0.0039	0.1217	0.0186
	200	0.2	−0.0727	0.0082	0.1410	0.0228

of power is due to missing lower trait values. The results also show that $\widetilde{F}_1(1/2|y^*)$ can be most powerful among the tests under consideration when $\rho_{01} = 0$ or 0.5.

Finally, we discuss the estimation problem. In simulations, we only considered $a_0 = a_1 = a$ under the additive model with $r = 500$ cases fixed, and MAF $= 0.4$ and $\rho_{01} = 0.9$ were also fixed. We find MLEs from the likelihoods $\widetilde{L}_{11}$ and $\widetilde{L}_1(y^*)$, respectively, where y^* was the minimum observed trait values of cases. Both bias and mean square error (MSE) were estimated from the simulations. The results are reported in Table 3.5. There is no clear pattern for the bias and MSE as ρ_{01} changes. However, the results show that using mixing likelihoods $\widetilde{L}_1(y^*)$ leads to larger bias and MSE compared to using $\widetilde{L}_{11}$. When s increases, the bias and MSE increase as well. Further, the bias and MSE increase with a using either likelihood.

Table 3.6 *MAFs, the estimates of log odds ratio* $\widehat{\beta}$, *and the p-values of F-test, using* $\widetilde{L}_{11}$ *and* $\widetilde{L}_1(y^*)$, *for the three SNPs in chromosome 1 that are associated with RA only (SNP rs2476601), with anti-CCP only (SNP rs12095496), and with both RA and anti-CCP (SNP rs2454170). Results are based on the additive model.*

SNP	MAF	$\widetilde{L}_{11}$		$\widetilde{L}_1(y^*)$	
		$\widehat{\beta}$	$\widetilde{F}_1$	$\widehat{\beta}$	$\widetilde{F}_1(y^*)$
rs2476601	0.084	0.063	0.600	−0.595	2.56×10^{-8}
rs12095496	0.180	0.452	0.000077	0.133	0.146
rs2454170	0.180	0.356	0.0053	0.428	2.90×10^{-6}

3.4   Application

Zheng *et al.* (2012*a*) scanned 38,739 SNPs of chromosome 1 in the GWAS of RA in GAW16. We consider three SNPs of chromosome 1 that are tested to have association with RA only, with anti-CCP only, and with both RA and anti-CCP, respectively. Zheng *et al.* (2012*a*) focused on testing associations. We focus on the estimation based on the likelihoods $\widetilde{L}_{11}$ and $\widetilde{L}_1(y^*)$. For the SNP associated with RA only, we consider SNP rs2476601, which is in *PTPN22*, a gene known to have association with RA among those with positive anti-CCP (Chen *et al.*, 2009). For the SNP associated with anti-CCP only, we use SNP rs12095496, which has the largest F-test statistic $\widetilde{F}_1(1/2)$ for association with anti-CCP (*p*-value is 0.000077) and is not significant for association with RA based on the trend test $Z(1/2)$ (*p*-value is 0.44). For the SNP with joint association, we consider SNP rs2454170, which was identified to have association with anti-CCP among those with RA (Suzuki *et al.*, 2003). The *p*-values for $Z(1/2)$ and $\widetilde{F}_1(1/2)$ are 0.00001 and 0.0053, respectively.

The GAW16 data contained 868 individuals with RA (cases), whose anti-CCP measures were also obtained, and 1,194 individuals without RA (controls), whose anti-CCP measures were not obtained. Anti-CCP is linked to RA with high sensitivity and specificity (Huizinga *et al.*, 2005). Among cases with anti-CCP, the minimum observed anti-CCP value in GAW16 was 20.053. Hence, like Zheng *et al.* (2012*a*), we use $y^* = \log(20.053)$. The log-transformation is applied to all anti-CCP values. The results of the analysis are reported in Table 3.6. Although the estimates of the log OR are likely biased, based on the discussion in Section 3.3, they are consistent with the *p*-values of the F-tests $\widetilde{F}_1(1/2)$ and $\widetilde{F}_1(1/2|y^*)$. For example, for SNP rs2476601, $\widehat{\beta} \approx 0$ when $\widetilde{L}_{11}$ is used and the corresponding *p*-value of $\widetilde{F}_1(1/2)$ is only about 0.6. But, when $\widetilde{L}_1(y^*)$ is used, $\widehat{\beta} = -0.595$ is much smaller, as the *p*-value of $\widetilde{F}_1(1/2|y^*)$ is also very significant.

3.5   Discussion

In this chapter, we discussed genetic association studies with either a binary trait or a quantitative trait. Some basic test statistics were reviewed. Analyses of mixed binary and quantitative traits were studied. We demonstrated through simulation studies that joint analysis can be more powerful than single trait analysis, especially when the genetic marker is associated with both types of traits. When quantitative trait values are only measured on a subset of population defined by case-control status, an approach by mixing likelihoods produced a more powerful test when the correlation of two traits are low to moderate. While the Type I error rates for testing association using the mixed likelihoods were close to the nominal levels in the simulations, larger bias, however, was produced in estimating the genetic effects. In Table 3.5, when $a = 0$ (for the null hypothesis), the bias and MSE are smaller than when $a > 0$ (for the alternative hypothesis). One possible explanation of this is the genetic model for the binary and quantitative traits. We used the additive model for both traits. However, the same additive model is defined differently for the binary and quantitative traits. This affects how the genotype is coded in the analysis. The data were simulated following the procedures described in Zheng *et al.* (2012*a*). For the case-control data, we first simulated the quantitative trait data under the additive model and then used a threshold model to obtain cases and controls. The additive model defined for the quantitative trait does not correspond to the additive model for the case-control data after applying the threshold model. Under the null hypothesis, however, the genetic model is not relevant. Hence, the bias is smaller.

The discussions in this chapter were based on a known genetic model (the additive model). A true genetic model is rarely known in practice. We did not examine the performance of the tests under a misspecified genetic model. Robust tests have been developed for testing single trait association (e.g., Freidlin *et al.*, 2002; Zheng *et al.*, 2009, 2012*b*). How some of those robust tests can be extended to the mixed types of traits has not been studied. Even for the same genetic marker, the genetic models for the binary and quantitative traits are often different. Thus, how to analyze mixed types of traits with different genetic models and different genetic effects (log odds ratios) requires further investigation.

Chapter 4

Bias in factor score regression and a simple solution

Takahiro Hoshino and Peter M. Bentler

4.1 Introduction

In social sciences, the main interest of research often lies in the quantitative expression of relationships between latent variables that represent (psychological or sociological) constructs. Structural equation modeling (SEM) is suitable for inference and description of these relationships (Bentler, 1983; Bentler and Weeks, 1980). As an important extension of Spearman's early ideas on factor analysis (Bartholomew, 2007), and built especially upon the many early contributions of Jöreskog (1977), today SEM includes a variety of statistical models that are highly relevant to psychological research (Bartholomew *et al.*, 2011; Lubke, 2010; Lee, 2007; Yuan and Bentler, 2007; Bollen and Curran, 2006; Bauer and Curran, 2004; Bollen, 2002). It is well known that SEM is frequently applied in the psychological (MacCallum and Austin, 2000) and related sciences (Hays *et al.*, 2005), but it also is becoming a useful methodology in many other fields. Gates *et al.* (2011), Aburatani (2011), Mulaik (2009), and Grace (2006) provide a general overview of the literature.

A standard estimation methodology for SEM is the method of maximum likelihood estimation (MLE). In contrast to simultaneous MLE of measurement and structural relations in SEM, however, some researchers prefer to estimate the latent variable scores, and then use traditional regression methods to determine their relations. This approach has still been used because standard statistical methodologies can be applied instead of requiring the use of specialized SEM programs, and it also allows the relative standing of individuals (cases, observations) to be studied. Unfortunately, it has the disadvantage of leading to problematic parameter estimates. In this chapter, we investigate the reason why estimates of parameters regarding the relationships between latent variables (i.e., correlations between latent variables) are biased when the estimated latent variable scores are used as if they are observed. We also propose a simple, valid estimation method using the estimated latent variable scores.

We consider a situation where two or more blocks of variables are observed on the same subjects. For example, consider a very simple SEM model depicted in Figure 4.1 (i.e., two blocks of variables and four variables for each block). There are several variants of estimation methods using estimated factor scores, but generally a stepwise estimation method can be divided into three steps:

Step 1. Estimate the parameters in the measurement part of the model (i.e., submodel 1 and 2 in Figure 4.1).

Step 2. Estimate the factor scores in exploratory (or confirmatory) factor analysis using the estimated parameters obtained in the step 1.

Step 3. Perform regression analysis, path analysis or factor analysis (i.e., submodel 3), where the variables of the model are the estimated factor scores.

The above procedure to estimate regression coefficients is usually called "factor score regression" because the estimated factor scores are used in a regression analysis, or a series of regressions, as if they were observed variables. For examples, see Zammuner (1998) and Gass (1996). In a later section, we consider the case where the observed variables are discrete and the latent variable scores are estimated using item response theory (IRT; see Baker and Kim, 2004).

Although simultaneous estimation in SEM can avoid the need for factor score regression, this approach may not be feasible. Many researchers have pointed out that there remain some problems in the practical application of SEM to data, especially when mixed continuous and ordinal variables are used, models are large, and simultaneous estimation methods, such as MLE or least squares methods are used. The main problem is that simultaneous estimation methods can be infeasible when continuous and discrete variables are mixed. If all observed variables are continuous, then simultaneous estimation methods only require estimates of moments, i.e., the sample mean vector and sample covariance matrix. However, including binary or ordinal variables also requires considering response patterns, as is typically done in IRT. A wedding of SEM and IRT methods is being developed. The generalized linear and nonlinear methodologies described in de Boeck and Wilson (2004) provide a partial approach because they allow the prediction of IRT item parameters and person parameters by external variables. In theory, simultaneous estimation such as that based on MLE is applicable to SEM with mixed continuous and nominal observed variables (Moustaki and Knott, 2000; Sammel et al., 1997). The generalized linear latent and mixed (GLLAMM) modeling framework of Skrondal and Rabe—Hesketh (2004) (see also Rabe–Hesketh et al., 2004) is also promising, since it allows IRT/SEM combinations theoretically by a unification and extension of multilevel and latent variable models to allow latent variable SEMs in the context of measurement models that permit a wide range of link functions and variable types (see Section 1.3 of Chapter 1; see also Chapter 8). However, their approach requires numerical integration and calculation of the likelihood, which is difficult to impossible when the model is complex or the number of variables is large. Thus, "estimation can be quite slow, especially if there are several random effects" (Rabe—Hesketh and Skrondal, 2005, p. 128). Since computational time is proportional to number of cases and the square of number of parameters, this methodology is not yet useful for larger models. Although there has been an important recent effort to joining IRT and SEM, the summary of Moustaki et al. (2004, p. 507) still holds: "On the other hand, IRT models have been developed recently and there is no flexible software available for fitting those models. If one wants to fit a model with many factors, one will probably have to use LISREL, Mplus or EQS."

Unfortunately, while methods such as Muthèn's (1984) and Lee et al.'s (1995) approaches based on polychoric and polyserial correlations are applicable to the case where there are continuous and ordered categorical observed variables, and are easy to implement in Mplus and EQS; and related methods such as pairwise likelihood (de Leon, 2005; Liu, 2007) are promising alternatives, these methods are not applicable when nominal observed variables are included in the model.

It is also well known that unless a sufficient number of good indicators of each factor is available, improper solutions (e.g., negative variance estimates, also known as Heywood cases) can occur frequently. Jöreskog (1967) reports that 9 out of 11 classical data sets possessed improper solutions, and Anderson and Gerbing (1984) report that with correct models, their simulation study found that 24.9% of replications had improper solutions. An important consequence is that test statistics no longer have their assumed distributions, and model evaluation becomes difficult (Savalei and Kolenikov, 2008; Stoel et al., 2006). There is an extensive literature on the reasons for improper solutions (Chen et al., 2001), but it is also known that factor score regression can reduce the occurrence of improper solutions through its division and separate treatment of the two parts of a complete model, namely, the measurement model and the structural relations model.

In order to avoid the problems stemming from simultaneous estimation in SEM, a frequently used method is a kind of stepwise estimation using estimated factor scores. A methodology such as this is encouraged by statistical packages such as SPSS FACTOR that make it easy to save "factor scores" or component scores for use in subsequent analyses (see Bentler and de Leeuw, 2011, for the mathematics of factor vs. component analysis). However, since such "factor scores" are linear combinations of variables, they contain error and are biased. Hence, factor score regression procedures using such scores produce biased estimates, usually underestimating the relationship between factors (see Section 4.3).

In an important paper, Skrondal and Laake (2001) point out the bias problem and propose a modified version of the above estimation procedure, especially for regression analysis between latent variables. Their proposed method (*i*) estimates the factor scores for dependent factors by the Bartlett method, (*ii*) estimates the factor scores for independent factors by the regression method (we review these factor score estimates below; see also Yanai and Ichikawa, 2007, pp. 287–289), and (*iii*) uses these estimated factor scores as if they were observed variables. They demonstrate the consistency of their proposed method, but the method has some disadvantages: (1) the method is not available for models that have more than three groups of factors (and indicators); (2) independent factors and dependent factors must be pre-specified before the analysis; (3) their method underestimates the correlations between factors; (4) their method is not available when there are discrete variables in the model; (5) their method neglects the effect of estimation of parameters of the factor models; and (6) they have not extended the theory to deal with higher-order factors. Their paper provides a solution for a certain (but restricted) case, but they do not clarify the cause of bias arising from using estimated factor scores.

A second and popular method related to factor score regression is the item parceling methodology. Item parceling involves summing or averaging item scores from two or more items and using these parcel scores (or scale score in personality psychology) as observed latent variable scores to estimate the relationships between latent variables (Bandalos, 2002). These composite scores are fixed and not iteratively updated. The rationale for the use of item parcels in SEM is as follows (Bandalos and Finney, 2001): (1) the reliability of item parcels will be greater than the raw scales (Kishton and Widaman, 1994; Cattell and Burdsall, 1975); (2) even when the data contain raw items that are nonnormally distributed or/and coarsely categorized, item parcels based on a large number of items often can be regarded as normally distributed, and normal theory MLE and generalized least squares estimation techniques are applicable to such data; (3) item parceling can reduce the number of variables in the analysis, thus also reducing the ratio of variables to subjects, which will lead to more stable estimates; and (4) item parceling typically leads to better model fit than estimation using the raw items (Thompson and Melancon, 1996). In spite of these advantages, item parceling has been criticized. There are at least two problems: (1) the resulting parameter estimates are sometimes biased, and then typically they are underestimated (Bandalos, 2002); and (2) the item parceling method does not always produce stable estimates (MacCallum *et al.*, 1999; Marsh *et al.*, 1998). Although there are some theoretical results (e.g., Yuan *et al.*, 1997), most conclusions on this methodology are mainly due to simulation studies. Further theoretical analyses on item parceling and factor score regression are still needed.

This chapter is organized as follows. In Section 4.2, the model assumptions are made. These assumptions appear to be slightly restrictive; however, they are the same for factor score regression or item parceling. In Section 4.3, we discuss the theoretical investigation of the sources of biases in factor score regression (also in the item parceling method). For notational convenience, we restrict our attention to a relatively simple model; however, the results apply to general cases. In Section 4.4, we propose an alternative estimation method based on the estimated latent variables. In Section 4.5, we provide simulation studies in various model setups to justify the validity of the proposed method when the number of subjects and observed variables are finite. We also provide an illustrative data analysis regarding the relationships between web browsing behavior and purchasing intent using the proposed method. Details of theoretical results are given in Section 4.7. Concluding remarks and discussions are provided in Section 4.8.

4.2 Model

We assume the model setup usually made in factor score regression and the item parceling method. Consider $J > 1$ measurement equations, in which each equation measures different latent variables.

In the jth measurement model, each $P_j \times 1$ observed variable vector $\mathbf{Y}_j$ is independently defined in terms of the Q_j-component factor vector $\mathbf{f}_j$, and the P_j-component error vector $\mathbf{e}_j$ by the jth measurement model,

$$\mathbf{Y}_j \;=\; g_j(\mathbf{f}_j) + \mathbf{e}_j, \quad j = 1, \cdots, J, \tag{4.1}$$

where $g_j(\cdot)$ is a linear or nonlinear function. The distribution of $\mathbf{Y}_j$ in the jth measurement model is also defined independently of the other measurement models and the structural model.

Measurement models vary with the level of measurement. If each element of $\mathbf{Y}_j$ is a continuous variable, the jth measurement model (4.1) is usually expressed as a linear factor analysis model, as follows:

$$\mathbf{Y}_j \;=\; \boldsymbol{\alpha}_j + \boldsymbol{\Lambda}_j \mathbf{f}_j + \mathbf{e}_j, \quad E(\mathbf{e}_j) = 0 \quad \text{and} \quad var(\mathbf{e}_j) = \boldsymbol{\Psi}_j, \tag{4.2}$$

where $\mathbf{e}_j$ follows the multivariate normal distribution with mean $\mathbf{0}$ and covariance matrix $\boldsymbol{\Psi}_j$, and $\boldsymbol{\Lambda}_j$ is the factor loading matrix.

If $\mathbf{Y}_j$ is dichotomous, the three-parameter logistic item response model (Embretson and Reise, 2000; Lord and Novick, 1968) is employed:

$$P(Y_{jk} = 1) \;=\; c_{jk} + \frac{c_{jk}}{1 + \exp\{-Da_{jk}(f_j - b_{jk})\}}, \tag{4.3}$$

where a_{jk}, b_{jk}, and c_{jk} are the kth item parameters of $\mathbf{Y}_j$, and $D = 1.702$. Sometimes, the probit item response model is also employed.

If $\mathbf{Y}_j$ is nominal or polytomous, the nominal response model proposed by Bock (1972) or the graded response model proposed by Samejima (1969) is also available.

We further assume that the joint distribution of factor vectors, $p(\mathbf{f}_1, \cdots, \mathbf{f}_J; \boldsymbol{\xi}_F)$ is a multivariate normal distribution, where $\boldsymbol{\xi}_F$ denotes the parameter vector. Therefore, the joint distribution of $\mathbf{Y}_1, \cdots, \mathbf{Y}_J$, is

$$p_{\mathbf{Y}_1, \cdots, \mathbf{Y}_J}(\mathbf{y}_1, \cdots, \mathbf{y}_J; \boldsymbol{\xi}_1, \cdots, \boldsymbol{\xi}_J, \boldsymbol{\xi}_F) = \int \cdots \int \prod_{J=1}^{J} p_j(\mathbf{y}_j | \mathbf{f}_j; \boldsymbol{\xi}_j) p(\mathbf{f}_1, \cdots, \mathbf{f}_J; \boldsymbol{\xi}_F) d\mathbf{f}_1 \cdots d\mathbf{f}_J, \tag{4.4}$$

where $p_j(\mathbf{y}_j | \mathbf{f}_j; \boldsymbol{\xi}_j)$ denotes the conditional distribution of $\mathbf{Y}_j$, given $\mathbf{f}_j$, and $\boldsymbol{\xi}_j$ is the parameter vector in the conditional distribution.

If $J = 2$, the entire model (4.4) is equivalent to the LISREL model proposed by Jöreskog (1970). If $J > 2$, the entire model can be considered as a "multiple indicator model," a submodel of SEM.

Parameter $\boldsymbol{\xi}_F$ contains the mean vector and the covariance matrix of factors,

$$\boldsymbol{v} = \begin{pmatrix} \boldsymbol{v}_1 \\ \boldsymbol{v}_2 \\ \vdots \\ \boldsymbol{v}_J \end{pmatrix}, \quad \boldsymbol{\Phi} = \begin{pmatrix} \boldsymbol{\Phi}_{11} & \boldsymbol{\Phi}_{12} & \cdots & \boldsymbol{\Phi}_{1J} \\ \boldsymbol{\Phi}_{21} & \boldsymbol{\Phi}_{22} & \cdots & \boldsymbol{\Phi}_{2J} \\ \vdots & \vdots & \ddots & \vdots \\ \boldsymbol{\Phi}_{J1} & \boldsymbol{\Phi}_{J2} & \cdots & \boldsymbol{\Phi}_{JJ} \end{pmatrix}, \tag{4.5}$$

where $\boldsymbol{v}_j$ is the mean vector, and $\boldsymbol{\Phi}_{jk}$ is the covariance matrix of $\mathbf{f}_j$ and $\mathbf{f}_k$.

Usually, the concern is not the covariance matrix of factors, but the parameters $\boldsymbol{\tau}$ of the structural equation. In this study, the objective of inference is to estimate parameters in the structural part of the model. The mean vector and the covariance matrix of factors, $\boldsymbol{v}$ and $\boldsymbol{\Phi}$, are structured by $\boldsymbol{\tau}$, as $\boldsymbol{v}(\boldsymbol{\tau})$ and $\boldsymbol{\Phi}(\boldsymbol{\tau})$.

Random effect and fixed effect models

For clarification, we define "random effect model" and "fixed effect model" as follows. The jth measurement model is called a "random effect model" if the factor scores follow a multivariate normal distribution. The distribution of $\mathbf{Y}_j$ can be expressed as follows:

$$p_{\mathbf{Y}_j}(\mathbf{y}_j; \boldsymbol{\xi}_j, \boldsymbol{\xi}_{F_j}) \;=\; \int p_j(\mathbf{y}_j | \mathbf{f}_j; \boldsymbol{\xi}_j) p(\mathbf{f}_j; \boldsymbol{\xi}_{F_j}) d\mathbf{f}_j, \tag{4.6}$$

where $\boldsymbol{\xi}_{F_j}$ is the parameter vector of the marginal distribution of $\mathbf{f}_j$ (i.e., the factor mean vector $\boldsymbol{\nu}_j$ and factor covariance matrix $\boldsymbol{\Phi}_{jj}$).

The jth measurement model is called a "fixed effect model" if the factor scores are not random variants but incidental parameters (Neyman and Scott, 1948). The distribution of $\mathbf{Y}_j$ can then be expressed as

$$p_{\mathbf{Y}_j}(\mathbf{y}_j; \mathbf{f}_j, \boldsymbol{\xi}_j). \tag{4.7}$$

Parameters $\boldsymbol{\xi}_1, \cdots, \boldsymbol{\xi}_J, \boldsymbol{\xi}_F$ are usually called the structural parameters, as distinguished from incidental parameters, $\mathbf{f}_1, \cdots, \mathbf{f}_J$. If we employ the random effect model, the objects of the inference are not incidental parameters but structural parameters. In the fixed effect model, both the structural and incidental parameters are usually estimated simultaneously.

It should be noted that if there are a large number of subjects, and the incidental parameters follow some distribution, these two models are virtually the same (Lindsay et al., 1991; Kiefer and Wolfowitz, 1956).

Further, we employ the fixed effect model and assume that there are a large number of subjects and that the incidental parameter vector $\mathbf{f}_j$ follows a multivariate normal distribution. Henceforth, we refer to the employed model as the "fixed effect Kiefer–Wolfowitz-type model."

In this chapter, it is also assumed that each random effect measurement model is identifiable, and the structural parameters can be estimated.

4.3   Bias due to estimated factor scores: Factor analysis model

This section discusses why the estimates obtained using factor score regression are generally biased. To be more concrete, we restate the factor score regression procedure as follows:

Step 1. Employ the random effect model (4.6) and obtain MLEs of structural parameters $\boldsymbol{\xi}_j, \boldsymbol{\xi}_{F_j}$, in each measurement model. For example, in the model depicted in Figure 4.1, regard submodels 1 and 2 as random effect models and estimate the parameters in these models.

Step 2. Fix the parameters at the estimates, and then estimate the factor scores for each subject in each measurement model (i.e., estimate factor scores f_X and f_Y independently.)

Step 3. Regard the estimated factor scores as the observed variables, then estimate the parameters regarding the relationships between the factors (i.e., regression coefficients, correlations, or factor loadings for higher-order factors). For example, estimate parameters in submodel 3 by regarding factor scores f_X and f_Y as if they were observed.

This procedure includes two sources of bias: (*i*) neglect of the uncertainty of the estimates of the structural parameters in the first step, and (*ii*) overestimation of the variances of factors in the third step. In this section, we focus our attention on the latter source of bias. Let the parameters of the measurement part be known. The effects of the estimated structural parameters are not considered in this section; please see Section 4.7 for this (i.e., proof of Proposition 3).

For the purpose of our demonstration, we assume two measurement models in which each factor vector is measured by some observed continuous indicators. Each measurement model is expressed as a factor analysis model (4.2):

$$\mathbf{Y}_j = \boldsymbol{\Lambda}_j \mathbf{f}_j + \mathbf{e}_j, \quad \mathbf{f}_j \sim N_{Q_j}(\boldsymbol{\nu}_j, \boldsymbol{\Phi}_{jj}) \text{ and } \mathbf{e}_j \sim N_{P_j}(\mathbf{0}, \boldsymbol{\Psi}_j), \quad j = 1, 2. \tag{4.8}$$

For notational simplicity, $\boldsymbol{\alpha}_j$ is fixed at zero for each measurement model. Let $\boldsymbol{\Phi}_{12}$ be the covariance between $\mathbf{f}_1$ and $\mathbf{f}_2$.

There are several estimation methods for factor scores in the factor analysis model. For illustrative purposes, two representative estimates are considered here, namely, Bartlett's method and the regression method. The former can be considered the MLE of the factor score vector in the fixed effect factor analysis, while the latter can be regarded as the Bayes posterior mean estimate. These two estimates, as well as other estimates, are expressed as the product of a matrix and the observed variable vector.

Let $\widehat{\mathbf{f}}_1 = \mathbf{A}_1\mathbf{Y}_1$ be the estimate of $\mathbf{f}_1$, and $\widehat{\mathbf{f}}_2 = \mathbf{A}_2\mathbf{Y}_2$ be the estimate of $\mathbf{f}_2$. Therefore, $cov(\mathbf{Y}_1, \mathbf{Y}_2) = \boldsymbol{\Lambda}_1\boldsymbol{\Phi}_{12}\boldsymbol{\Lambda}_2^\top$, and the joint distribution of the estimated factors is

$$\begin{pmatrix} \widehat{\mathbf{f}}_1 \\ \widehat{\mathbf{f}}_2 \end{pmatrix} \sim N_{Q_1+Q_2} \left(\begin{pmatrix} \mathbf{A}_1\boldsymbol{\mu}_1 \\ \mathbf{A}_2\boldsymbol{\mu}_2 \end{pmatrix}, \begin{pmatrix} \mathbf{A}_1(\boldsymbol{\Lambda}_1\boldsymbol{\Phi}_{12}\boldsymbol{\Lambda}_1^\top + \boldsymbol{\Psi}_1)\mathbf{A}_1^\top & \mathbf{A}_1(\boldsymbol{\Lambda}_1\boldsymbol{\Phi}_{12}\boldsymbol{\Lambda}_2^\top)\mathbf{A}_2^\top \\ \mathbf{A}_2(\boldsymbol{\Lambda}_2\boldsymbol{\Phi}_{21}\boldsymbol{\Lambda}_1^\top)\mathbf{A}_1^\top & \mathbf{A}_2(\boldsymbol{\Lambda}_2\boldsymbol{\Phi}_{22}\boldsymbol{\Lambda}_2^\top + \boldsymbol{\Psi}_2)\mathbf{A}_2^\top \end{pmatrix} \right) . \quad (4.9)$$

Using the well known relationship between the regression model and the multivariate normal distribution, the expectation of the regression coefficient is

$$cov(\widehat{\mathbf{f}}_1, \widehat{\mathbf{f}}_2)\mathbf{V}^{-1}(\widehat{\mathbf{f}}_2) = \mathbf{A}_2(\boldsymbol{\Lambda}_2\boldsymbol{\Phi}_{21}\boldsymbol{\Lambda}_1^\top)\mathbf{A}_1^\top \left\{ \mathbf{A}_2(\boldsymbol{\Lambda}_2\boldsymbol{\Phi}_{22}\boldsymbol{\Lambda}_2^\top + \boldsymbol{\Psi}_2)\mathbf{A}_2^\top \right\}^{-1}, \quad (4.10)$$

where $\mathbf{V}(\widehat{\mathbf{f}}_2) = var(\widehat{\mathbf{f}}_2)$. The expectation of the correlation matrix between $\widehat{\mathbf{f}}_1$ and $\widehat{\mathbf{f}}_2$ is

$$\left\{ \mathbf{A}_1(\boldsymbol{\Lambda}_1\boldsymbol{\Phi}_{12}\boldsymbol{\Lambda}_1^\top + \boldsymbol{\Psi}_1)\mathbf{A}_1^\top \right\}^{-1/2} \mathbf{A}_2(\boldsymbol{\Lambda}_2\boldsymbol{\Phi}_{21}\boldsymbol{\Lambda}_1^\top)\mathbf{A}_1^\top \left\{ \mathbf{A}_2(\boldsymbol{\Lambda}_2\boldsymbol{\Phi}_{22}\boldsymbol{\Lambda}_2^\top + \boldsymbol{\Psi}_2)\mathbf{A}_2^\top \right\}^{-1/2}. \quad (4.11)$$

If Bartlett's estimates are used to calculate each factor score,

$$\mathbf{A}_j = (\boldsymbol{\Lambda}_j^\top\boldsymbol{\Psi}_j^{-1}\boldsymbol{\Lambda}_j)^{-1}\boldsymbol{\Lambda}_j^\top\boldsymbol{\Psi}_j^{-1}, \quad j = 1, 2. \quad (4.12)$$

If regression estimates are used to calculate each factor score,

$$\mathbf{A}_j = \boldsymbol{\Phi}_{jj}\boldsymbol{\Lambda}_j^\top\boldsymbol{\Sigma}_j^{-1} = (\boldsymbol{\Phi}_{jj}^{-1} + \boldsymbol{\Lambda}_j^\top\boldsymbol{\Psi}_j^{-1}\boldsymbol{\Lambda}_j)^{-1}\boldsymbol{\Lambda}_j^\top\boldsymbol{\Psi}_j^{-1}, \quad j = 1, 2. \quad (4.13)$$

The expectations of the estimates of the means, the regression coefficient, and the covariance matrix are given in Table 4.1.

Table 4.1 *Estimates by various estimation methods using factor scores.*

Parameter	True	Bartlett
Mean of $\mathbf{f}_1$	$\boldsymbol{v}_1$	$\boldsymbol{v}_1$
Mean of $\mathbf{f}_2$	$\boldsymbol{v}_2$	$\boldsymbol{v}_2$
Regression coefficient	$\boldsymbol{\Phi}_{12}\boldsymbol{\Phi}_{22}^{-1}$	$\boldsymbol{\Phi}_{12}\{\boldsymbol{\Phi}_{22} + (\boldsymbol{\Lambda}_2^\top\boldsymbol{\Psi}_{22}^{-1}\boldsymbol{\Lambda}_2)^{-1}\}^{-1}$
Covariance matrix	$\boldsymbol{\Phi}_{12}$	$\boldsymbol{\Phi}_{12}$

Parameter	Regression	Skrondal and Laake (2001)
Mean of $\mathbf{f}_1$	$\boldsymbol{\Phi}_{11}\boldsymbol{\Lambda}_1^\top\boldsymbol{\Sigma}_1^{-1}\boldsymbol{v}_1$	$\boldsymbol{\Phi}_{11}\boldsymbol{\Lambda}_1^\top\boldsymbol{\Sigma}_1^{-1}\boldsymbol{v}_1$
Mean of $\mathbf{f}_2$	$\boldsymbol{\Phi}_{22}\boldsymbol{\Lambda}_2^\top\boldsymbol{\Sigma}_2^{-1}\boldsymbol{v}_2$	$\boldsymbol{v}_2$
Regression coefficient	$\boldsymbol{\Phi}_{11}(\boldsymbol{\Lambda}_1^\top\boldsymbol{\Sigma}_1^{-1}\boldsymbol{\Lambda}_1)\boldsymbol{\Phi}_{12}\boldsymbol{\Phi}_{22}^{-1}$	$\boldsymbol{\Phi}_{12}\boldsymbol{\Phi}_{22}^{-1}$
Covariance matrix	$\boldsymbol{\Phi}_{11}\boldsymbol{\Lambda}_1^\top\boldsymbol{\Sigma}_1^{-1}\boldsymbol{\Lambda}_1\boldsymbol{\Phi}_{12}\boldsymbol{\Lambda}_2^\top\boldsymbol{\Sigma}_2^{-1}\boldsymbol{\Lambda}_2\boldsymbol{\Phi}_{22}$	$\boldsymbol{\Phi}_{11}\boldsymbol{\Lambda}_1^\top\boldsymbol{\Sigma}_1^{-1}\boldsymbol{\Lambda}_1\boldsymbol{\Phi}_{12}\boldsymbol{\Lambda}_2^\top\boldsymbol{\Sigma}_2^{-1}\boldsymbol{\Lambda}_2\boldsymbol{\Phi}_{22}$

Note that in this section we neglect the fact that the true values of the structural parameters such as factor loadings are unknown and estimated. Therefore, the equivalence of the parameter and its expectation does not mean unbiasedness, but consistency (see proof of Proposition 1). As seen in Table 4.1, the bias does not disappear even when the number of subjects goes to infinity. See simulation studies in Section 4.5.

The regression method does not always produce underestimated regression coefficients or correlations, but it usually does. For example, suppose $\boldsymbol{\Phi}_{11} = \boldsymbol{\Phi}_{22} = \mathbf{I}$. Then, the expectation of the estimated regression coefficient

$$cov(\widehat{\mathbf{f}}_1, \widehat{\mathbf{f}}_2)\mathbf{V}^{-1}(\widehat{\mathbf{f}}_2) = \left\{ \mathbf{I} - (\mathbf{I} + \boldsymbol{\Lambda}_1^\top\boldsymbol{\Psi}_1\boldsymbol{\Lambda}_1)^{-1} \right\}\boldsymbol{\Phi}_{12}, \quad (4.14)$$

is not "greater than" the true regression coefficient Φ_{12}.

This section addresses only Bartlett's method, the regression method, and the method of Skrondal and Laake (2001), but it should be noted that estimated factor scores using the other methods yield inconsistent estimates. The estimated factor score is the sum of the true factor score and the error due to estimation, no matter what estimation method we use for factor scores. As shown in this section, the sample variance matrix of estimated factor scores is a biased estimate of the true variance matrix of factors. This fact results in the bias of factor score regression.

In this section, we explain why the bias occurs when estimated factor scores are used, but the degree of bias must be investigated to know whether this problem is of practical importance. The degree of bias due to the familiar three step estimation method using factor scores is examined by simulation studies in Section 4.5.

4.4 Proposed estimation method

To resolve the problem mentioned in the previous section, we propose a modified stepwise estimation method using estimated factor scores. The method is divided into four steps:

Step 1. Employ the random effect model and estimate the parameters in each measurement model (4.6). Then, obtain the MLEs of the structural parameters $\boldsymbol{\xi}_j$ (e.g., $\boldsymbol{\xi}_j = (\boldsymbol{\Lambda}_j, \boldsymbol{\Psi}_j)$, if $\mathbf{Y}_j$ is continuous), and $\boldsymbol{\xi}_{F_j} = (\boldsymbol{v}, \boldsymbol{\Phi})$, $\widetilde{\boldsymbol{\xi}}_j$ and $\widetilde{\boldsymbol{\xi}}_{F_j}$. The parameter estimation in the random effect model in this step is usually called marginal MLE in psychometrics (Bock and Aitkin, 1981). Henceforth, we term these marginal MLEs. If factor rotation is necessary, it should also be executed in this step.

Step 2. Employ the fixed effect model (4.7) for each measurement equation and fix $\boldsymbol{\xi}_j$ at $\widetilde{\boldsymbol{\xi}}_j$ obtained in the first step. Subsequently, calculate factor scores $\widehat{\mathbf{f}}_j^B$ by maximum likelihood (if observed variables are continuous, the method is simply Bartlett's method.)

Step 3. Estimate the factor mean vector and factor covariance matrix in the following manner. Let $\widehat{\boldsymbol{v}} = \sum_{i=1}^{N} \widehat{\mathbf{f}}_i^B / N$ be the sample mean of the estimated factor scores, where $\mathbf{f}_i$ is the latent variable vector $\mathbf{f} = (\mathbf{f}_1^{\top}, \cdots, \mathbf{f}_J^{\top})^{\top}$ for the ith subject. Let also $\mathbf{f}_{ij}$ be the value of $\mathbf{f}_j$ for the ith subject. Let the sample covariance matrix $\widehat{\boldsymbol{\Phi}}$ of the estimated factor scores be

$$\widehat{\boldsymbol{\Phi}} = \frac{1}{N} \sum_{i=1}^{N} (\widehat{\mathbf{f}}_i^B - \widehat{\boldsymbol{v}})(\widehat{\mathbf{f}}_i^B - \widehat{\boldsymbol{v}})^{\top}. \tag{4.15}$$

Further, use the sample mean of the estimated factor scores in the second step as the estimate of the factor mean vector $\widehat{\boldsymbol{v}}$. Use the estimate $\widetilde{\boldsymbol{\Phi}}_{jj}$ in the first step as the estimate of $\boldsymbol{\Phi}_{jj}$, instead of the sample covariance matrix of the estimated factor scores. Use the sample covariance matrix between $\mathbf{f}_j$ and $\mathbf{f}_k$ as the estimate of factor covariance $\boldsymbol{\Phi}_{jk}$ for $j \neq k$. Subsequently, the resulting estimate of $\boldsymbol{\Phi}$, $\boldsymbol{\Phi}^P$ can be expressed as

$$\widehat{\boldsymbol{\Phi}}^P = \begin{pmatrix} \widetilde{\boldsymbol{\Phi}}_{11} & \widehat{\boldsymbol{\Phi}}_{12} & \cdots & \widehat{\boldsymbol{\Phi}}_{1J} \\ \widehat{\boldsymbol{\Phi}}_{21} & \widetilde{\boldsymbol{\Phi}}_{22} & \cdots & \widehat{\boldsymbol{\Phi}}_{2J} \\ \vdots & \vdots & \ddots & \vdots \\ \widehat{\boldsymbol{\Phi}}_{J1} & \widehat{\boldsymbol{\Phi}}_{J2} & \cdots & \widetilde{\boldsymbol{\Phi}}_{JJ} \end{pmatrix} \tag{4.16}$$

Step 4. Estimate the parameters of the structural part in the model, using the generalized least squares method. Let $\boldsymbol{\tau}$ be the structural parameter vector of the structural equation part; $\boldsymbol{\sigma}$, the vector of the non-redundant elements of $\boldsymbol{v}, \boldsymbol{\Phi}$; and $\boldsymbol{\sigma}(\boldsymbol{\tau})$, the function of $\boldsymbol{\tau}$. Moreover, let $\widehat{\boldsymbol{\sigma}}$ be the corresponding estimate obtained in the third step, and let the estimate of $\boldsymbol{\tau}$ be the value that minimizes the following generalized least squared error function:

$$Q(\boldsymbol{\tau}) = \frac{1}{2} \{\widehat{\boldsymbol{\sigma}} - \boldsymbol{\sigma}(\boldsymbol{\tau})\}^{\top} \mathbf{W}^{-1} \{\widehat{\boldsymbol{\sigma}} - \boldsymbol{\sigma}(\boldsymbol{\tau})\}, \tag{4.17}$$

where $\mathbf{W}$ is the covariance matrix of $\widehat{\boldsymbol{\sigma}}$. See Section 4.7 for details.

4.4.1   Theoretical justification

There are several criteria for evaluating an estimation method in statistics, such as invariance, unbiasedness, efficiency, and so on. The most important issue is the consistency of the estimate, which implies that as the number of observations increases, the estimate converges to the true value of the parameter. We next investigate the consistency of the proposed estimate.

4.4.1.1   When observed variables are continuous

We show the consistency of the proposed estimate when the observed variables are continuous. Without loss of generality, we consider the case when the number of measurement models is two.

We use $\widetilde{\boldsymbol{\Phi}}_{jj}$, $j = 1, 2$, obtained in Step 1 as the consistent estimate of $\boldsymbol{\Phi}_{jj}$, instead of the sample variance matrices of the estimated factor scores. We can also show the following result.

Proposition 1 *The sample covariance matrix of the estimated factor scores* $\widehat{\boldsymbol{\Phi}}_{12} = \sum_{i=1}^{N} \big(\mathbf{f}_{i1}^B - \widehat{\mathbf{v}}_1\big)\big(\mathbf{f}_{i2}^B - \widehat{\mathbf{v}}_2\big)^{\top}/N$, *is the consistent estimate of* $\boldsymbol{\Phi}_{12}$, *considering the effect of estimation of structural parameters.*

See Hoshino and Bentler (2012) for the proof.

Therefore, $\widehat{\boldsymbol{\Phi}}^P$ in Step 3 is the consistent estimate of $\boldsymbol{\Phi}$. Following the properties of GLS estimates and the consistency of $\widehat{\boldsymbol{\sigma}}$, the proposed estimates of the parameters in structural equation obtained in Step 4 are consistent.

4.4.1.2   General case

We show that the proposed method has a kind of "consistency" (i.e., consistency at large) for general cases other than continuous observable variables. By "consistent at large," we mean that, with probability one, the estimation method finds the true values of structural parameters when the number of subjects and observed variables in each measurement model goes to infinity (i.e., $N \to \infty$ and $P_j \to \infty$). To prove the consistency of the proposed stepwise estimation method under a general model setup, the following additional propositions must be proved.

Proposition 2 *The estimates of* $\boldsymbol{\xi}_j$ *and* $\boldsymbol{\xi}_{F_j}$ *in the first step,* $\widetilde{\boldsymbol{\xi}}_j$ *and* $\widetilde{\boldsymbol{\xi}}_{F_j}$, *which are obtained under the random effect measurement model (4.6), are consistent, although we employ the fixed effect Kiefer–Wolfowitz-type model.*

Proposition 3 *The estimate* $\widehat{\boldsymbol{\Phi}}^P$ *is consistent at large.*

In the fixed effect model, factor scores are the incidental parameters in that the number of factor scores increases as the number of subjects increases. Further, the joint MLEs of $\boldsymbol{\xi}_j$ and $\boldsymbol{\xi}_{F_j}$ are not consistent (Neyman and Scott, 1948). However, as Kiefer and Wolfowitz (1956) point out, marginal MLEs are shown to be "consistent" in that the estimates converge to the true value, when the number of the indicator variables is large. Hence, Proposition 2 is true in the model setup. Then, it is sufficient to prove Proposition 3. The sketch of the proof of Proposition 3 is given in Section 4.7.

4.4.2   Alternative estimation method

In this section, by using simulation studies, we consider the influence of the number of subjects and variables. Kano (1983) and Shapiro (1984) show that when the variables are continuous, the estimate of $\boldsymbol{\tau}$ that minimizes the maximum likelihood discrepancy function (MLDF), where the mean vector and the covariance matrix are replaced by their consistent estimates, is consistent.

In this section, it is also shown that $\widehat{\boldsymbol{v}}$ and $\widehat{\boldsymbol{\Phi}}^P$ are consistent estimates of $\boldsymbol{v}$ and $\boldsymbol{\Phi}$, respectively. This indicates that the estimate of $\boldsymbol{\tau}$ that minimizes MLDF, where the mean and covariance are replaced by $\widehat{\boldsymbol{v}}$ and $\widehat{\boldsymbol{\Phi}}^P$, is also consistent for $\boldsymbol{\tau}$. Therefore, instead of the proposed fourth step, we can consistently estimate the structural parameter vector $\boldsymbol{\tau}$ using prevailing softwares such as SAS/STAT (*SAS/STAT User's Guide*, 1999), by considering $\widehat{\boldsymbol{\Phi}}^P$ as the sample covariance matrix of factors and modeling the structural equation part. It should be noted that the correct standard errors (SEs) cannot be evaluated by this method.

4.5 Simulation studies

In the previous section, the consistency of the proposed stepwise estimation method was proved. However, in order to show the validity of the proposed method in a moderate sample size and in a moderate number of indicators, the degree of bias must be investigated. In this section, some simulation studies that compare the proposed method with familiar methods are discussed.

4.5.1 Study 1: Two-dimensional factor analysis model

We modeled a situation in which two factors are assumed and each factor is measured by four observed indicators. The model we consider here is the factor analysis model

$$
\mathbf{Y} = \begin{pmatrix} \lambda_1 & \lambda_2 & \lambda_3 & \lambda_4 & 0 & 0 & 0 & 0 \\ 0 & 0 & 0 & 0 & \lambda_5 & \lambda_6 & \lambda_7 & \lambda_8 \end{pmatrix}^{\top} \mathbf{f} + \mathbf{e}, \tag{4.18}
$$

where $\mathbf{Y}$, $\mathbf{f}$, and $\mathbf{e}$ are the 8×1 observable variable vector, the 2×1 latent variable vector, and the 8×1 error vector, respectively. The first element of $\mathbf{f}$ is f_X, while the second one is labeled f_Y.

The structural equation is assumed as follows:

$$
f_Y = \beta f_X + \zeta, \tag{4.19}
$$

and the main point of interest in this study is β.

The model we assume here is depicted in Figure 4.1, with the true values of the factor loadings. The variances of the errors for the observed variables were all fixed at $var(e_j) = V(e_j) = 0.5$. The true value of β is 0.7; $var(f_X) = V(f_X)$ and $var(\zeta) = V(\zeta)$ were fixed at 1 and 0.51, respectively, so that $var(f_Y) = 1$. Five estimation methods are compared in this study: (1) factor score regression using the estimated factor scores by Bartlett's method (Bart), (2) factor score regression using the estimated factor scores by the regression method (Reg), (3) the modified estimation proposed by Skrondal and Laake (2001), (4) the proposed method (Prop), and (5) the MLE method in SEM. It should be noted that in methods (1)-(4), the factor scores of f_X and f_Y are independently estimated. Moreover, the factor variances are fixed at 1 in each method. The variances of the estimated factor scores in methods (1)-(4) are not equal to one; see Section 4.3.

Data generation and estimation were performed using the SAS package by combining SAS/IML program with proc CALIS in SAS/STAT. To calculate the mean and mean squared error (MSE) of estimates of β, 10,000 data sets were generated; the results are reported in Table 4.2.

These results indicate that even when the number of subjects is very large (i.e., $N = 10,000$), the factor score regression by Bartlett's method and the regression method underestimate β to a large extent. On the other hand, the proposed method and the MLE method produce valid estimates, even for relatively small sample sizes (i.e., $N = 300$).

4.5.2 Study 2: Two-level factor analysis model

In Study 2, we assume that three factors (i.e., level-one factor f_1, f_2, f_3) exist, and each factor is measured by four observed continuous indicators; hence, the total number of observed variables is

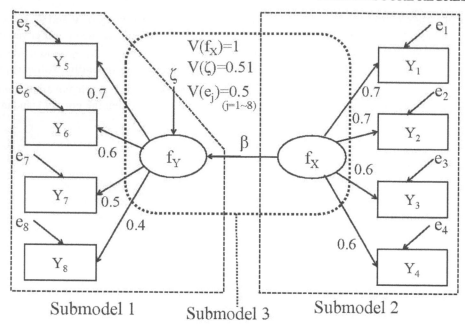

Figure 4.1 *Path diagram representation of model in Study 1.*

Table 4.2 *Resulting estimates from the five methods for Study 1.*

N	Method	Correlation (true value = 0.7)		Regression Coefficient (true value = 0.7)	
		Est	MSE	Est	MSE
	Bart	0.5190	0.0346	0.4995	0.0425
	Reg	0.5190	0.0346	0.5391	0.0285
300	Skro	0.5190	0.0346	0.6977	0.0045
	Prop	0.6953	0.0078	0.6953	0.0078
	MLE	0.6989	0.0028	0.6989	0.0028
	Bart	0.5206	0.0323	0.5010	0.0397
	Reg	0.5206	0.0323	0.5408	0.0254
10,000	Skro	0.5408	0.0254	0.6998	0.0001
	Prop	0.6997	0.0002	0.6997	0.0002
	MLE	0.7000	0.0000	0.7000	0.0000

12. Moreover, the three factors are assumed to measure a common factor (i.e., level-two factor f_4). Therefore, the structural equation is assumed to be as follows:

$$f_j = \gamma_j f_4 + \zeta_j, \quad j = 1, 2, 3. \tag{4.20}$$

The model assumed here is shown in Figure 4.2.

The true values of factor loadings are also shown in the figure. The variance of errors for the observed variables is 0.5; $var(f_4)$ was fixed at 1, and $var(\zeta_1) = 0.51$, $var(\zeta_2) = 0.64$, and $var(\zeta_3) = 0.75$, so that $var(f_1) = var(f_2) = var(f_3) = 1$.

In this model setup, estimation using Bartlett factor scores, the method using regression factor

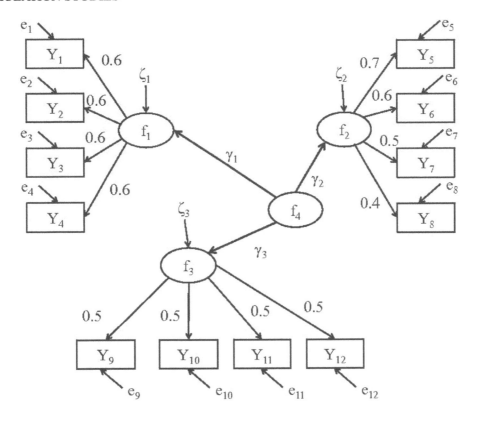

Figure 4.2 *Path diagram representation of model in Study 2.*

scores, and the method proposed by Skrondal and Laake (2001), all produce the same estimates, so the following three methods are compared: (1) factor analysis using estimated factor scores by Bartlett's method (Bart), (2) the proposed method (Prop), and (3) the MLE method in SEM.

To calculate the mean and the MSE of estimates of λ_j, $j = 1, 2, 3$, $N = 300$ and $10,000$ data sets were generated; the true values and the corresponding results are reported in Table 4.3.

Table 4.3 *Resulting estimates from the three methods for Study 2.*

N	Method	$\gamma_1 = 0.7$	$\gamma_2 = 0.6$	$\gamma_3 = 0.5$	MSE
	Bart	0.6085	0.5115	0.4058	0.0544
300	Prop	0.7038	0.6036	0.4969	0.0429
	MLE	0.7051	0.6038	0.4971	0.0388
	Bart	0.6035	0.5074	0.4089	0.0269
10,000	Prop	0.7005	0.5995	0.5007	0.0012
	MLE	0.7005	0.5998	0.5008	0.0010

These results indicate that even when the number of subjects is very large (i.e., $N = 10,000$), the factor score regression by the Bartlett method (i.e., the regression method, Skrondal and Laake, 2001) underestimates γ_j to a large extent. On the other hand, the proposed method and the MLE method produce valid estimates even for relatively small sample sizes (i.e., $N = 300$).

54   FACTOR SCORE REGRESSION

4.5.3   Study 3: Factor analysis and two-parameter logistic models

In Study 1, each measurement model is a factor analysis model. In this study, one factor is measured by continuous variables (in this case, the measurement model is a factor analysis), and the indicators of the other factor are binomial (in this case, the measurement model is the two-parameter logistic item response model (4.3)).

The true value of β was changed from -0.8 to 0.8, with increments of 0.2, and the number of subjects is $N = 200$ and 500. The number of items (i.e., indicators) is set to 20 and 50. For each model setup, we generated $2,200$ data sets, in order to obtain $2,000$ data sets that do not yield improper solutions by the three methods for each setup; hence, the resulting number of data sets was $79,200$. The data sets that yield improper solutions are discarded and not included in the study. The resulting number of data sets in each setup does not come up to $2,000$, when the number of items is 50 and $N = 200$, because the proportion of improper solutions in estimating polychoric and polyserial correlations is large. The total number of data sets included in the study was $67,358$.

The following three methods are carried out: (1) the proposed method (Prop), (2) the ordinary factor score regression (Old), and (3) the generalized least square estimation using estimated polychoric and polyserial correlations (poly-GLS).

Data generation was carried out by SAS/IML, and the estimation was performed using SAS/IML, Bilog-MG, and M-plus. The resulting mean and MSE of each estimate calculated using the $1,000$ data sets are listed in Table 4.4.

There are five points that should be noted. (1) The estimates of the proposed method are more accurate than the ordinary factor score regression in most of the model setups. (2) As the number of subjects increases, the estimates from the proposed method move closer to the true values. This is not the case in ordinary factor score regression. (3) As the number of subjects increases, both the estimates from the proposed method and ordinary factor score regression move closer to the true value. However, even when the number of items is 50, the MSEs of the ordinary method are very large compared to those of proposed method. This result is consistent with the theoretical investigation in Section 4.7; see (4.29). (4) The proposed method is sometimes biased compared to GLS using polyserial/polychoric correlations; however, it is not always biased. (5) GLS using polyserial/polychoric correlations yields improper solutions at a high rate, while the proposed method scarcely yields improper solutions (see Table 4.5). With regard to (4) and (5), the bias of the proposed method versus the polyserial/polychoric approach may be overstated due to the elimination of problematic samples from only the latter approach. Recently developed methods of dealing with polyserial/polychoric correlations could possibly minimize problems in achieving proper solutions (Yuan et al., 2011).

4.5.4   Study 4: Factor analysis and nominal response models

In Study 1, each measurement model is a factor analysis model. In this study, one factor is measured by continuous variables and hence, the measurement model is a factor analysis. The indicators of the other factor are nominal responses (Bock, 1972). In theory, simultaneous estimation such as that based on MLE is applicable to SEM with mixed continuous and nominal observed variables (Moustaki and Knott, 2000; Sammel et al., 1997). However, simultaneous estimation methods are generally very difficult for researchers to use in applied areas because the evaluation of the likelihood requires numerical integration, which becomes impractical in reasonably sized models. There is no prevailing software to deal with this model.

The true value of β is changed from -0.8 to 0.8, with increments of 0.2, and the number of subjects is $N = 200$ or 500. The number of items (indicators) is set to 25.

For each model setup, $2,000$ data sets were generated; hence, the resulting number of data sets was $36,000$. Data sets that yield improper solutions are discarded and not included in this study. The following two methods are carried out: (1) the proposed method (Prop) and (2) the ordinary

Table 4.4 *Resulting estimates from the three methods for Study 3.*

N	Item	True	Prop Est	Prop MSE	Old Est	Old MSE	Poly-GLS Est	Poly-GLS MSE
		0.8	0.8078	0.0042	0.6940	0.0143	0.8340	0.0050
		0.6	0.6096	0.0044	0.5233	0.0089	0.6341	0.0063
		0.4	0.4101	0.0042	0.3509	0.0054	0.4280	0.0059
		0.2	0.2076	0.0041	0.1778	0.0035	0.2175	0.0048
	50	0.0	0.0042	0.0040	0.0040	0.0031	0.0540	0.0047
		−0.2	−0.2001	0.0039	−0.1715	0.0038	−0.2190	0.0048
		−0.4	−0.4027	0.0040	−0.3461	0.0060	−0.4280	0.0063
		−0.6	−0.6048	0.0040	−0.5202	0.0095	−0.6337	0.0063
		−0.8	−0.8034	0.0040	−0.6896	0.0157	−0.8302	0.0075
200		0.8	0.8363	0.0063	0.6431	0.0304	0.8250	0.0032
		0.6	0.6339	0.0062	0.4874	0.0175	0.6331	0.0043
		0.4	0.4273	0.0057	0.3287	0.0090	0.4286	0.0047
		0.2	0.2198	0.0053	0.1690	0.0043	0.2223	0.0050
	20	0.0	0.0111	0.0050	0.0083	0.0033	0.0572	0.0051
		−0.2	−0.1987	0.0049	−0.1535	0.0055	−0.2145	0.0046
		−0.4	−0.4080	0.0048	−0.3130	0.0121	−0.4210	0.0045
		−0.6	−0.6166	0.0049	−0.4751	0.0203	−0.6251	0.0039
		−0.8	−0.8231	0.0051	−0.6341	0.0334	−0.8217	0.0032
		0.8	0.8050	0.0015	0.6909	0.0141	0.8121	0.0012
		0.6	0.6088	0.0016	0.5222	0.0078	0.6144	0.0015
		0.4	0.4099	0.0016	0.3522	0.0037	0.4135	0.0016
		0.2	0.2109	0.0016	0.1813	0.0016	0.2110	0.0016
	50	0.0	0.0106	0.0016	0.0091	0.0013	0.0302	0.0017
		−0.2	−0.1904	0.0016	−0.1650	0.0026	−0.2050	0.0016
		−0.4	−0.3931	0.0015	−0.3371	0.0055	−0.4080	0.0015
		−0.6	−0.5955	0.0014	−0.5099	0.0102	−0.6120	0.0013
		−0.8	−0.7959	0.0015	−0.6891	0.0161	−0.8125	0.0011
500		0.8	0.8133	0.0020	0.6497	0.0263	0.8069	0.0012
		0.6	0.6173	0.0021	0.4911	0.0143	0.6103	0.0014
		0.4	0.4160	0.0020	0.3313	0.0064	0.4096	0.0016
		0.2	0.2131	0.0019	0.1708	0.0022	0.2105	0.0017
	20	0.0	0.0996	0.0018	0.0084	0.0013	0.0350	0.0018
		−0.2	−0.1940	0.0018	−0.1543	0.0035	−0.2031	0.0017
		−0.4	−0.3992	0.0017	−0.3171	0.0088	−0.4063	0.0016
		−0.6	−0.6034	0.0018	−0.4823	0.0170	−0.6076	0.0014
		−0.8	−0.8079	0.0018	−0.6430	0.0284	−0.8082	0.0012

factor score regression (Old). Data generation was carried out by SAS/IML, and the estimation was performed using SAS/IML and MULTILOG.

The resulting mean and MSE of each estimate calculated using the 1,000 data sets are listed in Table 4.6. The simulation study shows that estimates from the proposed method are finer than those from ordinary factor score regression in most of the model setups. It is also observed that as the number of subjects increases, the estimates of the proposed method move closer to the true value. This is not the case for ordinary factor score regression.

Table 4.5 *Number of improper solutions for Study 3.*

N	Item	True	Number of data sets with improper solutions		Number of valid data sets
			Poly-GLS	Prop/Old	
200	50	0.8	725	14	1,472
		0.6	710	13	1,494
		0.4	705	12	1,488
		0.2	703	12	1,490
		0.0	706	13	1,487
		−0.2	705	12	1,488
		−0.4	703	11	1,484
		−0.6	708	10	1,482
		−0.8	718	7	1,473
	20	0.8	4	12	2,000
		0.6	4	13	2,000
		0.4	1	12	2,000
		0.2	1	12	2,000
		0.0	5	10	2,000
		−0.2	1	13	2,000
		−0.4	3	7	2,000
		−0.6	3	9	2,000
		−0.8	4	7	2,000
500	50	0.8	27	0	2,000
		0.6	25	0	2,000
		0.4	23	0	2,000
		0.2	25	0	2,000
		0.0	22	0	2,000
		−0.2	23	0	2,000
		−0.4	24	0	2,000
		−0.6	23	0	2,000
		−0.8	23	0	2,000
	20	0.8	1	0	2,000
		0.6	0	0	2,000
		0.4	0	0	2,000
		0.2	0	0	2,000
		0.0	0	0	2,000
		−0.2	0	0	2,000
		−0.4	0	0	2,000
		−0.6	1	0	2,000
		−0.8	0	0	2,000

4.6   Application

As an illustrative analysis, we consider the relationships between purchase intention of cars and the web browsing duration of the corresponding car makers. We use internet audience data provided by Video Research Interactive/Nielsen-Net Ratings Ltd. in Japan. The internet audience data are panel clickstream data on webpage access of approximately about 6,000 panel participants, and the data set includes all URLs, dates, duration of website visits, and referrer information with which panel participants view the website. In addition, the internet audience data include demographic characteristics and lifestyle data given by a questionnaire survey conducted annually. The panel

Table 4.6 *Resulting estimates from the two methods for Study 4.*

N	True	Prop		Old	
		Est	MSE	Est	MSE
200	0.8	0.7950	0.0059	0.5030	0.0903
	0.6	0.6020	0.0052	0.3750	0.0521
	0.4	0.4045	0.0048	0.2491	0.0246
	0.2	0.2012	0.0044	0.1235	0.0075
	0.0	−0.0012	0.0041	−0.0012	0.0017
	−0.2	−0.2046	0.0041	−0.1284	0.0068
	−0.4	−0.4111	0.0041	−0.2549	0.0231
	−0.6	−0.6098	0.0046	−0.3823	0.0490
	−0.8	−0.7975	0.0054	−0.5065	0.0890
500	0.8	0.7662	0.0033	0.4933	0.0949
	0.6	0.5776	0.0024	0.3670	0.0545
	0.4	0.3877	0.0018	0.2435	0.0251
	0.2	0.1992	0.0016	0.1241	0.0064
	0.0	0.0042	0.0015	0.0030	0.0006
	−0.2	−0.1950	0.0016	−0.1210	0.0068
	−0.4	−0.3962	0.0016	−0.2465	0.0242
	−0.6	−0.5948	0.0018	−0.3719	0.0526
	−0.8	−0.7875	0.0020	−0.5003	0.1029

participants were selected by the random digit dialing sampling method; therefore we can regard this data as a nationwide random sampling of consumers in Japan.

In web-marketing practice, it is important to know why some visitors stay longer on their websites or web services. By understanding this, marketing managers can determine the contents and design of their websites. In this application, we use the factors expressing intention to buy cars — measured in the questionnaire survey — as independent variables, and the logarithm of the website durations (in seconds) of the top five car makers in Japan (i.e., Toyota, Nissan, Honda, Mazda, Mitsubishi).

We restrict our analysis to the respondents who remained panel participants for the three months and those who viewed the target websites at least once during the observation period (i.e., October 2006 to March 2007). The resulting number of the respondents was $N = 805$. The purchase intents were asked by the survey in October 2006, as dichotomous (pick-any) items in which the respondents are asked whether they are willing to purchase a particular car or not. The number of cars considered here varies according to the companies: 23 for Toyota, 12 for Nissan, 10 for Honda, 6 for Mazda, and 10 for Mitsubishi. We use IRT modeling to consider maker-specific latent variable f_b of purchase intent for the cars made by car maker b. For example, the purchase intent factor for Toyota is measured by 15 dichotomous items (see Figure 4.3, where Y_{bj} represents the item for purchase intent for car $j = 1, \cdots, p_b$, of car maker $b = 1, \cdots, 5$).

Let v_b be logarithm of the sum of the website durations of car maker b during the observation period. The only information available here is maker-level aggregated duration (e.g., the browsing duration for Toyota is the sum of duration of the whole Toyota website pages), because it is very difficult to specify the specific car brand website, and some brand-specific websites are not constructed by the car makers. We can consider the following car-maker specific regression relationships:

$$v_b = \mu_b + \beta_b f_b + e_b, \quad b = 1, \cdots, B, \tag{4.21}$$

but it is more natural to assume the product-category level factor f_{car} behind the car-maker-specific factors $f_1, \cdots, f_B$, and it is also useful to separate the influence, by product-category level, intent to

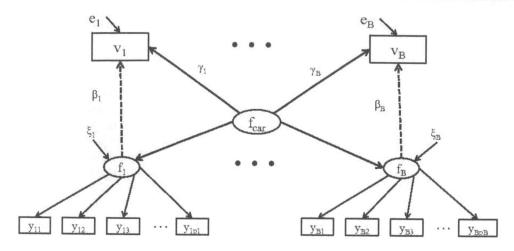

Figure 4.3 *Path diagram representation of model for study on web browsing duration and purchase intent.*

Table 4.7 *Results for the study on web browsing duration and purchase intent.*

Maker	Category Level		Maker-specific	
	Est	SE	Est	SE
Toyota	0.3093	0.1063	0.1705	0.0905
Nissan	0.3182	0.1308	0.1530	0.0894
Honda	0.1821	0.1349	0.2755	0.1094
Mazda	0.1322	0.1119	0.2091	0.1101
Mitsubishi	0.1409	0.1285	0.1605	0.1296

purchase from that by maker-specific intent, as follows:

$$v_b = \mu_b + \beta_b f_b + \gamma_b f_{car} + e_b, \quad b = 1, \cdots, B, \tag{4.22}$$

to understand the effect of loyalty to specific car maker on web browsing behavior.

The whole model contains 61 dichotomous variables, 5 continuous variables, five factors, and one two-level factor. Unfortunately, we could not estimate parameters by using full information MLE, and the model size is similarly problematic for polyserial/polychoric methodology. Therefore, we calculate the estimates by using the stepwise method proposed here. First, the latent variable scores for $f_1, \cdots, f_5$, are estimated separately using the five IRT models. The parameters of the two-level factor analysis model are calculated as conducted in Study 2.

The standardized coefficients and their SEs are shown in Table 4.7. It is easy to see that the influence of the product-category level factor f_{car} is very strong for Toyota and Nissan, which implies that the individuals who stay longer have a higher product-category level need for new cars, and browse these websites to search information useful in deciding what car they might purchase, not due to loyalty to these makers.

On the other hand, the influence of the car-maker-specific factor is stronger for Honda, which might be due to higher loyalty to Honda. From the results, one could recommend that the marketing managers of Honda should design their website to make an appeal for their originality, while managers of Toyota and Nissan should optimize their website to help consumers to compare and find advantages of their cars over competing cars.

4.7   Theoretical details

4.7.1   Generalized least squares as the fourth step

Following results shown in Lee *et al.* (1990), the resulting estimate of $\boldsymbol{\tau}$ at the fourth step can be shown to follow a multivariate normal distribution with mean $\boldsymbol{\tau}$ and covariance matrix $\left\{(\partial\boldsymbol{\Sigma}(\boldsymbol{\tau})/\partial\boldsymbol{\tau})^{\top}\mathbf{W}^{-1}(\partial\boldsymbol{\Sigma}(\boldsymbol{\tau})/\partial\boldsymbol{\tau})\right\}^{-1}$ asymptotically. Also, since it has been shown that $2Q(\boldsymbol{\tau})$ asymptotically follows a χ^2 distribution, we can execute hypothesis testing in the usual way (see Lee *et al.*, 1990, for details).

To use the above results, we must calculate the covariance matrix of $\widehat{\boldsymbol{\Sigma}}$ and $\mathbf{W}$. Let $\widehat{\boldsymbol{\phi}} = (\widehat{\mathbf{v}}^{\top}, vec(\widehat{\boldsymbol{\Phi}})^{\top})^{\top}$ and $\widetilde{\boldsymbol{\phi}}_D = vec(\widetilde{\boldsymbol{\Phi}}_D)$, where $\boldsymbol{\phi}_D$ is the vector containing all elements of $\boldsymbol{\Phi}_{jj}$, $j = 1\cdots,J$, and $vec(\mathbf{X})$ is the "rolled out" vector of columns of $\mathbf{X}$. Also, let $vec(\mathbf{f}, \boldsymbol{\phi}_D) = (\mathbf{f}^{\top}, \boldsymbol{\phi}_D^{\top})^{\top}$ and $vec(\widehat{\boldsymbol{\Phi}}, \widetilde{\boldsymbol{\phi}}_D) = (\widehat{\boldsymbol{\phi}}^{\top}, \widetilde{\boldsymbol{\phi}}_D^{\top})^{\top}$. For simplicity, we suppress the dimensions of the multivariate normal distributions in what follows.

Following pseudo-MLE theory (Parke, 1986; Gong and Samaniego, 1981),

$$vec(\widehat{\mathbf{f}}^B, \widetilde{\boldsymbol{\phi}}_D) \quad \sim \quad N(vec(\mathbf{f}, \boldsymbol{\phi}_D), \boldsymbol{\Sigma}_{\mathbf{f}, \boldsymbol{\phi}_D}), \tag{4.23}$$

where $\boldsymbol{\Sigma}_{\mathbf{f}, \boldsymbol{\phi}_D}$ is estimated as the inverse of the Fisher information matrix when the fixed effect model (4.7) is employed. Then, we obtain

$$vec(\widehat{\boldsymbol{\Phi}}, \widetilde{\boldsymbol{\phi}}_D) \quad \sim \quad N(vec(\boldsymbol{\Phi}, \boldsymbol{\phi}_D), \mathbf{A}\boldsymbol{\Sigma}_{\mathbf{f}, \boldsymbol{\phi}_D}\mathbf{A}^{\top}), \tag{4.24}$$

where

$$\mathbf{A} \quad = \quad \begin{pmatrix} \mathbf{I} \otimes \mathbf{a}^{\top} & 2\mathbf{N}_p(\mathbf{F}^{\top}\mathbf{R} \otimes \mathbf{I}_p) & \mathbf{0} \\ \mathbf{0} & \mathbf{0} & \mathbf{I} \end{pmatrix},$$

$\mathbf{N}_p = \frac{1}{2}(\mathbf{I}_{p^2} + \mathbf{K}_{pp})$, $\mathbf{a} = \frac{1}{N}\mathbf{1}$, $\mathbf{K}_{pp}$ is a commutation matrix, p is the dimension of $\boldsymbol{\Phi}$, and "$\otimes$" is the Kronecker product operator (Magnus and Neudecker, 1999).

Then $\mathbf{W}$ is expressed as $\mathbf{W} = \mathbf{P}\mathbf{A}\boldsymbol{\Sigma}_{\mathbf{f}, \boldsymbol{\phi}_D}\mathbf{A}^{\top}\mathbf{P}^{\top}$, where $\mathbf{P}$ is the appropriate permutation matrix.

4.7.2   Sketch of proof of Proposition 3

Let $\widehat{\mathbf{f}}_j^B$ be the estimate of the factor score vector in the jth measurement model. Also, let

$$\begin{pmatrix} \mathbf{I}_{\mathbf{f}_j} & \mathbf{I}_{\mathbf{f}_j\boldsymbol{\xi}_j} \\ \mathbf{I}_{\boldsymbol{\xi}_j, \mathbf{f}_j} & \mathbf{I}_{\boldsymbol{\xi}_j} \end{pmatrix}$$

be the Fisher information matrix for factor scores and structural parameters of the fixed effect model (4.7). Using the results of pseudo-MLE (Yuan and Bentler, 2007; Gong and Samaniego, 1981),

$$\begin{pmatrix} \widehat{\mathbf{f}}_j^B(\widetilde{\boldsymbol{\xi}}_j) \\ \widetilde{\boldsymbol{\xi}}_j \end{pmatrix} \quad \sim \quad N\left(\begin{pmatrix} \mathbf{f}_j \\ \boldsymbol{\xi}_j \end{pmatrix}, \begin{pmatrix} \mathbf{I}_{\mathbf{f}_j}^{-1} + \mathbf{I}_{\mathbf{f}_j}^{-1}\mathbf{I}_{\mathbf{f}_j\boldsymbol{\xi}}\boldsymbol{\Sigma}_{\boldsymbol{\xi}_j}\mathbf{I}_{\boldsymbol{\xi}_j, \mathbf{f}_j}\mathbf{I}_{\mathbf{f}_j}^{-1} & -\mathbf{I}_{\mathbf{f}_j}^{-1}\mathbf{I}_{\mathbf{f}_j\boldsymbol{\xi}}\boldsymbol{\Sigma}_{\boldsymbol{\xi}_j} \\ -\boldsymbol{\Sigma}_{\boldsymbol{\xi}_j}\mathbf{I}_{\boldsymbol{\xi}_j, \mathbf{f}_j}\mathbf{I}_{\mathbf{f}_j}^{-1} & \boldsymbol{\Sigma}_{\boldsymbol{\xi}_j} \end{pmatrix}\right), \tag{4.25}$$

where $\widetilde{\boldsymbol{\xi}}_j$ is the marginal MLE in the first step with variance $\boldsymbol{\Sigma}_{\boldsymbol{\xi}_j}$ (calculated by the inverse of the Fisher information matrix in the random effect model), and $\widehat{\mathbf{f}}_j^B(\widetilde{\boldsymbol{\xi}}_j)$ is the estimate of factor scores in the second step with $\widetilde{\boldsymbol{\xi}}_j$ given.

Because the mean of the marginal distribution is equal to the mean of the conditional distribution, it follows that

$$\underset{P_j \to \infty}{\mathrm{plim}}\, E(\widehat{\mathbf{f}}_j^B) \quad = \quad \underset{P_j \to \infty}{\mathrm{plim}}\, E\{E(\widehat{\mathbf{f}}_j^B|\mathbf{f}_j)\} \quad = \quad \mathbf{v}_j. \tag{4.26}$$

From the relationship between the variance of the marginal distribution and that of the conditional distribution, it is also shown that

$$\operatorname*{plim}_{P_j \to \infty} var(\widehat{\mathbf{f}}_j^B) = \operatorname*{plim}_{P_j \to \infty} \left[var\{E(\widehat{\mathbf{f}}_j^B | \mathbf{f}_j)\} + E\{var(\widehat{\mathbf{f}}_j^B | \mathbf{f}_j)\} \right] = var(\mathbf{f}_j) + \operatorname*{plim}_{P_j \to \infty} E\{var(\widehat{\mathbf{f}}_j^B | \mathbf{f}_j)\}. \quad (4.27)$$

From the definition, the covariance matrix of $\mathbf{f}_j$ is the true variance of factor $\mathbf{\Phi}_j$; $E\{var(\widehat{\mathbf{f}}_j^B | \mathbf{f}_j)\}$ is the variance of the factor score estimate $\widehat{\mathbf{f}}_j^B$, with the true factor score $\mathbf{f}_j$ given.

If the structural parameters in the measurement models are known, $E\{var(\widehat{\mathbf{f}}_j^B | \mathbf{f}_j)\}$ is the inverse matrix of the Fisher information matrix $\mathbf{I}_{\mathbf{f}_j}$ (or "test information matrix" in the area of educational statistics) with the structural parameters given in the jth fixed effect measurement model (4.7).

However, if we do not know the true values of structural parameters and if we evaluate the influence of the estimation in the first step, the second term in the right side of (4.27) can be expressed as follows, in the case of a large number of indicators:

$$\operatorname*{plim}_{P_j \to \infty} E\{var(\widehat{\mathbf{f}}_j^B | \mathbf{f}_j)\} = \mathbf{I}_{\mathbf{f}_j}^{-1} + \mathbf{I}_{\mathbf{f}_j}^{-1} \mathbf{I}_{\mathbf{f}_j \boldsymbol{\xi}_j} \boldsymbol{\Sigma}_{\boldsymbol{\xi}_j} \mathbf{I}_{\boldsymbol{\xi}_j \mathbf{f}_j} \mathbf{I}_{\mathbf{f}_j}^{-1}; \quad (4.28)$$

see Hoshino and Shigemasu (2008) for concrete expression.

It is shown that the sample covariance matrix of $\mathbf{f}_j$ using the factor scores (calculated in the third step) is inconsistent for estimating $\mathbf{\Psi}_{jj}$, because the following equation holds:

$$\operatorname*{plim}_{N, P_j \to \infty} \frac{1}{N} \sum_{i=1}^{N} (\widehat{\mathbf{f}}_{ji}^B - \overline{\mathbf{f}}_j^B)(\widehat{\mathbf{f}}_{ji}^B - \overline{\mathbf{f}}_j^B)^\top = \mathbf{\Phi}_{jj} + \mathbf{G}, \quad (4.29)$$

where $\overline{\mathbf{f}}_j^B = \frac{1}{N} \sum_{i=1}^{N} \widehat{\mathbf{f}}_j^B$ and $\mathbf{G} = \operatorname*{plim}_{N, P_j \to \infty} \frac{1}{N} \sum_{i=1}^{B} (\mathbf{I}_{\mathbf{f}_j}^{-1} + \mathbf{I}_{\mathbf{f}_j}^{-1} \mathbf{I}_{\mathbf{f}_j \boldsymbol{\xi}_j} \boldsymbol{\Sigma}_{\boldsymbol{\xi}_j} \mathbf{I}_{\boldsymbol{\xi}_j \mathbf{f}_j} \mathbf{I}_{\mathbf{f}_j}^{-1})$.

On the other hand, the sample covariance matrix between $\mathbf{f}_j$ and $\mathbf{f}_k$ using estimated factor scores is consistent for estimating $\mathbf{\Phi}_{jk}$:

$$\operatorname*{plim}_{N, P_j \to \infty} \widehat{\mathbf{\Phi}}_{jk} = \operatorname*{plim}_{N, P_j \to \infty} \frac{1}{N} \sum_{i=1}^{N} (\widehat{\mathbf{f}}_{ji}^B - \overline{\mathbf{f}}_j^B)(\widehat{\mathbf{f}}_{ki}^B - \overline{\mathbf{f}}_k^B)^\top = cov(\widehat{\mathbf{f}}_{ji}^B, \widehat{\mathbf{f}}_{ki}^B) = cov(\mathbf{f}_{ji}, \mathbf{f}_{ki}) = \mathbf{\Phi}_{jk}. \quad (4.30)$$

It follows that $\widetilde{\boldsymbol{\xi}}_j$ and $\widetilde{\boldsymbol{\xi}}_k$ are mutually independent for $j \neq k$, because each $\boldsymbol{\xi}_s$ is estimated for each measurement equation. Therefore, we obtain that $cov(\widehat{\mathbf{f}}_j^B, \widehat{\mathbf{f}}_k^B | \mathbf{f}_j, \mathbf{f}_k) = \mathbf{0}$, for $\forall j \neq k$. Therefore, we observe that $\widehat{\mathbf{\Phi}}^P$ is a consistent estimate of $\mathbf{\Phi}$ when the number of indicators is large.

4.8 Discussion

In this paper, we resolved the reason why estimates of the parameters related to the relationship between the latent variables, using estimated factor scores, are biased, even when the number of subjects is large. We also proposed a modified stepwise estimation method using estimated factor scores and proved some relevant asymptotic properties. The simulation studies indicated that the proposed method is valid under several model setups, while ordinary factor score regression is biased even for very large sample sizes. The proposed methodology has the following advantages:

1. The analysis becomes practical when the measurement scale of variables differs with the measurement models. For example, assume that the indicators of factor A in the first measurement model are continuous, and the indicators of factor B in the second measurement model are nominal. The MLE in such a model is impractical; however, it is very easy to infer the relationship between factors A and B by the proposed method, because most of available programs, such as SAS/STAT, solve the first measurement model. The second measurement model can be solved by using software such as MULTILOG. We can estimate the relationships between factors using the outputs of the available software.

2. The proposed method can eliminate the interpretational confounding problem in which the model setup of the structural equation part of a model affects the estimation of the measurement part, because the proposed method separates the estimation of parameters in the measurement model (i.e., the first step) from the parameters in the structural relations (i.e., the third and fourth parts).

3. Data sets with very a large number of variables can be modeled. The original raw data are not necessary to estimate the parameters in the structural part. The estimated factor scores, the estimates of the measurement parts, and their variance matrices are sufficient. With 30 sets of 10 variables, each measuring one latent factor, there are 300 observed variables but only 30 factors. Simultaneous analysis of all 300 variables is most likely impossible, but the proposed method makes the analysis for relationships between factors to be easily executed.

4. The proposed method can include variations in the estimates caused by factor rotation. With the exception of a recently developed method (Asparouhov and Muthèn, 2009), simultaneous estimation methods fail to incorporate factor rotation into the whole analysis, which leads them to underestimate the SEs of the parameter estimates in the structural part, if the rotated factor loadings are set as constants. The proposed method can evaluate the variation due to factor rotation by calculating the correct variance matrix of the rotated factor loadings (Ogasawara, 1998).

5. The proposed method diminishes improper solutions as compared to the previously proposed simultaneous estimation methods.

As discussed previously, some type of factor score regression is frequently used to analyze large scale data. The item parceling method that is most frequently used in psychology and the social and behavioral sciences, can be considered as a coarse variant of factor score regression. Several simulation studies have shown that item parceling causes the parameters in the structural part to be underestimated (Bandalos, 2002; MacCallum *et al.*, 1999; Marsh *et al.*, 1998). The theoretical investigations in Section 4.4 and the appendices also show that correlations or regression coefficients are underestimated in studies in which scale scores (i.e., parcel scores) are used to estimate factor scores. The item parceling method using scale scores can be expected to produce a coarser estimate than the factor score regression using Bartlett's or regression estimates; thus, they are also expected to cause a more serious bias. It is clear that this issue requires further research.

In our development of IRT models, we emphasized the classical unidimensional latent trait model based on binary or nominal data. However, our approach is in no way limited to unidimensional measurement models. A multidimensional IRT model, possibly based on polytomous items, can certainly be utilized. Estimates of the underlying factor scores can subsequently be obtained and utilized in our methodology. However, when the variables are not factorially complex, i.e., they do not have loadings on more than one factor, then the simpler approach of proceeding factor by factor is an attractive alternative, since it can avoid integrals associated with marginal MLE that can become intractable as the number of factors increases. Although computational improvements in handling multidimensional IRT models are being developed (An and Bentler, 2012; Wu and Bentler, 2012; An and Bentler, 2011; Cai, 2010), it is possible that they will not be competitive with the factor by factor approach when this is feasible. Research will have to evaluate these alternatives in realistic contexts.

The measurement models discussed in this chapter, regardless of the scale level of the variables involved, are of the standard factor analytic type, where variation in the factor generates variation in observed variables. Another type of measurement model, sometimes called a formative or causal indicator model, also exists. In such models, variation in the observed variables is presumed to generate variation in the latent variable, i.e., the latent variable is generated by the observed variables (Bollen and Bauldry, 2011; Hardin *et al.*, 2011). While formative models can be specially constructed so that standard factor analytic measurement models can be applied (Treiblmaier *et al.*, 2011), and hence, they become amenable to our approach, research is needed to determine whether factor score regression could be implemented more directly with standard approaches to formative measurement.

Chapter 5

Joint modeling of mixed count and continuous longitudinal data

Jian Kang and Ying Yang

5.1 Introduction

The main motivating data set comes from the Interstitial Cystitis Data Base (ICDB) (Propert *et al.*, 2000). Interstitial cystitis (IC) is a painful condition due to inflammation of tissues of the bladder wall. The cause remains unknown and there are no effective diagnostic tools. It is usually diagnosed by ruling out other conditions such as sexually transmitted disease, bladder cancer, and bladder infections. Some studies (Kirkemo *et al.*, 1997) suggest that IC can be characterized by at least one of the following symptoms: pain in the pelvic or bladder area, urgency (pressure to urinate), and frequency of urination. It has been shown that the different symptom patterns are caused by different underlying disease mechanisms. Thus, to get a better understanding of this chronic disease, it is important to determine whether symptoms tend to co-fluctuate together (suggesting a single underlying aetiology) or vary independently (suggesting multiple mechanisms at work). The ICDB kept track of pain, urgency, and urinary frequency, for a prevalent cohort of patients with IC over a couple of months. For each patient, the pain score, urinary urgency, and urinary frequency were observed as three major longitudinal outcomes. Also, we have the demographic characteristics and clinical biomarkers of patients which were also recorded over time. The pain score and urinary urgency take ordinal/continuous values, while the urinary frequency is an integer that counts the number of urinary episodes during a given period. It is thus natural to assume the urinary frequency to follow Poisson distribution. The goal of the analysis is to use regression models to investigate the association between demographic/clinical biomarkers and IC-symptom variables. Also, it is of interest to study the change in correlation over time between the continuous and Poisson outcomes which will provide implications on the disease mechanism. There are missing values in the longitudinal outcomes because some patients drop out of the study. This motivates the need for developing statistical methods to deal with missing data.

In this chapter, we review different regression models developed to jointly analyze mixed Poisson and continuous longitudinal outcomes. We discuss how to handle missing data using these regression models. In addition, we propose a new imputation method for bivariate longitudinal data with monotone missing data patterns. There are two types of models we consider: semiparametric and parametric models. Yang *et al.* (2007) develop several semiparametric regression methods by modeling the marginal expectations of Poisson and continuous outcomes and then exploring the changes in the correlations between the two outcomes. More recently, Yang and Kang (2010) proposed a fully parametric model using an extension of the multivariate Poisson model (Karlis and Meligkotsidou, 2005; Karlis, 2003), taking into account the over-dispersion in the data. They assume an explicit covariance structure for the Poisson longitudinal outcomes observed at different time points, which then determines the joint distribution of the multiple Poisson outcomes. Given the Poisson outcomes, they model the conditional distribution of continuous longitudinal outcomes by the multivariate normal distribution. For parameter estimation, they propose an expectation-conditional maximization (ECM) algorithm (McLachlan and Krishnan, 1997; Meng and Rubin, 1993) to obtain the maximum likelihood estimate (MLE). A likelihood ratio testis proposed as well, to assess the significance of effects of covariates on the longitudinal outcomes. This method is different from those based on the generalized estimating equations (GEE) approach, which only needs the mean and working covariance functions rather than a specific joint distribution for the Poisson outcomes.

Missing data are very common in biomedical and clinical research, and many statistical methods have been developed motivated by different problems (Ibrahim *et al.*, 2005; Little and Rubin, 2002; Roy and Lin, 2002; Ibrahim *et al.*, 2001; Troxel *et al.*, 1998). The ICDB also involves some missing values, and it is unclear what the true missing data mechanism is in this data set. Yang and Kang (2010) assumed that the missing data mechanisms for the two outcomes are independent but nonignorable, namely, dependent on both the observed and missing data for the two outcomes. A logistic regression is then used to model the missing data mechanism, with the full likelihood specified from a selection model. A Monte Carlo EM (MCEM) algorithm (Zhu *et al.*, 2007; Booth and Hobert, 1999; Delyon *et al.*, 1999; Wei and Tanner, 1990) is proposed to obtain the MLE. This algorithm is developed especially to solve intractable multidimensional integrations. In addition to the likelihood approach, imputation based methods are also popular for modeling missing data. To the best of our knowledge, there are no imputation methods developed especially for bivariate longitudinal data analysis. In this chapter, we introduce a sequential imputation method for this purpose.

This chapter is organized as follows. In Section 5.2, we discuss several models for joint analysis of mixed Poisson and continuous outcomes, including both semiparametric and parametric models, assuming no missing data; these models are factorization models, a brief survey of which is provided in Chapter 1. We then present methods to model missing data in Section 5.3. The models and methods are then applied to ICDB data in Section 5.4. A discussion is provided in Section 5.5.

5.2   Complete data model

In this section, we review models proposed by Yang *et al.* (2007) and Yang and Kang (2010). We first assume that there is no missing data. Let n be the number of subjects, T be the number of measurements over time, and P be the number of covariates in the model. For $i = 1, \cdots, n$, and $j = 1, \cdots, T$, let X_{ij}, Y_{ij} and $\mathbf{z}_{ij}$, respectively denote the Poisson outcome, continuous outcome, and the vector of covariates measured at time t_j, where $\mathbf{z}_{ij} = (Z_{ij1}, \cdots, Z_{ijP})^\top$. We also set $\mathbf{X}_i = (X_{i1}, \cdots, X_{iT})^\top$, $\mathbf{Y}_i = (Y_{i1}, \cdots, Y_{iT})^\top$, and $\mathbf{Z}_i = (\mathbf{z}_{i1}, \cdots, \mathbf{z}_{iT})^\top$.

5.2.1   GEE approach

To use the GEE approach, it is necessary to specify the mean and covariance structure of the bivariate longitudinal data. For the mean structure, the marginal mean of the Poisson outcome is specified based on a log-linear regression model, and given the Poisson outcome, the conditional mean of the continuous outcome is modeled by a linear regression model, where the Poisson outcome is included as a covariate, i.e.,

$$\begin{aligned}
\log\{E(\mathbf{X}_i)\} &= \mathbf{Z}_i^\top \boldsymbol{\beta}_1, \\
E(Y_{ij}|\mathbf{X}_i) &= \mathbf{z}_{ij}^\top \boldsymbol{\beta}_2 + \gamma_1(X_{ij} - \lambda_{ij}) + \gamma_2 S_i,
\end{aligned}$$

where $\lambda_{ij} = \exp(\mathbf{z}_{ij}^\top \boldsymbol{\beta}_1)$ and $S_i = \sum_{j=1}^{T}(X_{ij} - \lambda_{ij})$. Parameters γ_1 and γ_2 characterize the association between $\mathbf{X}_i$ and $\mathbf{Y}_i$, where γ_1 represents the effect of X_{ij} on Y_{ij}, and γ_2 reflects the average effect of $\mathbf{X}_i$ on Y_{ij}. For simplicity, let $\mathbf{W}_i = (\mathbf{W}_{i1}, \cdots, \mathbf{W}_{iT})$, with $\mathbf{W}_{ij} = (\mathbf{z}_{ij}^\top, X_{ij} - \lambda_{ij}, S_i)^\top$, and $\boldsymbol{\alpha} = (\boldsymbol{\beta}_2^\top, \gamma_1, \gamma_2)^\top$. Then, $E(Y_{ij}|\mathbf{X}_i) = \mathbf{W}_{ij}^\top \boldsymbol{\alpha}$. The covariance for Poisson and continuous outcomes can be approximated by a compound symmetry covariance structure, i.e.,

$$cov(\mathbf{X}_i) = \mathbf{V}_{1i} \approx \boldsymbol{\Delta}_i^{1/2}\{(1-\rho_X)\mathbf{I} + \rho_X \mathbf{J}\}\boldsymbol{\Delta}_i^{1/2},$$

$$cov(\mathbf{Y}_i|\mathbf{X}_i) = \mathbf{V}_{2i} \approx \sigma^2\{(1-\rho_Y)\mathbf{I}+\rho_Y\mathbf{J}\},$$

where ρ_X and ρ_Y represent the within subject correlation for Poisson and continuous outcomes, respectively; $\mathbf{\Delta}_i$ is a diagonal matrix with elements $var(X_{ij}) = \lambda_{ij}$, $\mathbf{I}$ is a $T \times T$ identity matrix, and $\mathbf{J}$ is a $T \times T$ matrix of 1s. According to the above mean and covariance specifications, the following estimating equations can be derived based on routine calculations:

$$S(\mathbf{\Theta}) = \sum_{i=1}^n \begin{pmatrix} \mathbf{\Delta}_i\mathbf{Z}_i & -(\gamma_1+\gamma_2)\mathbf{\Delta}_i\mathbf{Z}_i \\ \mathbf{0} & \mathbf{W}_i \end{pmatrix} \begin{pmatrix} \mathbf{V}_{1i}^{-1} & \mathbf{0} \\ \mathbf{0} & \mathbf{V}_{2i}^{-1} \end{pmatrix} \begin{pmatrix} \mathbf{X}_i - \mathbf{\lambda}_i \\ \mathbf{Y}_i - \mathbf{W}_i^\top\mathbf{\alpha} \end{pmatrix} = \mathbf{0}, \quad (5.1)$$

where $\mathbf{\lambda}_i = (\lambda_{i1},\cdots,\lambda_{iT})^\top$ and $\mathbf{\Theta} = (\mathbf{\beta}_1^\top,\mathbf{\alpha}^\top,\rho_X,\rho_Y,\sigma^2)^\top$. To solve the above estimating equations, an iterative algorithm is suggested, see Yang et al. (2007) for details. The basic idea is to partition the parameters into two parts, $\{\mathbf{\beta}_1,\mathbf{\alpha}\}$ and $\{\rho_X,\rho_Y,\sigma^2\}$. Note that given estimates of $\mathbf{\beta}_1$ and $\mathbf{\alpha}$, we can obtain the regression residuals, based on which we can get estimates of ρ_X, ρ_Y, and σ^2. On the other hand, given ρ_X, ρ_Y, and σ^2, (5.1) can be easily solved to obtain estimates of $\mathbf{\beta}_1$ and $\mathbf{\alpha}$. Following the standard theory of GEE (Liang and Zeger, 1986), it can be shown that this algorithm produces consistent asymptotically normal estimates of $\mathbf{\beta}_1$ and $\mathbf{\alpha}$. A robust "sandwich" estimate for the covariance matrix of $\widehat{\mathbf{\beta}}_1$ and $\widehat{\mathbf{\alpha}}$ is given by

$$\mathbf{\Sigma}_e = \mathbf{I}_0^{-1}\mathbf{I}_1\mathbf{I}_0^{-1}, \quad (5.2)$$

where $\mathbf{I}_0 = \sum_{i=1}^n \mathbf{A}_i\mathbf{B}_i\mathbf{A}_i^\top$, $\mathbf{I}_1 = \sum_{i=1}^n \mathbf{A}_i\mathbf{B}_i\widehat{cov(\mathbf{X}_i,\mathbf{Y}_i)}\mathbf{B}_i\mathbf{A}_i^\top$,

$$\mathbf{A}_i = \begin{pmatrix} \mathbf{\Delta}_i\mathbf{Z}_i & -(\gamma_1+\gamma_2)\mathbf{\Delta}_i\mathbf{Z}_i \\ \mathbf{0} & \mathbf{W}_i \end{pmatrix}, \quad \mathbf{B}_i = \begin{pmatrix} \mathbf{V}_{1i}^{-1} & \mathbf{0} \\ \mathbf{0} & \mathbf{V}_{2i}^{-1} \end{pmatrix},$$

and $\widehat{cov(\mathbf{X}_i,\mathbf{Y}_i)} = \{\mathbf{X}_i-\mathbf{\lambda}_i(\widehat{\mathbf{\beta}}_1),\mathbf{Y}_i-\mathbf{W}_i\widehat{\mathbf{\alpha}}\}\{\mathbf{X}_i-\mathbf{\lambda}_i(\widehat{\mathbf{\beta}}_1),\mathbf{Y}_i-\mathbf{W}_i\widehat{\mathbf{\alpha}}\}^\top$. To test the hypothesis

$$H_0: \mathbf{L}\mathbf{\Theta} = 0 \quad \text{vs.} \quad H_1: \mathbf{L}\mathbf{\Theta} \neq \mathbf{0}, \quad (5.3)$$

where $\mathbf{L}$ is a user-defined $k \times (2P+2)$ matrix. The number k suggests the number of restrictions on $\mathbf{\Theta}$ in null hypothesis H_0, and the number $2P+2$ is the dimension of the $\mathbf{\Theta}$. The following test statistic has been shown to have a good performance (Rotnitzky and Jewell, 1990),

$$T = S(\mathbf{\Theta})\mathbf{I}_0^{-1}\mathbf{L}^\top(\mathbf{L}\mathbf{\Sigma}_e\mathbf{L}^\top)^{-1}\mathbf{L}\mathbf{I}_0^{-1}S^\top(\mathbf{\Theta}),$$

which asymptotically follows a chi-square distribution with k degrees of freedom. We reject the null hypothesis if T is larger than a critical value at a given level.

5.2.2 Modeling correlations between outcomes

In this section, we discuss regression models to evaluate the effect of covariates on the sample correlation coefficient between the two longitudinal outcomes. Specifically, we consider the following log-linear model:

$$\log\left(\frac{1+r_j}{1-r_j}\right) = \overline{\mathbf{Z}}_j^\top\mathbf{b}_1 + t_j b_2 + \varepsilon_j, \quad (5.4)$$

for $j = 1,\cdots,T$, where $\overline{\mathbf{Z}}_j$ is the average of the continuous covariate at time t_j, i.e., $\overline{\mathbf{Z}}_j = \sum_{i=1}^n \mathbf{Z}_{ij}/n$; r_j is the sample correlation computed using the Poisson and continuous outcomes measured at time j, and that is,

$$r_j = \frac{\sum_{i=1}^n (X_{ij}-\overline{X}_j)(Y_{ij}-\overline{Y}_j)}{\sqrt{\sum_{i=1}^n (X_{ij}-\overline{X}_j)^2}\sqrt{\sum_{i=1}^n (Y_{ij}-\overline{Y}_j)^2}}, \quad (5.5)$$

where $\overline{X}_j = \sum_{i=1}^{n} X_{ij}/n$ and $\overline{Y}_j = \sum_{i=1}^{n} Y_{ij}/n$. The estimates of coefficients b_1 and b_2 can be obtained by method of least squares.

Alternatively, we can fit the following nonlinear correlation regression model:

$$r_j = \frac{\exp(\overline{\mathbf{Z}}_j^{\top} b_1 + t_j b_2) - 1}{\exp(\overline{\mathbf{Z}}_j^{\top} b_1 + t_j b_2) + 1} + \varepsilon_j, \tag{5.6}$$

for $j = 1, \cdots, T$, model (5.4) or (5.6) is not applicable when some $\mathbf{z}_{ij}$ are categorical random variables. Even if all $\mathbf{z}_{ij}$s are continuous, the averaging might not be the best method to summarize all the information. Yang et al. (2007) suggest an alternative choice which assumes a parametric distribution of $\mathbf{z}_{ij}$ and uses a sufficient statistic of $\mathbf{z}_{ij}$.

5.2.3 Varying-coefficients model

Yang et al. (2007) also propose a varying-coefficients model to address the question of how the covariance between the two longitudinal outcomes changes over time. Specifically, it is assumed that

$$\log(X_{ij} + 1) = \mathbf{z}_{ij}^{\top} \boldsymbol{\beta}_1(t_j) + \varepsilon_{ij1} \quad \text{and} \quad Y_{ij} = \mathbf{z}_{ij}^{\top} \boldsymbol{\beta}_2(t_j) + \varepsilon_{ij2}, \tag{5.7}$$

where the independent random errors $\boldsymbol{\varepsilon}_{ij} = (\varepsilon_{ij1}, \varepsilon_{ij2})^{\top} \sim N_2(\mathbf{0}, \boldsymbol{\Sigma}(t_j))$, with

$$\boldsymbol{\Sigma}(t) = \begin{pmatrix} \sigma_1^2(t) & \rho(t)\sigma_1(t)\sigma_2(t) \\ \rho(t)\sigma_1(t)\sigma_2(t) & \sigma_2^2(t) \end{pmatrix}.$$

It is of primary interest to test whether the correlation function $\rho(t)$ and variance functions $\sigma_1^2(t)$ and $\sigma_2^2(t)$ change over time. The covariance matrix $\boldsymbol{\Sigma}(t)$ can be estimated using the following two-step method.

Step 1 Given time point t_j, the ordinary least-squares method is used to estimate $\boldsymbol{\beta}_1(t_j)$ and $\boldsymbol{\beta}_2(t_j)$. Based on the residuals $\widehat{\boldsymbol{\varepsilon}}_{ij}$, for $j = 1, \cdots, T$, from observations from all individuals at a given time, the estimate of $\boldsymbol{\Sigma}(t_j)$ is given by

$$\widehat{\boldsymbol{\Sigma}}(t_j) = \frac{1}{n - p - 1} \sum_{i=1}^{T} \widehat{\boldsymbol{\varepsilon}}_{ij} \widehat{\boldsymbol{\varepsilon}}_{ij}^{\top}.$$

Step 2 Given the estimates of $\boldsymbol{\Sigma}(t)$ on $t_1, \cdots, t_T$, the components $\sigma_1^2(t)$, $\sigma_2^2(t)$ and $\rho(t)$ of $\boldsymbol{\Sigma}(t)$ are separately smoothed using the standard smoothing techniques (e.g., kernel estimate, smoothing spline, or local polynomial).

The problem of interest is to test the hypothesis

$$H_0 : \boldsymbol{\Sigma}(t_j) = \boldsymbol{\Sigma}_0 \quad \text{for all} \quad j = 1, \cdots, T,$$

where

$$\boldsymbol{\Sigma}_0 = \begin{pmatrix} \sigma_{01}^2 & \rho_0 \sigma_{01} \sigma_{02} \\ \rho_0 \sigma_{01} \sigma_{02} & \sigma_{02}^2 \end{pmatrix} \tag{5.8}$$

is a constant positive definite matrix, against the alternative

$$H_1 : \boldsymbol{\Sigma}(t_{j_1}) \neq \boldsymbol{\Sigma}(t_{j_2}) \ \exists j_1 \neq j_2 \in \{1, \cdots, T\}. \tag{5.9}$$

For this testing problem, a likelihood ratio testis suggested. The test statistic is given by

$$LR = \frac{\sup_{\boldsymbol{\Theta} \in \widetilde{\boldsymbol{\Theta}}_0} L(\boldsymbol{\Theta}; \mathbf{X}, \mathbf{Y}, \mathbf{Z})}{\sup_{\boldsymbol{\Theta} \in \widetilde{\boldsymbol{\Theta}}_1} L(\boldsymbol{\Theta}; \mathbf{X}, \mathbf{Y}, \mathbf{Z})} = \frac{\widehat{\sigma}_{01}^2 \widehat{\sigma}_{02}^2 (1 - \widehat{\rho}_0^2)^{-nT/2}}{\prod_{j=1}^{T} \{\widehat{\sigma}_1^2(t_j) \widehat{\sigma}^2(t_j)(1 - \widehat{\rho}^2(t_j))\}^{-n/2}},$$

where $L(\boldsymbol{\Theta}; \mathbf{X}, \mathbf{Y}, \mathbf{Z})$ is the likelihood function, $\boldsymbol{\Theta} = (\boldsymbol{\theta}(t_1), \cdots, \boldsymbol{\theta}(t_M))^\top$, $\boldsymbol{\theta}(t_j) = (\boldsymbol{\beta}_1(t_j)^\top, \boldsymbol{\beta}_2(t_j)^\top$, $\sigma_1(t_j), \sigma_2(t_j), \rho(t_j))^\top$, $\widetilde{\boldsymbol{\Theta}}_1$ represents the whole parameter space, and $\widetilde{\boldsymbol{\Theta}}_0$ is the parameter space under H_0. Under H_0, testing statistic $-2 \log LR$ follows an asymptotic chi-square distribution with $3(T-1)$ degrees of freedom.

If the null hypothesis is rejected, a data transformation is suggested to account for variance change, to make inference on $\boldsymbol{\beta}_i(t_j)$, for $i = 1, 2$, and $j = 1, \cdots, T$. Specifically, the transformed data are

$$
\begin{pmatrix} X_{ij}^* \\ Y_{ij}^* \end{pmatrix} = \widehat{\Sigma}(t_j)^{-1/2} \begin{pmatrix} \log(X_{ij}+1) \\ Y_{ij} \end{pmatrix}, \quad \mathbf{z}_{ij}^* = \widehat{\Sigma}(t_j)^{-1/2} \mathbf{z}_{ij}.
$$

According to model (5.7), we have

$$
X_{ij}^* = \mathbf{z}_{ij}^\top \boldsymbol{\beta}_1(t_j) + \varepsilon_{ij1}^*, \quad Y_{ij}^* = \mathbf{z}_{ij}^\top \boldsymbol{\beta}_2(t_j) + \varepsilon_{ij2}^*, \tag{5.10}
$$

where the independent random error $\boldsymbol{\varepsilon}_{ij}^* = (\varepsilon_{ij1}^*, \varepsilon_{ij2}^*)^\top \sim N_2(\mathbf{0}, \mathbf{I}_2)$, $\mathbf{I}_2$ is a 2×2 identity matrix, and $\boldsymbol{\beta}_k(t) = (\beta_{k1}(t), \beta_{k2}(t), \cdots, \beta_{kP}(t))^\top$, $k = 1, 2$. Similar to testing the variance function, a likelihood ratio test is suggested to test the following null hypothesis

$$
H_0 : \beta_{kp}(t_j) = 0, \quad \forall j = 1, \cdots, T, \quad \text{vs.} \quad H_1 : \beta_{kp}(t_j) \neq 0, \quad \exists j \in \{1, \cdots, T\}. \tag{5.11}
$$

5.2.4   *Fully parametric model*

In this section, we introduce a fully parametric model for joint analysis of mixed Poisson and continuous longitudinal data (Yang and Kang, 2010). It explicitly models the joint distribution of the longitudinal data using an extended multivariate Poisson model accounting for overdispersion. It assumes

$$
X_{ij} = \widetilde{X}_{ij} + a\widetilde{X}_{i0}, \tag{5.12}
$$

where $\widetilde{X}_{ij}$, $j = 0, 1, \cdots, T$, are mutually independent Poisson random variables with mean λ_{ij} and $a \geq 0$ is a pre-specified integer. The conditional distribution of $\mathbf{Y}_i = (Y_{i1}, \cdots, Y_{iT})^\top$, given $\mathbf{X}_i = (X_{i1}, \cdots, X_{iT})^\top$, is assumed to be a multivariate normal distribution with mean $\boldsymbol{\mu}_i = (\mu_{i1}, \cdots, \mu_{iT})^\top$ and covariance $\boldsymbol{\Sigma}_i$. The following regression model is proposed to estimate the effect of covariate $\mathbf{z}_{ij}$:

$$
\log \lambda_{i0} = \omega, \quad \log \lambda_{ij} = \mathbf{z}_{ij}^\top \boldsymbol{\beta}_1, \quad \mu_{ij} = \mathbf{Z}_{ij}^\top \boldsymbol{\beta}_2 + \gamma_1(X_{ij} - a\lambda_{i0} - \lambda_{ij}) + \gamma_2 S_i, \tag{5.13}
$$

where $S_i = \sum_{j=1}^T (X_{ij} - a\lambda_{i0} - \lambda_{ij})$ represents the subject effect. The covariance function for $\mathbf{Y}_i$ can be specified based on the problem. Yang and Kang (2010) discuss a compound symmetry correlation function, i.e.,

$$
\boldsymbol{\Sigma}_i = \sigma^2 \{(1-\rho)\mathbf{I} + \rho\mathbf{J}\}, \quad i = 1, \cdots, n, \tag{5.14}
$$

which has been shown to be valid for the ICDB data.

Given the probability mass function for the multivariate mixed Poisson distribution, the likelihood for the data is complicated and computationally intensive. Yang and Kang (2010) propose an ECM algorithm to mitigate this problem. Write $\widetilde{\mathbf{X}}_i = (\widetilde{X}_{i0}, \widetilde{X}_{i1}, \cdots, \widetilde{X}_{iT})^\top$ as unobserved data. Note that $\mathbf{X} = (\mathbf{X}_1^\top, \cdots, \mathbf{X}_n^\top)^\top$ is observed, according to (5.12), and $\widetilde{X}_{ij}$ is completely determined by $\widetilde{X}_{i0}$ and X_{ij}. Thus, only $\widetilde{X}_{i0}$ is missing in this model. By the independence of $\widetilde{X}_{ij}$, $j = 0, 1, \cdots, T$, the complete data likelihood is now simple and easy to compute. Write $\widetilde{\mathbf{X}} = (\widetilde{\mathbf{X}}_1, \cdots, \widetilde{\mathbf{X}}_n)^\top$, $\mathbf{Y} = (\mathbf{Y}_1^\top, \cdots, \mathbf{Y}_n^\top)^\top$, and $\mathbf{Z} = (\mathbf{Z}_1, \cdots, \mathbf{Z}_n)^\top$, the complete data log-likelihood is given by

$$
\ell(\boldsymbol{\Theta}; \widetilde{\mathbf{X}}, \mathbf{Y}, \mathbf{Z}) = \sum_{i=1}^n \left(\sum_{j=1}^T \log \pi(\widetilde{X}_{ij}; \mathbf{Z}_i, \boldsymbol{\Theta}) + \log \pi(\mathbf{Y}_i; \widetilde{\mathbf{X}}_i, \mathbf{Z}_i, \boldsymbol{\Theta}) \right),
$$

where $\pi(\widetilde{X}_{ij}; \mathbf{Z}_i, \mathbf{\Theta})$ is the Poisson probability mass function. To make $\pi(\mathbf{Y}_i; \widetilde{\mathbf{X}}_i, \mathbf{Z}_i, \mathbf{\Theta})$ simple, it is reasonable to assume that $\widetilde{\mathbf{X}}_i$ does not provide extra information on the conditional distribution of $\mathbf{Y}_i$, given $\mathbf{X}_i$, i.e., $\mathbf{Y}_i$ and $\widetilde{\mathbf{X}}_i$ are conditionally independent given $\mathbf{X}_i$. This further implies that $\pi(\mathbf{Y}_i; \widetilde{\mathbf{X}}_i, \mathbf{Z}_i, \mathbf{\Theta}) = \pi(\mathbf{Y}_i; \mathbf{X}_i, \mathbf{Z}_i, \mathbf{\Theta})$, which is a multivariate normal density function with mean $\boldsymbol{\mu}_i$ in (5.13) and covariance $\boldsymbol{\Sigma}_i$ in (5.14). To find the MLE, the following ECM algorithm is suggested:

E-Step: Using $\mathbf{X}_i$ and the current estimates $\mathbf{\Theta}^{(k)}$ after the kth iteration, calculate the expectation of $\widetilde{\mathbf{X}}_{i0}$, $i = 1, \cdots, n$,

$$e_{i0}^{(k)} = E(\widetilde{X}_{i0} | \mathbf{X}_i, \mathbf{\Theta}^{(k)}), \quad e_{ij}^{(k)} = X_{ij} - a e_{i0}^{(k)},$$

and $Q(\mathbf{\Theta} | \mathbf{\Theta}^{(k)}) = \ell(\mathbf{\Theta}; \mathbf{e}^{(k)}, \mathbf{Y}, \mathbf{Z})$.

CM-Step 1: Given the estimate of $\{\boldsymbol{\beta}, \omega\}$ at kth iteration, update estimates of $\{\boldsymbol{\alpha}, \sigma^2, \rho\}$ by

$$(\boldsymbol{\alpha}^{(k+1)}, \sigma^{2,(k+1)}, \rho^{(k+1)}) = \underset{\boldsymbol{\alpha}, \sigma^2, \rho}{\arg\max} \, Q(\boldsymbol{\beta}^{(k)}, \omega^{(k)}, \boldsymbol{\alpha}, \sigma^2, \rho | \mathbf{\Theta}^{(k)}).$$

CM-Step 2: Given the estimates of $\{\boldsymbol{\alpha}, \sigma^2, \rho\}$ at kth iteration, update the estimate of $\{\boldsymbol{\beta}, \omega\}$ by

$$(\boldsymbol{\beta}^{(k+1)}, \omega^{(k+1)}) = \underset{\boldsymbol{\beta}, \omega}{\arg\max} \, Q(\boldsymbol{\beta}, \omega, \boldsymbol{\alpha}^{(k+1)}, \sigma^{2,(k+1)}, \rho^{(k+1)} | \mathbf{\Theta}^{(k)}).$$

Note that $Q(\mathbf{\Theta} | \mathbf{\Theta}^{(k)}) \neq E\{\ell(\mathbf{\Theta}; \widetilde{\mathbf{X}}, \mathbf{Y}, \mathbf{Z})\}$, but this scheme is equivalent to the standard EM algorithm, since $\widetilde{\mathbf{X}}$ is the sufficient statistics for the multivariate Poisson distribution. The standard errors (SEs) based on the inverse of the complete data information matrix and standard Wald statistics can be used to test parameter significance. See Yang and Kang (2010) for details on computing the expectation in E-step, as well as on how to select an appropriate tuning parameter a to model the overdispersion of the Poisson outcome.

5.3   Handling missing data problem

In this section, our emphasis is on how to deal with the missing data problem in mixed Poisson and continuous longitudinal data. We mainly focus on two approaches, namely, the method based on a parametric missing data model and imputation.

5.3.1   Parametric missing data model

It is straightforward to extend the fully parametric model described in Section 5.2.4 to handle missing data. In particular, Yang and Kang (2010) consider a nonignorable missing data mechanism. They use a fully likelihood-based approach and choose a selection model for the joint distribution of the missingness indicator and observed data. For subject i, let $\mathbf{X}_i^\top = (\mathbf{X}_{i,o}^\top, \mathbf{X}_{i,m}^\top)$ denote the partition of the Poisson outcome into the observed and missing components. Let $\mathbf{Y}_i^\top = (\mathbf{Y}_{i,o}^\top, \mathbf{Y}_{i,m}^\top)$ denote the same partition for the continuous outcome. Let $\mathbf{R}_i = (R_{i1}, \cdots, R_{iT})^\top$ and $\mathbf{U}_i = (U_{i1}, \cdots, U_{iT})^\top$, respectively, denote the missing indicators for the Poisson and continuous outcomes, where $R_{ij} = 1$, if X_{ij} is observed, and $R_{ij} = 0$, otherwise. Similarly, $U_{ij} = 1$, if Y_{ij} is observed, and $U_{ij} = 0$, otherwise. For simplicity, R_{ij} and U_{ij}, $j = 1, \cdots, T$, are assumed to be conditionally independent given (X_{ij}, Y_{ij}). For $r, u \in \{0, 1\}$,

$$
\begin{aligned}
f(r, u | X_{ij}, Y_{ij}, \tau) &= P(R_{ij} = r, U_{ij} = u | X_{ij}, Y_{ij}, \tau) = \pi_{ij}^r (1 - \pi_{ij})^{1-r} \phi_{ij}^u (1 - \phi_{ij})^{1-u}, \\
\text{logit}(\pi_{ij}) &= \tau_{10} + \tau_{11} X_{ij} + \tau_{12} Y_{ij}, \\
\text{logit}(\phi_{ij}) &= \tau_{20} + \tau_{21} X_{ij} + \tau_{22} Y_{ij},
\end{aligned}
\tag{5.15}
$$

where $\boldsymbol{\tau} = (\tau_{10}, \tau_{11}, \tau_{12}, \tau_{20}, \tau_{21}, \tau_{22})^{\top}$. The log-likelihood of complete data and missing indicators is

$$\log L(\boldsymbol{\Theta}, \boldsymbol{\tau}; \mathbf{X}, \mathbf{Y}, \mathbf{Z}, \mathbf{R}, \mathbf{U}) = \sum_{i=1}^{n} \left(\sum_{j=1}^{T} \log f(R_{ij}, U_{ij} | X_{ij}, Y_{ij}, \boldsymbol{\tau}) + \log \pi(\mathbf{X}_i, \mathbf{Y}_i | \mathbf{Z}_i, \boldsymbol{\Theta}) \right). \quad (5.16)$$

A Monte Carlo EM algorithm is suggested to find the MLE. Let $(\boldsymbol{\Theta}^{(k)}, \boldsymbol{\tau}^{(k)})$ denote the updated parameters at the kth iteration. Let $E^{(k)}(\cdot)$ represent the expectation operator with respect to the density of missing data given observed data, i.e., $\pi(\mathbf{X}_m, \mathbf{Y}_m | \mathbf{X}_o, \mathbf{Y}_o, \mathbf{R}, \mathbf{S}, \boldsymbol{\Theta}^{(k)}, \boldsymbol{\tau}^{(k)})$. Let $Q(\boldsymbol{\Theta}, \boldsymbol{\tau} | \boldsymbol{\Theta}^{(k)}, \boldsymbol{\tau}^{(k)}) = E^{(k)} \{ \log L(\boldsymbol{\Theta}, \boldsymbol{\tau} | \mathbf{X}, \mathbf{Y}, \mathbf{Z}, \mathbf{R}, \mathbf{U}) \}$ that needs to be computed in the E-step. However, there are no closed forms of this equation and the numerical integration is very computationally intensive. Thus, a Monte Carlo method is suggested to estimate $Q(\boldsymbol{\Theta}, \boldsymbol{\tau} | \boldsymbol{\Theta}^{(k)}, \boldsymbol{\tau}^{(k)})$ by

$$\begin{aligned}
\widehat{Q}(\boldsymbol{\Theta}, \boldsymbol{\tau} | \boldsymbol{\Theta}^{(k)}, \boldsymbol{\tau}^{(k)}) &= \frac{1}{m} \sum_{i=1}^{n} \sum_{t=1}^{m} \log f(\mathbf{R}_i, \mathbf{U}_i | \mathbf{X}_{i,m}^{(t)}, \mathbf{X}_{i,o}, \mathbf{Y}_{i,m}^{(t)}, \mathbf{Y}_{i,o}), \boldsymbol{\tau}) \quad (5.17) \\
&\quad + \log \pi(\mathbf{X}_{i,m}^{(t)}, \mathbf{X}_{i,o}, \mathbf{Y}_{i,m}^{(t)}, \mathbf{Y}_{i,o} | \mathbf{Z}_i, \boldsymbol{\Theta}).
\end{aligned}$$

The algorithm outline at the kth iteration, is as follows:

E-step: For subject i, $i = 1, \cdots, n$, draw M random samples from distribution of $bfX_{i,m}$ and $\mathbf{Y}_{i,m}$, from $\pi(\mathbf{X}_m, \mathbf{Y}_m | \mathbf{X}_o, \mathbf{Y}_o, \mathbf{R}, \mathbf{S}, \boldsymbol{\Theta}^{(k)}, \boldsymbol{\tau}^{(k)})$;

M-step: Set $(\boldsymbol{\Theta}^{(k+1)}, \boldsymbol{\tau}^{(k+1)}) = \mathrm{argmax}_{\boldsymbol{\Theta}, \boldsymbol{\tau}} \, \widehat{Q}(\boldsymbol{\Theta}, \boldsymbol{\tau} | \boldsymbol{\Theta}^{(k)}, \boldsymbol{\tau}^{(k)})$;

The key issue of this MCEM algorithm is how to sample $\{\mathbf{X}_{i,m}, \mathbf{Y}_{i,m}\}$. Note that

$$\pi(\mathbf{X}_{i,m}, \mathbf{Y}_{i,m} | \mathbf{X}_{i,o}, \mathbf{Y}_{i,o}, \mathbf{R}_i, \mathbf{U}_i) = \pi(\mathbf{Y}_{i,m} | \mathbf{X}_{i,o}, \mathbf{X}_{i,m}, \mathbf{Y}_{i,o}, \mathbf{R}_i, \mathbf{U}_i) \pi(\mathbf{X}_{i,m} | \mathbf{X}_{i,o}, \mathbf{Y}_{i,o}, \mathbf{R}_i, \mathbf{U}_i).$$

Thus, we can first draw $\mathbf{X}_{i,m}$, then draw $\mathbf{Y}_{i,m}$, given $\mathbf{X}_{i,m}$. Two sampling schemes have been suggested, namely, rejection sampling and importance sampling. For details, please see Yang and Kang (2010).

5.3.2 Imputation

In this section, we discuss an imputation method for the analysis of bivariate longitudinal data with ignorable missingness. The mixed Poisson and continuous longitudinal outcomes can be considered as a special case in this approach. For the Poisson longitudinal outcome, we can usually take logarithmic transformation as in model (5.7). To make the model more general and easier to present, we redefine the notations in this section. Let $\mathbf{Y}^{\top} = (\mathbf{Y}_1^{\top}, \mathbf{Y}_2^{\top})$ denote the bivariate longitudinal outcomes, where $\mathbf{Y}_v = (\mathbf{Y}_{v1}, \cdots, \mathbf{Y}_{vN})^{\top}$, $\mathbf{Y}_{vi} = (Y_{vi1}, \cdots, Y_{viT})^{\top}$, and Y_{vit} denotes response variable $v = 1, 2$, of individual $i = 1, \cdots, N$, at time t, where N is the number of individuals and T is the length of the time period. Let $\mathbf{X} = (\mathbf{X}_1, \cdots, \mathbf{X}_N)^{\top}$ be the predictor variable, where $\mathbf{X}_i = (\mathbf{X}_{i1}, \cdots, \mathbf{X}_{iT})^{\top}$, $\mathbf{X}_{it} = (X_{it1}, \cdots, X_{itP})^{\top}$, and X_{itp} denotes predictor variable p of individual i at time t. For simplicity, let $\bar{\mathbf{Y}}_{vit} = (Y_{vi1}, \cdots, Y_{vit})^{\top}$ and $\bar{\mathbf{X}}_{it} = (X_{i1}, \cdots, X_{it})^{\top}$ denote the observations before time t of $\mathbf{Y}_{vi}$ and $\mathbf{X}_i$, respectively. Our primary interest is in estimating $E(\mathbf{Y}_{vi} | \mathbf{X}_i) = \boldsymbol{\mu}_{vi} = (\mu_{vi1}, \cdots, \mu_{viT})^{\top}$, $v = 1, 2$, $i = 1, \cdots, N$.

We assume that $\mathbf{X}$ is fully observed but the components of $\mathbf{Y}$ are possibly missing. Let $\mathbf{R}^{\top} = (\mathbf{R}_1^{\top}, \mathbf{R}_2^{\top})$ denote the observed missingness indicator of $\mathbf{Y}$, where $\mathbf{R}_v = (\mathbf{R}_{v1}, \cdots, \mathbf{R}_{vN})^{\top}$, $\mathbf{R}_{vi} = (R_{vi1}, \cdots, R_{viT})^{\top}$, and $R_{vit} = 1$, when Y_{vit} is observed, and $R_{vit} = 0$, when Y_{vit} is missing. Furthermore, we assume the baseline of $\mathbf{Y}$ should be observed, i.e., $R_{vi1} = 1$, for all $v = 1, 2$, $i = 1, \cdots, N$.

Note that $R_{vit} = 1$, for all v, i, and t. This implies $\mathbf{Y}$ is fully observed. We have particularly discussed the GEE approach for the complete mixed Poisson and continuous longitudinal outcomes in the previous section. For the general bivariate longitudinal data, we propose the following GEE approach. First, μ_{vi} is specified to satisfy

$$\mu_{vi} = h_v^{-1}(\mathbf{X}_i \boldsymbol{\beta}_v), \quad (5.18)$$

where $h_v(\cdot)$, $v = 1,2$ are two known link functions, $\boldsymbol{\beta} = (\boldsymbol{\beta}_1^\top, \boldsymbol{\beta}_2^\top)^\top$, $\boldsymbol{\beta}_v = (\beta_{v1}, \cdots, \beta_{vP})^\top$, and the estimating equation for $\boldsymbol{\beta}$ is

$$U(\boldsymbol{\beta}, \boldsymbol{\rho}) = \sum_{i=1}^N \begin{pmatrix} \frac{\partial \mu_{1i}}{\partial \boldsymbol{\beta}_1} & \mathbf{0} \\ \mathbf{0} & \frac{\partial \mu_{2i}}{\partial \boldsymbol{\beta}_2} \end{pmatrix}^\top cov_\rho^{-1} \begin{pmatrix} \mathbf{Y}_{1i} \\ \mathbf{Y}_{2i} \end{pmatrix} \begin{pmatrix} \mathbf{Y}_{1i} - \boldsymbol{\mu}_{1i} \\ \mathbf{Y}_{2i} - \boldsymbol{\mu}_{2i} \end{pmatrix} = \mathbf{0}, \quad (5.19)$$

where $\partial \mu_{vi} / \partial \boldsymbol{\beta}_v = (\partial \mu_{vit} / \partial \boldsymbol{\beta}_{vp})_{T \times P}$, $v = 1,2$;

$$cov_\rho \begin{pmatrix} \mathbf{Y}_{1i} \\ \mathbf{Y}_{2i} \end{pmatrix} = \boldsymbol{\Delta}_i^{1/2} \begin{pmatrix} \boldsymbol{\Omega}_{1i}(\rho_1) & \boldsymbol{\Omega}_{12i}(\rho_{12}) \\ \boldsymbol{\Omega}_{12i}^\top(\rho_{12}) & \boldsymbol{\Omega}_{2i}(\rho_2) \end{pmatrix} \boldsymbol{\Delta}_i^{1/2},$$

$$\boldsymbol{\Delta}_i = \begin{pmatrix} diag\{var(\mathbf{Y}_{1i})\} & \mathbf{0} \\ \mathbf{0} & diag\{var(\mathbf{Y}_{2i})\} \end{pmatrix},$$

$\boldsymbol{\rho} = (\rho_1, \rho_2, \rho_{12})^\top$, $\boldsymbol{\Omega}_{vi}(\rho_v)$ is a "working" correlation matrix of Y_{vi}, and $\boldsymbol{\Omega}_{vui}(\rho_{vu})$ is also a "working" matrix which depicts the correlation between $\mathbf{Y}_{1i}$ and $\mathbf{Y}_{2i}$. One can choose $\boldsymbol{\Omega}_{vi}(\rho_v)$ as the identity matrix or the equal-correlation matrix, and the specified correlation structure does not need to be correct; $\boldsymbol{\Omega}_{vui}(\rho_{vu})$ has a similar property. The GEE estimate, say $\widehat{\boldsymbol{\beta}}$, is the root of $U(\boldsymbol{\beta}, \widehat{\boldsymbol{\rho}}) = \mathbf{0}$, where $\widehat{\boldsymbol{\rho}}$ is any consistent estimate.

We focus here on a monotone missing data pattern for bivariate outcomes. Let $\{R_{vit}, v = 1,2; i = 1, \cdots, N; t = 1, \cdots, T\}$ denote the observed missing data indicators, which satisfy the following $N \times (T-1)$ inequalities:

$$\begin{pmatrix} R_{1i,t+1} \\ R_{2i,t+1} \end{pmatrix} \leq \begin{pmatrix} R_{1it} \\ R_{2it} \end{pmatrix}, \quad (5.20)$$

for $t = 1, \cdots, T-1$; $i = 1, \cdots, N$. When data are under monotone missing pattern (5.20), we have that if $R_{vit} = 1$, then $R_{vi,t-1} = R_{vi,t-2} = \cdots = R_{vi1} = 1$, and if $R_{vit} = 0$, then $R_{vi,t+1} = R_{vi,t+2} = \cdots = R_{viT} = 0$, which implies that the missing data pattern of individual i can be expressed by the last observation time vector (a_{1i}, a_{2i}), where $a_{vi} = \{t : R_{vi1} = R_{vi2} = \cdots = R_{vit} = 1, R_{vi,t+1} = 0\} = \{t : R_{vit} = 1, R_{vi,t+1} = 0\}$.

We assume that the missingness of $Y_{v'it}$ only depends on the previous observation $\bar{Y}_{vit}$; then we consider the following missing mechanism:

$$P\left(R_{vit} = 1 \middle| \begin{array}{l} \mathbf{X}_i, \mathbf{Y}_{vi}, \mathbf{Y}_{v'i}, \\ R_{via_1} = R_{v'ia_2} = 1 \end{array}\right) = P\left(R_{vit} = 1 \middle| \begin{array}{l} \mathbf{X}_i, \bar{\mathbf{Y}}_{via_1}, \bar{\mathbf{Y}}_{v'ia_2}, \\ R_{via_1} = R_{v'ia_2} = 1 \end{array}\right), \quad (5.21)$$

for all $(v, v') = (1,2), (2,1)$; $t = 2, \cdots, T$; $r = 1, \cdots, t-1$; $s = 1, \cdots, t$; this mechanism is called missing at random (MAR). Then, we have

$$E\left(Y_{vit} \middle| \begin{array}{l} \mathbf{X}_i, \bar{\mathbf{Y}}_{via_1}, \bar{\mathbf{Y}}_{v'ia_2}, \\ R_{via_1} = R_{vit} = R_{v'ia_2} = 1 \end{array}\right) = E\left(Y_{vit} \middle| \begin{array}{l} \mathbf{X}_i, \bar{\mathbf{Y}}_{via_1}, \bar{\mathbf{Y}}_{v'ia_2}, \\ R_{via_1} = R_{v'ia_2} = 1, R_{vit} = 0, \end{array}\right). \quad (5.22)$$

This shows that $E(Y_{vit}|\mathbf{X}_i, \bar{\mathbf{Y}}_{via_1}, \bar{\mathbf{Y}}_{v'ia_2}, R_{via_1} = R_{v'ia_2} = 1, R_{vit} = 0)$ can be consistently estimated by the sample mean of the observed Y_{vit}s having the same history as $\bar{\mathbf{Y}}_{via_1}$ and $\bar{\mathbf{Y}}_{v'ia_2}$, i.e., the estimate is obtained by noting that $E(Y_{vit}|\mathbf{X}_i, \bar{\mathbf{Y}}_{via_1}, \bar{\mathbf{Y}}_{v'ia_2}, R_{via_1} = R_{vit} = R_{v'ia_2} = 1)$ is the regression of Y_{vit} on the predictors $\mathbf{X}_i$, $\bar{\mathbf{Y}}_{vir}$, and $\bar{\mathbf{Y}}_{v'is}$. Based on this property, we can use the regression imputation method to replace missing values with imputed values resulting from the prediction via the regression model constructed from the observed sample and imputed values. In order to establish the sequential regression imputation scheme, we first classify the bivariate longitudinal data Y_{vit} as cell (a_1, a_2, v, t), which includes the sample, both missing and observed, with missing data pattern (a_1, a_2) at time t for the variable $\mathbf{Y}_v$. We can easily find that data Y_{vit} in cell (a_1, a_2, v, t) is missing when $a_v < t$, and is observed when $a_v \geq t$. For example, we lay out the missing data cells in Table

5.1 for the case $T = 3$. Next, for each missing data cell, there are three types of data cells we should consider: one is the cell used to predict the missing data cell, called missing predictor cell (MPC); the other two are the cells used to construct the regression model, which include both predictor and response, and are called constructing predictor cell (CPC) and constructing response cell (CRC), respectively. Table 5.1 contains the illustration of the three types of cells for the regression imputation of missing data cell $(2,1,1,3)$.

Table 5.1 *Missing data patterns (a_1, a_2, v, t) for the bivariate longitudinal data $(T = 3)$ and illustration of regression imputation for cell $(2,3,1,3)$.*

| (a_1, a_2) | $t = 1$ | | $t = 2$ | | $t = 3$ | |
	$v = 1$	$v = 2$	$v = 1$	$v = 2$	$v = 1$	$v = 2$
$(1,1)$			$(1,1,1,2)$	$(1,1,2,2)$	$(1,1,1,3)$	$(1,1,2,3)$
$(1,2)$			$(1,2,1,2)$		$(1,2,1,3)$	$(1,2,2,3)$
$(1,3)$			$(1,3,1,2)$		$(1,3,1,3)$	
$(2,1)$				$(2,1,2,2)$	$(2,1,1,3)$	$(2,1,2,3)$
$(2,2)$					$(2,2,1,3)$	$(2,2,2,3)$
$(2,3)$	Δ^*	Δ^*	Δ^*	Δ^*	$(2,3,1,3)$	Δ^*
$(3,1)$				$(3,1,2,2)$		$(3,1,2,3)$
$(3,2)$						$(3,2,2,3)$
$(3,3)$	Δ	Δ	Δ	Δ	Λ	Δ

NOTE: (a) Δ^*: *missing predictor cell (MPC) for cell $(2,3,1,3)$; (b) Λ: constructing predictor cell (CPC) for cell $(2,3,1,3)$; (c) Δ: constructing response cell (CRC) for cell $(2,3,1,3)$.*

In general, the cells used to construct the regression model should be observed, but if we arrange an appropriate order to impute the data cells, it is possible to use the imputed data for the construction of the regression model. According to the missing mechanism (5.21) and the properties of conditional expections, for $t = 3, \cdots, T$, $a_1 = 2, \cdots, t - 1$, $a_2 = 2, \cdots, t$, $j = 1, \cdots, a_1 - 1$, and $k = 1, \cdots, a_2 - 1$, we have

$$E_{jk}\left\{ E\left(Y_{vit} \middle| \begin{array}{l} \bar{\mathbf{Y}}_{via_1}, \bar{\mathbf{Y}}_{v'ia_2}, R_{vit} = 0, \\ \mathbf{X}_i, R_{via_1} = R_{v'ia_2} = 1 \end{array} \right) \right\} = E\left(Y_{vit} \middle| \begin{array}{l} \bar{\mathbf{Y}}_{vi,a_1-j}, \bar{\mathbf{Y}}_{v'i,a_2-k}, R_{vit} = 0, \\ \mathbf{X}_i, R_{vi,a_1-j} = R_{v'i,a_2-k} = 1 \end{array} \right), \quad (5.23)$$

where $E_{jk}(\cdot)$ is the expectation operator with respect to the joint density of $Y_{vi,a_1-j+1}, \cdots, Y_{via_1}$, $Y_{v'i,a_2-k+1} \cdots, Y_{v'ia_2}$. This implies that the imputed value $\widehat{Y}_{vit}$ constructed from the estimate of $E(Y_{vit} | \mathbf{X}_i, \bar{\mathbf{Y}}_{vir}, \bar{\mathbf{Y}}_{v'ia_1}, R_{via_2} = R_{v'is} = 1, R_{vit} = 0)$ is the unbiased estimate of

$$E(Y_{vit} | \mathbf{X}_i, \bar{\mathbf{Y}}_{vi,r-j}, \bar{\mathbf{Y}}_{v'i,a_1-k}, R_{vi,a_2-j} = R_{v'i,s-k} = 1, R_{vit} = 0),$$

and the cell (a_1, a_2, v, t) with imputed data can be used as a CRC for the missing data cell $(a_1 - j, a_2 - k, v, t)$. It also suggests the sequential imputation scheme for all the missing data cells by the order indicator $Q(\cdot)$ of cell (a_1, a_2, v, t):

$$Q(a_1, a_2, v, t) \quad = \quad T + t - a_1 - a_2, \quad (5.24)$$

which implies that following the order defined by (5.24) to impute, we could use both observations and imputed values in CRC to construct the imputation regression for the corresponding missing data cell. Take $T = 4$ as an example, and the specified order of imputation could be found in Table 5.2.

Thus, for the each missing data cell (a_1, a_2, v, t) with $a_v < t$, the three types of data cells used to

Table 5.2 *Imputation order $Q(a_1,a_2,v,t)$ for the bivariate longitudinal data ($T = 4$), and an illustration of imputation with observations and imputed values for cell $(3,2,1,4)$.*

(a_1,a_2)	$t=1$ $v=1$	$v=2$	$t=2$ $v=1$	$v=2$	$t=3$ $v=1$	$v=2$	$t=4$ $v=1$	$v=2$
$(1,1)$			4th	4th	5th	5th	6th	6th
$(1,2)$			3rd		4th	4th	5th	5th
$(1,3)$			2nd		3rd		4th	4th
$(1,4)$			1st		2nd		3rd	
$(2,1)$				3rd	4th	4th	5th	5th
$(2,2)$					3rd	3rd	4th	4th
$(2,3)$					2nd		3rd	3rd
$(2,4)$					1st		2nd	
$(3,1)$				2nd		3rd	4th	4th
$(3,2)$	Δ^*	Δ^*	Δ^*	Δ^*	Δ^*	2nd	3rd	3rd
$(3,3)$	Δ	Δ	Δ	Δ	Δ		2nd$^{\#}$	2nd
$(3,4)$	Δ	Δ	Δ	Δ	Δ		1st$^{\#}$	
$(4,1)$				1st		2nd	Λ	3rd
$(4,2)$	Δ	Δ	Δ	Δ	Δ	1st	Λ	2nd
$(4,3)$	Δ	Δ	Δ	Δ	Δ		Λ	1st
$(4,4)$	Δ	Δ	Δ	Δ	Δ		Λ	

NOTE: (a) Δ^*: MPC for cell $(3,2,1,4)$; (b) Δ: CPC for cell $(3,2,1,4)$; (c) #: CRC with imputed value for cell $(3,2,1,4)$; (d) Λ: CRC with observations for cell $(3,2,1,4)$.

construct imputation regression could be expressed as

$$\text{MPC set} = \bigcup_{v'=1}^{2} S_{v'}(a_1,a_2;a_1,a_2,v,t),$$

$$\text{CPC set} = \bigcup_{(a_1',a_2'):a_v'\geq v} \bigcup_{v'=1}^{2} S_{v'}(a_1',a_2';a_1,a_2,v,t), \qquad (5.25)$$

$$\text{CRC set} = \bigcup_{(a_1',a_2'):a_v'\geq v} \{(a_1',a_2',v,t)\},$$

where

$$S_u(a_1',a_2';a_1,a_2,v,t) = \begin{cases} \{(a_1',a_2',v',s) : s \leq \min(t-1,a_u)\} & \text{, for } v' = v \\ \{(a_1',a_2',v',s) : s \leq \min(t,a_u)\} & \text{, for } v' \neq v \end{cases}.$$

The CRC set in (5.25) contains cells of both observations and imputed values. Table 5.2 illustrates how to construct the regression cells for missing cell $(3,2,1,4)$.

It is important to model $E(Y_{vit}|\mathbf{X}_i, \bar{\mathbf{Y}}_{via_1}, \bar{\mathbf{Y}}_{v'ia_2}, R_{via_1} = R_{v'ia_2} = 1, R_{vit} = 0)$, and make a prediction for the missing value. When $\mathbf{Y}$ arises from a normal distribution, the conditional expectation is equivalent to a linear model. Hence, we have

$$E\left(Y_{vit} \,\middle|\, \begin{matrix} \mathbf{X}_i, \bar{\mathbf{Y}}_{via_1}, \bar{\mathbf{Y}}_{uia_2}, \\ R_{via_1} = R_{v'ia_2} = 1, R_{vit} = 0 \end{matrix}\right) = \beta_0 + \bar{\mathbf{Y}}_{via_1}^{\top}\boldsymbol{\beta}_v + \bar{\mathbf{Y}}_{v'ia_2}^{\top}\boldsymbol{\beta}_{v'}, \qquad (5.26)$$

where $\beta_0 = E(Y_{vit})$, $\boldsymbol{\beta}_v = (\beta_{v1},\cdots,\beta_{vr})^{\top}$, $\boldsymbol{\beta}_{v'} = (\beta_{v'1},\cdots,\beta_{v's})^{\top}$, $(\boldsymbol{\beta}_v^{\top},\boldsymbol{\beta}_{v'}^{\top})^{\top}$ depicts the correlation between Y_{vit} and $(\bar{\mathbf{Y}}_{vir}, \bar{\mathbf{Y}}_{uis})^{\top}$. The estimate of parameters in (5.26) is easily obtained using the likelihood approach. When $\mathbf{Y}$ is not normal, the conditional expectation can be modeled as a generalized linear model. Specifically, we can fit a log-linear model for the Poisson outcome. Also, a nonparametric regression can be considered to fit such model. Since more than one predictor can be in the model, the direct and simple method is multivariate kernel regression, which usually requires a large sample and does not perform well in our problem. Additive and generalized additive models have been shown to be very useful as they naturally generalize the linear regression model and allow for an interpretation of marginal changes. Thus, we have

$$E\left(Y_{vit} \middle| \begin{array}{c} \mathbf{X}_i, \bar{\mathbf{Y}}_{via_1}, \bar{\mathbf{Y}}_{v'ia_2}, \\ R_{via_1} = R_{v'ia_2} = 1, R_{vit} = 0 \end{array}\right) = c + \sum_{j=1}^{r} g_{vj}(Y_{vij}) + \sum_{j=1}^{s} g_{v'j}(Y_{v'ij}). \quad (5.27)$$

This model can be fitted by a modified back-fitting algorithm (Hastie and Tibshirani, 1990).

5.4   Application

In this section, we illustrate the models discussed in Sections 5.2 and 5.3 by analyzing the ICDB data. We focus on the longitudinal outcomes urgency, $\mathbf{Y}$, and urinary frequency, $\mathbf{X}$. As discussed in the introduction, it is straightforward to assume urinary frequency as a Poisson outcome, and urgency as a continuous response. The data are cleaned by removing subjects with completely missing data or extreme values. The demographic results are summarized in Table 5.3. The total number of subjects is $N = 611$ and the total number of time points is $T = 48$. We create short names for the clinical biomarkers (i.e., covariates) that are included in the analysis. Specifically, the urine volume at the first sensation is indicated as urod_7; the urine volume at maximal capacity is urod_9; and whether or not the previous interstitial cystitis was diagnosed by a physician is indicated as shx_2; urod_7 and urod_9 are both continuous variables, while shx_2 is a binary variable taking values 1 for "yes" or 0 for "no." Covariate severity is a discrete variable with three categories, which are indicated as sev_1, sev_2, and sev_3, respectively.

Figure 5.1 shows that the correlation between longitudinal outcome Y_{it} and X_{it} increases over time. Also, there is strong evidence that the covariance significantly changes over time based on the formal statistical test (5.11) (p-value< 0.001) in Section 5.2.2. The solid lines in Figure 5.1 were obtained by kernel method in which the bandwidth was chosen based on the least-squares cross-validation technique. Figure 5.1 shows that the estimated standard deviations $\hat{\sigma}_1(t)$ and $\hat{\sigma}_2(t)$ decrease over time. The model in Section 5.2.3 could be used to study the correlation change over time, but it fails to provide a satisfactory statistical interpretation, if there exist discrete covariates in the model. Thus, it is not the best choice for the ICDB data, but it can potentially be applied to analyze other data.

5.4.1   Complete data model

In this section, by removing subjects with at least one missing value in the ICDB, we only use the complete cases to illustrate the methods we discussed in Section 5.2.

5.4.1.1   GEE approach

We apply the GEE approach in Section 5.2.1 to the ICDB data in this section. The GEE approach provides estimates of the time-independent covariate effects. Covariate severity is a three-level categorical variable which is converted to two dummy variables. Respective estimates of γ_1 and γ_2 are 0.258 and 0.001, and their associated p-values are 0.07 and 0.06, respectively. Estimates of $\boldsymbol{\beta}_1$ and $\boldsymbol{\beta}_2$, and their associated p-values are listed in Table 5.4. The results imply that gender, shx_2, urod_7, urod_9, and age are significantly associated with urgency. Also, patients with level 3 severity have significantly higher urgency than those with the other two levels. We can also see that income and severity are both significantly associated with frequency.

Table 5.3 *Demographic statistics of ICDB data.*

Characteristics	Number of Patients	Percentage
Sex		
M	54	8.8
F	557	91.2
Race		
White	569	93.1
Other	42	6.87
Marital status		
Partnered	430	70.4
Alone	181	29.6
Employment		
Employed	372	60.9
Unemployed	68	11.1
Home/Retired	171	28.0
Education		
High School or less	260	42.6
College or Advanced	351	57.5
Annual household income ($)		
Less than 30,000	175	28.6
30,000 or greater	436	71.4
Previous interstitial cystitis diagnosis by physician		
Yes	413	67.6
No	198	32.4

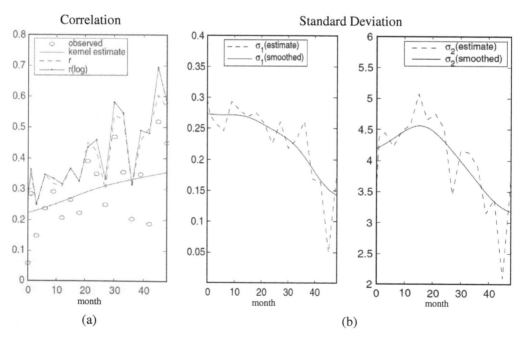

(a)   (b)

Figure 5.1 *Plots of (a) correlation between Y_{it} and X_{it} over time, and (b) standard deviations $\sigma_1(t)$ and $\sigma_2(t)$ for X_{it} and Y_{it}.*

Table 5.4 *Estimates of regression coefficients for ICDB data obtained by the GEE approach.*

Effect		Urgency ($k = 1$)			Frequency ($k = 2$)		
		Est	SE	p-value	Est	SE	p-value
Sex	β_{k1}	−0.438	0.118	<0.001	−0.380	0.257	0.070
Income	β_{k2}	0.003	0.002	0.960	−0.453	0.103	<0.001
shx_2	β_{k3}	0.217	0.051	<0.001	0.161	0.291	0.290
urod_7	β_{k4}	−0.001	<0.001	<0.001	−0.002	0.003	0.260
urod_9	β_{k5}	−0.002	<0.001	<0.001	−0.001	0.001	0.200
Age	β_{k6}	0.021	0.005	<0.001	0.003	0.020	0.560
sev_1	β_{k7}	−0.045	0.040	0.860	3.735	1.004	<0.001
sev_2	β_{k8}	0.436	0.295	0.070	5.406	1.268	<0.001
sev_3	β_{k9}	0.934	0.284	<0.001	6.656	1.281	<0.001

5.4.1.2 Varying-coefficients model

In contrast, we fit the transformed varying-coefficients model (5.10), which accounts for the effect of covariance structure changing over time. Figure 5.2 shows the estimates for outcomes urgency and frequency. The advantage of the varying-coefficients model is that a likelihood ratio test can be performed for the null hypothesis (5.11) to assess the significance of covariate effects over time. The p-values are also shown in Figure 5.2. Similar to the results from the GEE approach, covariates sex, shx_2, urod_7, urod_9, age, and severity are all significantly associated with urgency, while covariates income, shx_2, and severity are significant for frequency. There is only one major difference in results between the GEE approach and the varying-coefficients model: covariate shx_2, i.e., whether or not previous interstitial cystitis diagnosis is diagnosed by physician, is not significant in the GEE analysis but is significant in the varying-coefficients model. From Figure 5.2, the effect of shx_2 dramatically changed after 40 months; the GEE approach assumes a time-constant effect, which fails to capture this feature.

5.4.1.3 Fully parametric model

To apply the fully parametric model discussed in Section 5.2.4 to the ICDB data, it is important to check the model assumptions via exploratory data analysis. We can compare the marginal empirical distribution of X_{ij} with a Poisson distribution with mean $\overline{X} = \sum_{i,j} X_{ij} / \sum_i T_i$, e.g., see Figure 3 (Yang and Kang, 2010). The marginal cumulative distribution function of X_{ij} is quite close to that of Poisson distribution. Thus, it is reasonable to assume X_{ij} follows a Poisson distribution. Similarly, the histogram of Y_{ij} is plotted with a normal density function with mean $\overline{Y} = \sum_{i,j} Y_{ij} / \sum_i T_i$, and variance $\sum_{i,j} (Y_{ij} - \overline{Y})^2 / (\sum_i T_i - 1)$. The comparison indicates that the marginal distribution of Y_{ij} can be roughly assumed as a normal distribution.

To check the assumption of exchangeable covariance of multivariate Poisson data $\mathbf{X}_i$, and the exchangeable correlation of multivariate continuous data $\mathbf{Y}_i$, we compute the sample covariance and correlation. Due to missing values, the size of the last 18-month data is small, which might produce biased results when we estimate the sample covariance and correlation. We thus do not report their associated estimates here. The sample covariance of $\mathbf{X}_i$ ranges from 0.8 to 1.2, and the sample correlation of the multivariate continuous data $\mathbf{Y}_i$ is between 0.4 and 0.6 (see Figure 4 in Yang and Kang, 2010). Also, there are no strong patterns showing that the covariance or the correlation decreases as the observed time difference increases.

We apply the complete data model in Section 5.2.4 to obtain the parameter estimates $\{\widehat{\boldsymbol{\beta}}_1, \widehat{\boldsymbol{\beta}}_2, \widehat{\gamma}_1, \widehat{\gamma}_2\}$. The log-likelihood has a larger value when $a = 1$ compared to when $a = 2$. This indicates that the Poisson outcome does not have the overdispersion problem. The parameter estimates

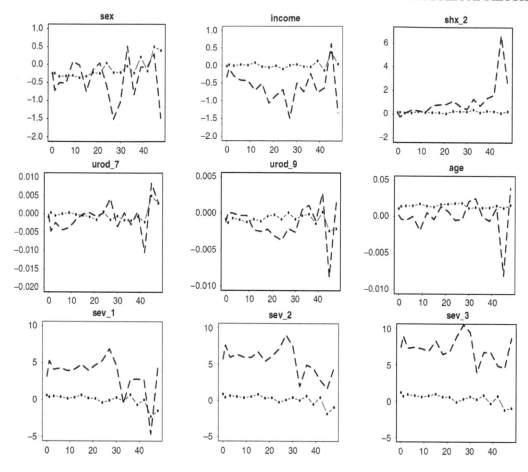

Figure 5.2 *Varying-coefficients estimates for $\boldsymbol{\beta}_1(t)$ (dotted line) and $\boldsymbol{\beta}_2(t)$ (dashed line) in model (5.10) with p-values for testing null hypothesis (5.11).*

$\widehat{\gamma}_1$ and $\widehat{\gamma}_2$ are 0.323 and 0.0247, and the associated p-values are 0.0012 and 0.2345, respectively. Table 5.5 summarizes the parameter estimates $\widehat{\boldsymbol{\beta}}_1$ and $\widehat{\boldsymbol{\beta}}_2$, and the associated p-values.

Table 5.5 shows that not all the covariates are significantly associated with urgency, and age is marginally significant with p-value 0.088. However, income, urod_7, and severity, are all significantly associated with frequency. We conclude that these covariates are important to get a good understanding of the disease. While estimates are similar to those of the GEE method, there are differences between the two concerning significance. Specifically, in the covariates for the Poisson outcome (i.e., frequency), sex, shx_2, urod_7, urod_9, and sev_3, are significant in the GEE analysis but not in the parametric model; in the covariates for the continuous outcome (i.e., urgency), urod_7 is significant in the parametric model but not in the GEE analysis.

5.4.2   Missing data model

Due to patients dropping out of the study, the number of observations of both longitudinal outcomes dramatically decreases over time. We find that the frequency of the observed outcome after 36 months is less than 20%. This implies that the useful information we could draw after 36 months is very limited. Therefore, in the following analysis, we only consider data observed before 36 months, when the frequency of the observed outcome is above 21.6%. The overall missing rate is increasing over time and the number of observations per patient is most likely around 4 to 6.

Table 5.5 *Estimates of regression coefficients for ICDB data obtained from the fully parametric model using available data.*

Effect		Urgency ($k=1$)			Frequency ($k=2$)		
		Est	SE	*p*-value	Est	SE	*p*-value
Sex	β_{k1}	−0.372	0.430	0.193	−0.082	0.264	0.378
Income	β_{k2}	−0.002	0.313	0.498	−0.488	0.157	0.001
shx_2	β_{k3}	0.187	0.347	0.295	0.072	0.145	0.311
urod_7	β_{k4}	−0.001	0.002	0.380	−0.002	0.001	0.035
urod_9	β_{k5}	−0.001	0.001	0.284	−0.001	0.001	0.202
Age	β_{k6}	0.012	0.009	0.088	−0.001	0.005	0.401
sev_1	β_{k7}	−0.135	1.220	0.456	3.440	0.730	<0.001
sev_2	β_{k8}	0.192	1.053	0.428	4.997	0.668	<0.001
sev_3	β_{k9}	0.447	1.059	0.336	6.037	0.670	<0.001

5.4.2.1 Fully parametric model

We use the missing data model in Section 5.3.1 to estimate parameters and test the significance of covariates. Respective estimates $\widehat{\gamma}_1$ and $\widehat{\gamma}_2$ are 0.341 and 0.112, and their corresponding *p*-values are 0.012 and 0.646. Results for $\widehat{\boldsymbol{\beta}}_1$ and $\widehat{\boldsymbol{\beta}}_2$ are listed in Table 5.6.

Table 5.6 *Estimates of regression coefficients for ICDB data obtained from fully parametric model, taking missing data into account.*

Effect		Urgency ($k=1$)			Frequency ($k=2$)		
		Est	SE	*p*-value	Est	SE	*p*-value
Sex	β_{k1}	−0.288	0.390	0.230	0.141	0.243	0.281
Income	β_{k2}	0.002	0.242	0.497	−0.444	0.141	0.001
shx_2	β_{k3}	0.171	0.258	0.254	0.045	0.131	0.365
urod_7	β_{k4}	−0.001	0.002	0.345	−0.001	0.001	0.136
urod_9	β_{k5}	−0.001	0.001	0.262	−0.001	0.001	0.091
Age	β_{k6}	0.013	0.007	0.039	−0.001	0.005	0.491
sev_1	β_{k7}	−0.335	1.089	0.379	2.876	0.667	<0.001
sev_2	β_{k8}	0.007	0.924	0.497	4.474	0.609	<0.001
sev_3	β_{k9}	0.272	0.933	0.385	5.526	0.612	<0.001

The covariate significance assessed by the missing data model is a little different from the results from the complete case analysis. First, age becomes significantly associated with urgency, and urod_7 is no longer significant. However, estimates from the two methods are relatively similar. This suggests that ignoring missing data may lead to efficiency loss and possibly biased inference.

Estimates in the missingness indicator model are also obtained, i.e., $\widehat{\tau}_{10} = \widehat{\tau}_{20} = 0.0164, \widehat{\tau}_{11} = \widehat{\tau}_{21} = -0.08, \widehat{\tau}_{12} = \widehat{\tau}_{22} = -0.0864$. This implies that the missingness mechanism of the two outcomes are the same. It also indicates that the data in our example are not completely missing at random (MCAR).

5.4.2.2 Imputation

We also apply the imputation method discussed in Section 5.3.2 to the ICDB data. We use a log-linear regression model as the imputation model for the Poisson outcome, and a linear regression model for the continuous outcome. After imputation, we use the GEE approach in Section 5.2.1.

Estimates $\widehat{\gamma}_1$ and $\widehat{\gamma}_2$ are 0.234 and 0.022, respectively, with associated p-values 0.12 and 0.08. Results for $\widehat{\boldsymbol{\beta}}_1$ and $\widehat{\boldsymbol{\beta}}_2$ are listed in Table 5.7.

Table 5.7 *Estimates of regression coefficients for ICDB data obtained by the GEE approach, with imputation.*

Effect		Urgency ($k=1$)			Frequency ($k=2$)		
		Est	SE	p-value	Est	SE	p-value
Sex	β_{k1}	−0.323	0.234	0.916	0.131	0.234	0.288
Income	β_{k2}	0.002	0.124	0.494	−0.345	0.153	0.012
shx_2	β_{k3}	0.241	0.344	0.242	0.067	0.165	0.342
urod_7	β_{k4}	−0.001	0.003	0.631	−0.001	0.002	0.691
urod_9	β_{k5}	−0.002	0.003	0.748	−0.001	0.001	0.158
Age	β_{k6}	0.018	0.008	0.012	0.003	0.006	0.309
sev_1	β_{k7}	−0.236	0.989	0.594	3.413	0.760	<0.001
sev_2	β_{k8}	0.009	0.934	0.496	5.245	0.668	<0.001
sev_3	β_{k9}	0.198	0.823	0.405	6.242	0.678	<0.001

The imputation results are quite similar to those from the fully parametric analysis with missing data.

5.5 Discussion

In this chapter, we reviewed several semiparametric and fully parametric regression models for jointly analyzing Poisson and continuous longitudinal outcomes. Among the complete data models, the GEE approach is relatively easy to implement. The varying-coefficients model can capture the change in the covariance between two longitudinal outcomes over time; in addition, it allows the covariate effect to change over time. This might have good statistical interpretations for some particular problems. The fully parametric approach explicitly models the joint distribution of longitudinal Poisson outcomes. A special compound symmetry covariance structure is discussed. By introducing a tuning parameter in the model representation, this model has the ability to mitigate the problem that the Poisson outcome is usually overdispersed. The computational cost of the fully parametric approach is higher than those of the other two methods, because it requires an iterative algorithm, i.e., ECM algorithm.

We also discussed two methods that handle missing data: the likelihood approach and an imputation method. The likelihood approach is an extension of the fully parametric model for complete case analysis. It can deal with nonignorable missing data mechanisms, where an assumption on modeling of the missing data mechanism needs to be made. The results might be sensitive to this assumption, however. The imputation method considers a monotone missing data pattern and assumes data are missing at random. Computationally, the likelihood approach uses an MCEM algorithm, which carries out the Monte Carlo simulation in the E-step to compute the intractable integration. This significantly increases the computational cost. In contrast, the imputation method only requires the fitting of a standard generalized linear regression. Also, the imputation method we proposed in this chapter is more general than the likelihood approach. It can be applied to any bivariate longitudinal outcomes with monotone missing data patterns.

There are several future directions that can be pursued on the methods we discussed in this chapter. For semiparametric modeling of mixed count and continuous outcomes, an alternative to the GEE method is the quadratic inference function (QIF) method. It overcomes some of the difficulties in the use of the GEE methods such as its sensitivity to outliers and its numerical instability. For the fully parametric approach, a more general parametric family is needed. The multivariate Poisson model we considered restricts correlations to be positive, has the restrictions on positive correlations, which might not be true in real data applications. For modeling of missing data using the likelihood approach, it is of interest to develop more efficient computing strategies. For the imputation method, it is worth building a robust method for non-monotone missingness patterns in bivariate longitudinal data.

Chapter 6

Factorization and latent variable models for joint analysis of binary and continuous outcomes

Armando Teixeira–Pinto and Jaroslaw Harezlak

6.1 Introduction

Many studies, particularly in the health sciences, involve measurements of multiple outcomes on each subject. When the outcomes are of the same type and measured on the same scale, such as continuous variables, classical tools of multivariate statistics can be used. However, multivariate methods to analyze outcomes measured on different scales or measuring different underlying variables, i.e., non-commensurate outcomes, are less common and rarely used in applied research.

A common approach used in the presence of non-commensurate outcomes is to analyze each outcome separately, ignoring the potential correlation between the outcomes. There are several disadvantages of this approach. One aspect that seems obvious is the loss of efficiency when the extra information contained in the correlation between the outcomes is ignored. Although this may seem intuitive, the loss of efficiency is not as straightforward as one may think. Drawing a parallel with multivariate linear regression and considering multiple correlated continuous outcomes being modeled by a set of covariates, the efficiency gained in modeling the outcomes in a multivariate fashion versus separate linear regressions for each outcome, is only observed if the equations for each outcome involve different covariates (Bartels and Fiebig, 1991). In other words, if all the outcomes share the same covariates, then the estimates of the regression parameters (and their standard errors) obtained in the multivariate linear model are exactly the same as the ones obtained from separate univariate linear regressions. In this case, the multivariate model produces more efficient estimates than the univariate approach (Zellner, 1963; Zellner, 1962). In practice, however, one should not expect to observe dramatic differences in the estimates produced by multivariate models when compared to univariate regressions, if the outcomes are modeled with the same covariates (Teixeira–Pinto and Normand, 2009).

Nevertheless, the advantages of modeling multiple outcomes with multivariate methods are not restricted to situations of having outcome-specific covariates. With separate analysis, it is harder to answer intrinsic multivariate questions such as jointly testing a covariate effect on all the outcomes. Also, if an outcome is missing for some individuals, separate models may produce biased estimates if the missing mechanism depends on the other outcomes. Chapter 13 describes in more detail the problem of incomplete data in the context of mixed outcomes.

The main obstacle for jointly modeling variables of different nature is the nonexistence of an obvious multivariate distribution that accommodates this situation. In this chapter, we present two likelihood-based approaches to jointly model binary and continuous outcomes. The first method considers a factorization of the joint likelihood and fits a model to each component of the factorization. The other approach uses a latent variable to model the correlation between the outcomes.

This chapter is organized as follows. We introduce data from a clinical trial comparing two different stents in Section 6.2 to motivate the discussion in succeeding chapters. In Sections 6.4 and 6.5, two classes of joint models are presented and illustrated using the stents clinical trial data. They are contrasted as well with results from separate analyses, which are discussed in Section 6.3. Section 6.6 includes SAS codes used in the analyses. Finally, Section 6.7 concludes the chapter with a summary and some remarks.

6.2   Clinical trial on bare-metal and drug-eluting stents

To motivate the models presented in the subsequent sections, we use a data set to illustrate and compare the methods. The data were collected in a clinical trial comparing bare-metal and drug-eluting stents (Holmes *et al.*, 2004). With aging, there is an accumulation of cholesterol plaques in the walls of the blood vessels that then narrows the diameter of vessels, leading to coronary disease. Stents are small metallic coils used to help keep a blocked artery open and reduce the chance of it narrowing. Recently, drug-eluting stents were introduced as an alternative to the bare-metal stents. The primary aim of this trial was to show the efficacy of drug-eluting stents. Another objective was to identify baseline risk factors predictive of poor outcome after stenting procedure.

Patients at risk for coronary restenosis were randomized to standard bare-metal stenting or drug-eluting stenting. Several baseline characteristics were recorded, but in this example we limit the analysis to patient's age and angiographic pre-stenting characteristics (length of lesion, percentage of diameter stenosis, and minimum lumen diameter). Additionally, the covariate group identifies the enrollment arm (i.e., bare-metal or drug-eluting stent) of each patient.

We consider two clinical outcomes usually assessed in patients undergoing the stenting procedure. These outcomes were measured 9 months after stenting and are defined as follows:

1. a binary outcome *restenosis*, which is an adverse outcome that indicates the renarrowing of the artery; It is defined as a stenosis with diameter greater than 50% at 9 months after stenting;

2. a continuous outcome *late loss*, defined as the difference between the diameters of the stented segment right after the stenting procedure compared with follow-up angiogram at 9 months.

A total of 1052 patients were initially enrolled in the trial. However, given the invasive nature of the exam, the angiography at 9 months was only performed in a random subsample of 699 patients. We use this subsample throughout the chapter. An important characteristic of these data is the strong correlation (i.e., 0.72) between *restenosis* and *late loss*.

6.3   Separate analyses

A common approach to analyze multiple outcomes in applied research is to carry out separate (i.e., univariate) regression analyses for each response variable and ignore the correlation between them. We first describe this framework to be used later for comparison with multivariate models, and also to introduce the notation used in this chapter. Let Y_{bi} and Y_{ci} denote measurements of a binary and continuous variables, respectively, obtained on each of N subjects in a particular study. In addition, each subject has a $r_b \times 1$ covariate vector $\mathbf{x}_{bi}$, believed to be associated with Y_{bi}, and another $r_c \times 1$ vector $\mathbf{x}_{ci}$ of covariates that may be associated with Y_{ci}. In this setting, we use a probit-regression model for the binary outcome Y_{bi} and a linear regression model for the continuous response Y_{ci} as

$$\text{probit}(\mu_{bi}) = \Phi^{-1}(\mu_{bi}) = \mathbf{x}_{bi}^{\top}\boldsymbol{\beta}_b \quad \text{and} \quad Y_{ci} = \mathbf{x}_{ci}^{\top}\boldsymbol{\beta}_c + \varepsilon_i, \tag{6.1}$$

where $\mu_{bi} = E(Y_{bi})$, $\boldsymbol{\beta}_b = (\beta_{b1}, \cdots, \beta_{br_b})^{\top}$, $\boldsymbol{\beta}_c = (\beta_{c1}, \cdots, \beta_{cr_c})^{\top}$, $\Phi^{-1}(\cdot)$ is the quantile function of the standard normal distribution $\Phi(\cdot)$, and $\varepsilon_i \sim N(0, \sigma_c^2)$. Note that (6.1) implies that $\mu_{bi} = \Phi(\mathbf{x}_{bi}^{\top}\boldsymbol{\beta}_b)$.

Although a probit regression model is not commonly used in health research, it appears often in other areas of applied statistics. Another choice for the link function associated with binary outcomes is the logit link, leading to a logistic model instead of the probit model. The main reason that leads us to favor the probit link here is the existence of closed-form solutions for some calculations that are introduced in the multivariate models. We should note however, that for practical purposes the results given by the probit and logistic regressions are very similar, with the estimates for the logistic regression being approximately 1.6 times those of the probit model. More details on this approximation can be found in (Demidenko, 2004).

The interpretation of the regression parameters for the models in (6.1) is the same as in univariate generalized linear regression models. For example, β_{ck} is the change in the mean of Y_{ci}, for a unit increase in the kth covariate, fixing the other covariates; as well, β_{bk} is the change in the probit of the mean of Y_{bi} for an increase of one unit in the covariate x_{bk}, conditional on the remaining covariates. Using the similarity between the probit and logistic models mentioned above, $\exp(1.6\beta_{bk})$ is a close approximation to the adjusted odds ratio associated with the covariate x_{bk}. To simplify the discussion, we omit the conditional notation on the covariates in the formulas, as it should be clear from the context when the statements are conditional on $\mathbf{x}_{bi}$ and $\mathbf{x}_{ci}$.

The univariate models were applied to the example introduced in Section 6.2. For the continuous outcome *late loss*, we used a linear regression model to compare the bare-metal and drug-eluting stents. The covariates age and lesion length were also included in the model. The percentage of stenosis and minimum lumen diameter before the procedure turned out to be not associated with *late loss*, after adjusting for the other variables, and were removed in the final model (see Table 6.1).

Table 6.1 *Estimates (and their SEs and p-values) obtained from separate univariate models, for each covariate and with all covariates, for outcomes restenosis and late loss from the stents clinical trial data, with a probit regression for restenosis and a linear regression model for late loss.*

Covariate	Separate analysis per covariate			With all covariates		
	Est	SE	p-value	Est	SE	p-value
Restenosis						
Group	1.496	0.149	< 0.001	1.552	0.155	< 0.001
Age	−0.003	0.005	0.500	—	—	—
% stenosis pre-stent	0.005	0.004	0.303	−0.029	0.012	0.012
Minimum lumen diameter	−0.271	0.141	0.055	−1.157	0.381	0.002
Lesion length	0.210	0.089	0.019	0.270	0.104	0.010
Late loss						
Group	0.577	0.044	< 0.001	0.570	0.044	< 0.001
Age (per 10 yrs.)	−0.035	0.023	0.130	−0.043	0.021	0.035
% stenosis pre-stent	0.003	0.002	0.151	—	—	—
Minimum lumen diameter	−0.110	0.062	0.078	—	—	—
Lesion length	0.077	0.042	0.068	0.073	0.038	0.051
Error SD (σ_c)	—	—	—	0.578	0.016	—

We applied the probit model for binary outcome *restenosis*. The covariates stent group (bare-metal vs. drug-eluting stent), percentage of stenosis before stenting, minimum lumen diameter before stenting, and lesion length, all had significant associations with *restenosis*. Age was not associated with the outcome after adjusting for the other covariates, and was not included in the final model. The results of the analysis are presented in Table 6.1. As expected, given the randomization of the treatment, the adjusted differences observed between the two arms are similar to the non-adjusted differences in both outcomes. Percentage of stenosis and minimum lumen diameter became significantly associated with *restenosis* in the model with all covariates. Also, age and lesion length became significantly associated with continuous outcome *late loss*.

In the next two sections, we present joint (i.e., multivariate) regression approaches that are comparable with the univariate models, but also take into account the potential correlation between the outcome variables.

6.4   Factorization models for binary and continuous outcomes

Consider the setting described in the previous section. One of the first approaches proposed to jointly model mixed outcomes has its roots in the general location model (GLOM) (Olkin and Tate, 1961). The main idea is to factorize the joint distribution and fit a univariate model to each component of the factorization. Fitzmaurice and Laird (1995) propose a model for correlated binary and continuous variables by considering one of the two possible factorizations of their joint distribution as $f_{Y_{bi},Y_{ci}}(y_{bi},y_{ci}) = f_{Y_{bi}}(y_{bi})f_{Y_{ci}|Y_{bi}}(y_{ci}|y_{bi})$. The expected values of the outcomes are related to the covariates $\mathbf{x}_{bi}$ and $\mathbf{x}_{ci}$. For example, using a probit regression for Y_{bi}, and a linear regression for Y_{ci}, conditional on Y_{bi}, we have

$$\text{probit}(\mu_{bi}) = \mathbf{x}_{bi}^\top \boldsymbol{\beta}_b \quad \text{and} \quad Y_{ci} = \mathbf{x}_{ci}^\top \boldsymbol{\beta}_c + \tau(Y_{bi} - \mu_{bi}) + \varepsilon_{ci}, \tag{6.2}$$

where ε_{ci} is assumed to follow a $N(0,\sigma_c^2)$, and τ is the parameter for the regression of Y_{ci} on Y_{bi}, which models the correlation between the variables.

The correlation that results from this model is given by

$$corr(Y_{bi},Y_{ci}) \quad = \quad \frac{\tau}{\sqrt{\tau^2 + \dfrac{\sigma_c^2}{\Phi(\mathbf{x}_{bi}^\top \boldsymbol{\beta}_b)\{1 - \Phi(\mathbf{x}_{bi}^\top \boldsymbol{\beta}_b)\}}}}. \tag{6.3}$$

Note that as the absolute size of τ increases, when compared to σ_c, the correlation approaches 1 or -1. Large absolute values of τ indicate a strong correlation between the two outcomes; if $\tau = 0$, the two outcomes are independent given the covariates.

A convenient property of this factorization is that the model parameters maintain a marginal interpretation in both regression equations, identical to the univariate models (6.1). This is obvious for the probit model associated with the binary outcome Y_{bi}, where β_{bk} is the change on the probit of its expected value, for a unit increase in the kth covariate, conditional on the other covariates in the regression. For the linear model, β_{ck} is both the change in the mean of Y_{ci}, given Y_{bi}, for an increase of one unit in the covariate x_{ck}, fixing the other covariates, as well as the change in the mean of Y_{ci}, because the residual $Y_{bi} - \mu_{bi}$ has expectation 0.

Despite the interpretations of the regression parameters for this model and those from the univariate analysis being the same, the two approaches have different distributional assumptions regarding the continuous outcomes Y_{ci}. For the linear model in (6.1), we have the usual normality assumption on the distribution of Y_{ci}. For the linear component of the factorization model (6.2), Y_{ci} is assumed to be normally distributed, conditional on Y_{bi} and the covariates, implying that marginalizing over Y_{bi} yields a mixture of two normal distributions for the marginal distribution of Y_{ci}. A simple calculation shows that the variance of Y_{ci} depends on $\mathbf{x}_{bi}$, i.e.,

$$var(Y_{ci}) \quad = \quad E\{var(Y_{ci}|Y_{bi})\} + var\{E(Y_{ci}|Y_{bi})\} \quad = \quad \sigma_c^2 + \tau^2 \Phi(\mathbf{x}_{bi}^\top \boldsymbol{\beta}_b)\{1 - \Phi(\mathbf{x}_{bi}^\top \boldsymbol{\beta}_b)\}.$$

We fitted the factorization model (6.2) to the stenting data, and compared it with separate analysis of each outcome. The part of the model corresponding to the binary outcome has the same structure as the corresponding univariate model (6.1). Thus, the estimated effects of the covariates on *restenosis* are the same in the two approaches, as well as the respective standard errors (SEs). However, for *late loss*, the estimates of the covariate effects and their SEs are different in the univariate and the factorization models. Typically, the SEs are smaller in the joint model. This difference in the SEs may be explained by the inclusion of the correlation between the outcomes in the joint model, but is also due to the different structure of the error term variance. As explained above, this factorization assumes a different variance for each level of the binary outcome, while in the univariate model, the variance of the error term does not depend on the binary outcome.

Table 6.2 *Estimates (and their SEs and p-values) obtained from separate models and from the factorization model, for mixed outcomes restenosis and late loss from the stents clinical trial data, with probit link for restenosis and identity link for late loss.*

Covariate	Separate models			Factorization model		
	Est	SE	*p*-value	Est	SE	*p*-value
Restenosis						
Group	1.552	0.155	<0.001	1.552	0.155	<0.001
% of stenosis pre-stent	−0.029	0.012	0.012	−0.029	0.012	0.012
Minimum lumen diameter	−1.157	0.381	0.002	−1.157	0.381	0.002
Lesion length	0.270	0.104	0.010	0.270	0.104	0.010
Late loss						
Group	0.570	0.044	<0.001	0.579	0.033	<0.001
Age (per 10 yrs.)	−0.043	0.021	0.035	−0.028	0.016	0.083
Lesion length	0.073	0.038	0.051	0.074	0.029	0.010
Error SD (σ_c)	0.578	0.016	—	—	—	—
Association (τ)	—	—	—	1.047	0.048	—

An alternative to model (6.2) is proposed in Catalano and Ryan (1992), and uses the reverse factorization of the joint distribution, i.e., $f_{Y_{bi},Y_{ci}}(y_{bi},y_{ci}) = f_{Y_{ci}}(y_{ci})f_{Y_{bi}|Y_{ci}}(y_{bi}|y_{ci})$. This factorization can be motivated by assuming that the binary outcome Y_{bi} is a dichotomization of an underlying latent continuous variable Y_{bi}^*,

$$Y_{bi} = \begin{cases} 0 & \text{, if } Y_{bi}^* \leq 0 \\ 1 & \text{, if } Y_{bi}^* > 0 \end{cases}. \tag{6.4}$$

Catalano and Ryan (1992) originally refer to this model as a latent variable model because of the underlying unobserved variable. However, we prefer the designation "reverse factorization," as a different latent variable model is considered in Section 6.5.

The underlying variable Y_{bi}^* is assumed, conditional on the covariates $\mathbf{x}_{bi}$, to follow $N(0,(\sigma_b^*)^2)$. We now consider the following bivariate linear regression model:

$$\begin{aligned} Y_{bi}^* &= \mathbf{x}_{bi}^\top \boldsymbol{\beta}_b^* + \varepsilon_{bi}^* \\ Y_{ci} &= \mathbf{x}_{ci}^\top \boldsymbol{\beta}_c + \varepsilon_{ci} \end{aligned}, \quad \text{with} \quad \begin{pmatrix} \varepsilon_{bi}^* \\ \varepsilon_{ci} \end{pmatrix} \sim N_2\left(\mathbf{0}, \begin{pmatrix} (\sigma_b^*)^2 & \tau^*\sigma_c\sigma_b^* \\ \tau^*\sigma_c\sigma_b^* & \sigma_c^2 \end{pmatrix}\right).$$

The joint bivariate normal distribution of Y_{bi}^* and Y_{ci} can be written as the product of the marginal distribution of $Y_{ci} \sim N(\mathbf{x}_{ci}^\top\boldsymbol{\beta}_c, \sigma_c^2)$ and the conditional distribution of Y_{bi}^*, given Y_{ci}, where

$$Y_{bi}^*|Y_{ci}=y_{ci} \sim N\left(\mathbf{x}_{bi}^\top\boldsymbol{\beta}_b^* + \frac{\sigma_b^*\tau^*}{\sigma_c}(y_{ci}-\mathbf{x}_{ci}^\top\boldsymbol{\beta}_c), (\sigma_b^*)^2\left\{1-(\tau^*)^2\right\}\right).$$

The above formulation implies the following model for the binary outcome, conditional on the continuous outcome:

$$E(Y_{bi}|Y_{ci}=y_{ci}) = P(Y_{bi}^*>0|Y_{ci}=y_{ci}) = \Phi\left(\frac{\mathbf{x}_{bi}^\top\boldsymbol{\beta}_b^* + \frac{\sigma_b^*\tau^*}{\sigma_c}(y_{ci}-\mathbf{x}_{ci}^\top\boldsymbol{\beta}_c)}{\sigma_b^*\sqrt{1-(\tau^*)^2}}\right).$$

Unfortunately, not all parameters in this model are estimable, so we need to reparametrize it as

$$E(Y_{bi}|Y_{ci}=y_{ci}) = \Phi\left(\mathbf{x}_{bi}^\top\boldsymbol{\beta}_b + \tau(y_{ci}-\mathbf{x}_{ci}^\top\boldsymbol{\beta}_c)\right), \tag{6.5}$$

where $\boldsymbol{\beta}_b = \boldsymbol{\beta}_b^* / \{\sigma_b^* \sqrt{1-(\tau^*)^2}\}$ and $\tau = \tau^* / \{\sigma_c \sqrt{1-(\tau^*)^2}\}$.

Finally, using the above construction to write the joint model for the binary and continuous outcomes, we have

$$\text{probit}(\mu_{bi}) = \mathbf{x}_{bi}^\top \boldsymbol{\beta}_b + \tau(Y_{ci} - \mu_{ci}) \quad \text{and} \quad Y_{ci} = \mathbf{x}_{ci}^\top \boldsymbol{\beta}_c + \varepsilon_{ci}. \tag{6.6}$$

In this case, $\boldsymbol{\beta}_b$ represents the effects of covariates on the probit of μ_{bi}, conditional on Y_{ci}. Therefore, they cannot be directly compared with the marginal effects obtained through the univariate probit model in (6.1), which are unconditional (i.e., marginal) effects. The advantage of using the probit over the logit link is the possibility to obtain closed-form expressions of the marginal effects. Because $(Y_{ci} - \mu_{ci})$ follows $N(0, \sigma_c^2)$, the marginal effects $\boldsymbol{\beta}_b^{\mathrm{m}}$, obtained by integrating over Y_{ci}, can be similarly obtained by rescaling the conditional effects. This yields $\boldsymbol{\beta}_b^{\mathrm{m}} = \boldsymbol{\beta}_b / \sqrt{1 + \sigma_c^2 \tau^2}$.

It is also possible to obtain an explicit expression for the correlation induced by (6.6). It should be noted that the correlation is bounded approximately by 0.8 (and -0.8), which corresponds to the maximum correlation between a normal and dichotomous variables (Gradstein, 1986). This is seen in the following:

$$corr(Y_{bi}, Y_{ci}) = \left(\frac{\tau \sigma_c}{\sqrt{1 + \sigma_c^2 \tau^2}} \right) \frac{\phi\left(\frac{\mathbf{x}_{bi}^\top \boldsymbol{\beta}_b}{\sqrt{1 + \sigma_c^2 \tau^2}} \right)}{\sqrt{\Phi\left(\frac{\mathbf{x}_{bi}^\top \boldsymbol{\beta}_b}{\sqrt{1 + \sigma_c^2 \tau^2}} \right) \left\{ 1 - \Phi\left(\frac{\mathbf{x}_{bi}^\top \boldsymbol{\beta}_b}{\sqrt{1 + \sigma_c^2 \tau^2}} \right) \right\}}} < \frac{0.4}{0.5}, \tag{6.7}$$

where $\phi(\cdot)$ is the standard normal density.

6.5   Latent variable models for binary and continuous outcomes

In this section, we discuss a different approach to building a joint model for mixed binary and continuous outcomes, by introducing a latent variable.

Sammel *et al.* (1997) discuss a latent variable model, where it is assumed that the two outcomes are physical manifestations of an unobserved (latent) continuous variable. Conditional on this latent variable, the outcomes are assumed to be independent, and the effects of the covariates of interest are modeled through the latent variable.

Let U_i denote the latent variable, and let x_i be a covariate of interest, such as treatment. If U_i were directly observable for an individual, a possible model would be the linear regression model $U_i = \gamma x_i + \delta_i$, with $\delta_i \sim N(0,1)$. Note that fixing the variance of δ_i to 1 does not represent a constraint, as we could assume that U_i is appropriately scaled. The parameter γ represents the association between the covariate and the unobserved latent variable. Now, the outcomes Y_{bi} and Y_{ci} are modeled as functions of the latent variable U_i. The conditional models are

$$\text{probit}\{E(Y_{bi}|U_i)\} = \beta_{b1} + \beta_{b2} U_i \quad \text{and} \quad Y_{ci} = \beta_{c1} + \beta_{c2} U_i + \varepsilon_{ci}, \tag{6.8}$$

where $\varepsilon_{ci} \sim N(0, \sigma_c^2)$. Here, β_{b2} and β_{c2} indicate the strength of the association between the observed outcomes and the latent variable. Conceptually, this model is very appealing because it quantifies the notion of the outcomes measuring an underlying construct. However, the drawback is that some covariance parameters are also parameters of the mean structure. For example, because $E(Y_{ci}) = \beta_{c1} + \beta_{c2} \gamma x_{ik}$ and $var(Y_{ci}) = \beta_{c2}^2 + \sigma_2^2$, the model is sensitive to misspecification of the correlation structure, as described by Sammel *et al.* (1997). See also Sammel and Ryan (2002) and Sammel *et al.* (1999).

Another approach based on latent variables is introduced in Dunson (2000). A major difference between this approach and Sammel *et al.*'s (1997) model lies in the association between the responses and the covariates. In Dunson's (2000) approach, the covariates are not included in the latent variable structure, but rather are introduced separately in the model. In the case of binary and continuous outcomes, Dunson's (2000) conditional model, given U_i, is

$$\text{probit}\{E(Y_{bi}|U_i)\} = \mathbf{x}_{bi}^{\top}\boldsymbol{\beta}_b + \lambda_b U_i \text{ and } Y_{ci} = \mathbf{x}_{ci}^{\top}\boldsymbol{\beta}_c + \lambda_c U_i + \varepsilon_{ci}, \tag{6.9}$$

where $\varepsilon_{ci} \sim N(0, \sigma_c^2)$, and $U_i \sim N(0, \sigma_u^2)$ is a subject-specific latent variable. The latent variable shared by both outcomes induces the correlation, and it is assumed that given the latent variable, the two outcomes are independent. However, λ_b, λ_c, σ_u, and σ_c are not identifiable. Fixing these parameters to any constant will result in a misspecification of the correlation between the outcomes (Teixeira–Pinto and Normand, 2009).

To better understand this argument, consider a similar idea for two correlated continuous outcomes, Y_1 and Y_2, and assume the conditional models (given U_i)

$$Y_{1i} = \mathbf{x}_{1i}^{\top}\boldsymbol{\beta}_1 + \lambda_1 U_i + \varepsilon_{1i} \text{ and } Y_{2i} = \mathbf{x}_{2i}^{\top}\boldsymbol{\beta}_2 + \lambda_2 U_i + \varepsilon_{2i}, \tag{6.10}$$

where $\varepsilon_{1i} \sim N(0, \sigma_1^2)$, $\varepsilon_{2i} \sim N(0, \sigma_2^2)$, and $U_i \sim N(0, \sigma_u^2)$. The parameters associated with the variance components of the outcomes (i.e., $\lambda_1, \lambda_2, \sigma_u, \sigma_1$, and σ_2) are again not identifiable. We have to restrict at least two parameters to obtain an identifiable model. The correlation induced by the model is given by

$$\frac{\lambda_1 \lambda_2 \sigma_u^2}{\sqrt{(\lambda_1^2 \sigma_u^2 + \sigma_1^2)(\lambda_2^2 \sigma_u^2 + \sigma_2^2)}}.$$

If we constrain the parameters λ_1 and λ_2 to be 1, for example, the correlation becomes $\sigma_u^2 / \sqrt{(\sigma_u^2 + \sigma_1^2)(\sigma_u^2 + \sigma_2^2)}$. It is easy to build a case where such model fails to induce the correct correlation. Suppose that $var(Y_1) = \sigma_u^2 + \sigma_1^2 = 0.5$, $var(Y_2) = \sigma_u^2 + \sigma_2^2 = 5$, and $corr(Y_1, Y_2) = 0.8$. It follows that $\sigma_u^2 < 0.5$ and the correlation induced by model (6.9) becomes

$$corr(Y_1, Y_2) < \frac{0.5}{\sqrt{0.5 \times 5}} \approx 0.32,$$

which is incorrect. Fixing the variances of the error terms, or that of the latent variable, will lead to similar inconsistencies. A similar argument can be given for model (6.9). Although there is one fewer parameter than in model (6.10), there is less information to estimate the parameters because $var(Y_b)$ is fully determined by $E(Y_b)$.

Teixeira–Pinto and Normand (2009) show how to constrain the parameters in model (6.9) using a similar idea to the scaled multivariate mixed model proposed by Lin et al. (2000). Let Y_{1i} and Y_{2i} be two continuous normally distributed outcomes associated with covariates $\mathbf{x}_{1i}$ and $\mathbf{x}_{2i}$, respectively. Given the covariates, we assume that the two outcomes are correlated. We define $Y_{1i}^* = Y_{1i}/\sigma_1$ and $Y_{2i}^* = Y_{2i}/\sigma_2$, and the conditional model, given U_i, is then

$$Y_{1i}^* = \frac{Y_{1i}}{\sigma_1} = \mathbf{x}_{1i}^{\top}\boldsymbol{\beta}_1^* + U_i + \varepsilon_{1i}^* \text{ and } Y_{2i}^* = \frac{Y_{2i}}{\sigma_2} = \mathbf{x}_{2i}^{\top}\boldsymbol{\beta}_2^* + U_i + \varepsilon_{2i}^*, \tag{6.11}$$

where σ_1 and σ_2 are scaling parameters, such that $\varepsilon_{1i}^* \sim N(0, 1)$ and $\varepsilon_{2i}^* \sim N(0, 1)$, and $U_i \sim N(0, \sigma_u^2)$ is a latent variable that induces the correlation between Y_{1i}^* and Y_{2i}^*. We can rewrite (6.11) and obtain the final expression for the conditional joint latent model (given U_i) for the two continuous outcomes, as

$$Y_{1i} = \mathbf{x}_{1i}^{\top}\boldsymbol{\beta}_1 + \sigma_1 U_i + \varepsilon_{1i} \text{ and } Y_{2i} = \mathbf{x}_{2i}^{\top}\boldsymbol{\beta}_2 + \sigma_2 U_i + \varepsilon_{2i}, \tag{6.12}$$

where $\boldsymbol{\beta}_1 = \sigma_1 \boldsymbol{\beta}_1^*$, $\boldsymbol{\beta}_2 = \sigma_2 \boldsymbol{\beta}_2^*$, $\varepsilon_{1i} \sim N(0, \sigma_1^2)$, $\varepsilon_{2i} \sim N(0, \sigma_2^2)$, and $U_i \sim N(0, \sigma_u^2)$. The correlation between the two outcomes induced by the model is $corr(Y_1, Y_2) = \sigma_u^2/(1 + \sigma_u^2)$, so that the range of correlations that we can model is $[0,1)$. This model still constrains the correlation to be positive; however, if the two outcomes are negatively correlated, a simple transformation of one of the outcomes will reverse the sign of the correlation.

These considerations motivate the constraints for model (6.9) as follows. Let Y_{bi} and Y_{ci} be binary and continuous outcomes associated with covariates $\mathbf{x}_{bi}$ and $\mathbf{x}_{ci}$, respectively. We want to develop a joint model that takes into account the potential correlation between Y_{bi} and Y_{ci}. The outcome Y_{ci} is assumed to be normally distributed, given the covariates $\mathbf{x}_{ci}$. Using the same construction of the reverse factorization, suppose there is an underlying variable Y_{bi}^*, which is normally distributed given the covariates $\mathbf{x}_{bi}$, and is associated with binary outcome Y_{bi}, as described in model (6.4). Define $Y_{ci}^* = Y_{ci}/\sigma_c$, where σ_c is a scale parameter for the continuous outcome. The conditional regression equations (given U_i) for the two variables can be written as

$$Y_{bi}^* = \mathbf{x}_{bi}^\top \boldsymbol{\beta}_b^* + U_i + \varepsilon_{bi}^* \quad \text{and} \quad Y_{ci}^* = \mathbf{x}_{ci}^\top \boldsymbol{\beta}_c^* + U_i + \varepsilon_{ci}^*, \tag{6.13}$$

with $\varepsilon_{bi}^* \sim N(0,1)$, $\varepsilon_{ci}^* \sim N(0,1)$, and $U_i \sim N(0,\sigma_u^2)$. The variances of the error terms are fixed at 1 by design. This is just a convenient standardization to obtain a common variance and does not represent a restriction of the model. The latent variable U_i is introduced in both equations to induce the correlation between the outcomes. It is assumed that given U_i, Y_{bi}^* and Y_{ci}^* are independent, and consequently Y_{bi} and Y_{ci} are also independent, given U_i.

Because $E(Y_{ci}) = \sigma_c E(Y_{ci}^*)$, we can write the equation for Y_{ci}, given U_i, as $Y_{ci} = \mathbf{x}_{ci}^\top \boldsymbol{\beta}_c + \sigma_c U_i + \varepsilon_{ci}$, where $\boldsymbol{\beta}_c = \sigma_c \boldsymbol{\beta}_c^*$ and $\varepsilon_{ci} \sim N(0,\sigma_c^2)$. The correlation between Y_{bi}^* and Y_{ci} is a function of σ_u only, and is given by $\sigma_u^2/(1+\sigma_u^2)$. However, Y_{bi}^* is not observed. We can write the conditional regression equation for the binary outcome Y_{bi}, given U_i, as $P(Y_{bi}=1|U_i) = P(Y_{bi}^*>0|U_i) = \Phi(\mathbf{x}_{bi}^\top \boldsymbol{\beta}_b^* + U_i)$. The final model is then

$$\text{probit}\{P(Y_{bi}=1|U_i)\} = \mathbf{x}_{bi}^\top \boldsymbol{\beta}_b' + U_i \quad \text{and} \quad Y_{ci} = \mathbf{x}_{ci}^\top \boldsymbol{\beta}_c + \sigma_c U_i + \varepsilon_{ci}. \tag{6.14}$$

The correlation between Y_{bi} and Y_{ci} that results from this model can be calculated as

$$corr(Y_{bi}, Y_{ci}) = \left(\frac{\sigma_u^2}{1+\sigma_u^2}\right) \frac{\phi\left(\frac{\mathbf{x}_{bi}^\top \boldsymbol{\beta}_b'}{\sqrt{\sigma_u^2+1}}\right)}{\sqrt{\Phi\left(\frac{\mathbf{x}_{bi}^\top \boldsymbol{\beta}_b'}{\sqrt{\sigma_u^2+1}}\right)\left\{1-\Phi\left(\frac{\mathbf{x}_{bi}^\top \boldsymbol{\beta}_b'}{\sqrt{\sigma_u^2+1}}\right)\right\}}}. \tag{6.15}$$

The parameters $\boldsymbol{\beta}_b'$ in model (6.14) have to be interpreted conditional on U_i. Given U_i, β_{bk}' is the change on the probit scale of the expected value of Y_{bi} for an increase of one unit in the covariate x_{bk}. For this reason, the parameters $\boldsymbol{\beta}_b'$ of the latent model cannot be directly compared with the regression parameters of the marginal models, such as (6.1) and (6.2). To obtain the marginal effects that can be compared with the other models, we have to average over the U_is as follows:

$$P(Y_{bi}=1) = \int_{-\infty}^{+\infty} P(Y_{bi}=1|u_i) f_{U_i}(u_i) du_i = \Phi\left(\frac{\mathbf{x}_{bi}^\top \boldsymbol{\beta}_b'}{\sqrt{1+\sigma_u^2}}\right).$$

Thus, $\boldsymbol{\beta}_b = \boldsymbol{\beta}_b'/\sqrt{1+\sigma_u^2}$ are the marginal effects associated with the covariates. For the continuous outcome, $\boldsymbol{\beta}_c$ can be interpreted either as conditional or marginal covariate effects.

The log-likelihood for the model is written as

$$\log \prod_{i=1}^n f_{Y_{bi},Y_{ci}}(y_{bi}, y_{ci}) = \log \prod_{i=1}^n \int_{-\infty}^{+\infty} f_{Y_{bi}|U_i}(y_{bi}|u_i) f_{Y_{ci}|U_i}(y_{ci}|u_i) f_{U_i}(u_i) du_i$$

$$= \log \prod_{i=1}^n \int_{-\infty}^{+\infty} \{\Phi(\mu_{bi}+u_i)\}^{y_{bi}} \{1-\Phi(\mu_{bi}+u_i)\}^{1-y_{bi}}$$

$$\times \frac{1}{2\pi\sigma_c\sigma_u} \exp\left\{-\frac{1}{2\sigma_c^2}(y_{ci}-\mu_{ci}-\sigma_c u_i)^2 - \frac{1}{2\sigma_u^2}u_i^2\right\} du_i, \tag{6.16}$$

where $\mu_{bi} = \mathbf{x}_{bi}^\top \boldsymbol{\beta}_b'$ and $\mu_{ci} = \mathbf{x}_{ci}^\top \boldsymbol{\beta}_c$. Estimates $\widehat{\boldsymbol{\beta}}_b$ of the marginal effects can be obtained from $\widehat{\boldsymbol{\beta}}_b'/\sqrt{1 + \widehat{\sigma}_u^2}$. The SEs for $\widehat{\boldsymbol{\beta}}_b$ can be approximated using the delta method.

The properties of the probit link allow a simplification of the likelihood for the latent variable model (6.16), since the integral in (6.16) has a closed-form solution. Solving this integral, we show that this is the same model as the reverse factorization but with a different parameterization, as follows:

$$\log \prod_{i=1}^n f_{Y_{bi}, Y_{ci}}(y_{bi}, y_{ci}) = \log \prod_{i=1}^n \{\Phi(p_i)\}^{y_{bi}} \{1 - \Phi(p_i)\}^{1 - y_{bi}} \phi\left(\frac{y_{ci} - \mu_{ci}}{\sigma_c \sqrt{1 + \sigma_u^2}}\right), \quad (6.17)$$

where

$$p_i = \frac{1}{\sqrt{\frac{2\sigma_u^2 + 1}{\sigma_u^2 + 1}}} \left\{\mu_{bi} + \frac{\sigma_u^2}{\sigma_c(1 + \sigma_u^2)}(y_{ci} - \mu_{ci})\right\}. \quad (6.18)$$

Letting the reverse factorization coefficient $\boldsymbol{\beta}_b^{\mathrm{rf}} = \boldsymbol{\beta}_b' \sqrt{(\sigma_u^2 + 1)/(2\sigma_u^2 + 1)}$, and $\tau = \sigma_u^2/\sqrt{2\sigma_u^2 + 1}$, we get

$$\log \prod_{i=1}^n f_{Y_{bi}, Y_{ci}}(y_{bi}, y_{ci}) = \log \prod_{i=1}^n \phi\left(\frac{y_{ci} - \mathbf{x}_{ci}^\top \boldsymbol{\beta}_c}{\sigma_c \sqrt{1 + \sigma_u^2}}\right) \left\{\Phi\left(\mathbf{x}_{bi}^\top \boldsymbol{\beta}_b^{\mathrm{rf}} + \tau(y_{ci} - \mathbf{x}_{ci}^\top \boldsymbol{\beta}_c)\right)\right\}^{y_{bi}}$$

$$\times \left\{1 - \Phi\left(\mathbf{x}_{bi}^\top \boldsymbol{\beta}_b^{\mathrm{rf}} + \tau(y_{ci} - \mathbf{x}_{ci}^\top \boldsymbol{\beta}_c)\right)\right\}^{1 - y_{bi}}. \quad (6.19)$$

The likelihood (6.19) takes the same form as the reverse factorization likelihood in model (6.6); that is, both approaches result in the same model with different parameterizations.

We fit the latent variable model (6.14) to the stenting example, using the same covariates as the approach that fits separate regressions for the outcomes *restenosis* and *late loss*. We should note that the distributional assumption for the error term associated with the linear model for *late loss* is the same for the latent variable model and the univariate linear regression. However, the coefficients associated with *restenosis* have different interpretations in the two approaches. For the latent variable model, the effects of group, percent of stenosis, minimum lumen diameter, and lesion length are adjusted for the covariates in the model, and are conditional on the latent variable. We can obtain the marginal effects (with respect to the latent variable), by dividing the coefficients by $\sqrt{1 + \sigma_u^2}$. The results presented in Table 6.3 are already converted to the marginal effects and are directly comparable to the univariate regressions.

The comparison between the joint model using a latent variable, and the univariate regressions for each outcome (see Table 6.3) brings a number of interesting observations. For the binary outcome *restenosis*, the effects of the covariates are slightly different in both approaches, and the corresponding SEs are, in general, smaller for the joint model. For the continuous outcome *late loss*, the coefficient estimates for group and lesion length are identical (the small differences are explained by numerical errors) and their SEs are similar. These two covariates are shared by both outcomes. The coefficient estimate for the non-shared covariate age is identical in both approaches, but its SE is smaller in the latent variable model than in the univariate regression (i.e., 0.017 vs. 0.021). In summary, we observed gains in efficiency for the estimates in the joint model associated with outcome-specific covariates, and identical efficiency for common covariates (i.e., group and lesion length).

6.6   Software

In this section, we present SAS code for fitting the latent variable model (6.14) using the stents clinical trial data in Section 6.2. We use PROC NLMIXED to program the likelihood for the model.

Table 6.3 *Estimates (and their SEs and p-values) obtained from separate models and from the joint model, i.e., model (6.14), using a latent variable, for mixed outcomes restenosis and late loss from the stents clinical trial data, with probit link for restenosis and identity link for late loss.*

Covariate	Separate models			Latent variable model		
	Est	SE	*p*-value	Est	SE	*p*-value
Restenosis						
Group	1.552	0.155	< 0.001	1.414	0.143	< 0.001
% stenosis pre-stent	−0.029	0.012	0.012	−0.036	0.009	< 0.001
Minimum lumen diameter	−1.157	0.381	0.002	−1.219	0.304	< 0.001
Lesion length	0.270	0.104	0.010	0.260	0.106	0.014
Late loss						
Group	0.570	0.044	< 0.001	0.579	0.044	< 0.001
Age (per 10 yrs.)	−0.043	0.021	0.035	−0.040	0.017	0.002
Lesion length	0.073	0.038	0.051	0.073	0.038	0.051
Error SD (σ_c)	0.578	0.016	—	0.223	0.016	—
Latent SD (σ_u)	—	—	—	2.379	0.206	—

From our experience, the convergence of PROC NLMIXED is highly dependent on the starting values for model coefficients. One strategy to obtain good starting values is to fit separate regressions to each outcome.

```
/*SAS code to fit the model to the clinical trial data comparing bare-metal
and drug-eluting stents.*/

/*PROC NLMIXED options - MAXITER, QPOINTS and GCONV - may require some tuning
for convergence.*/

/*Outcome restenosis (REST) is modeled using a probit link and covariates trial arm
(GROUP), percentage of stenosis pre-stenting (PCTDS), minimum lumen diameter (MLD),
and lesion length(LL); outcome late loss (LS) is modeled using the identity link
and covariates trial arm (GROUP), age (AGE), and lesion length (LL).*/

PROC NLMIXED DATA=trialdata MAXITER=400 QPOINTS=25 GCONV=10E-15;
PARMS  a1=1.4  b1=1.6  c1=-0.03  d1=-1.2  e1=0.3
    a2=0.4  b2=0.6  e2=0.1  f2=-0.04
    sigma_u=2  sigma_c=0.5;  *initial values

BOUNDS  sigma_c>0.001,sigma_u>0.001;

      prob1 = a1 + b1*GROUP + c1*PCTDS + d1*MLD + e1*LL + u;
      mean2 = a2 + b2*GROUP + e2*LL + f2*AGE + sigma_c*u;

      logl1 = REST*log(PROBNORM(prob1)) + (1-REST)*log(PROBNORM(-prob1));
      logl2 = - log(sigma_c*sqrt(2*3.14)) - 0.5*((LL-mean2)/sigma_c)**2;
  ll = logl1 + logl2; *log-likelihood for the data

  MODEL REST ~ GENERAL(ll);
  RANDOM u ~ NORMAL(0,sigma_u**2)
        SUBJECT=PATIENT;

    *Estimate the marginal effects for REST;
ESTIMATE a1mar = a1/sqrt(1+sigma_u**2);
```

```
ESTIMATE b1mar = b1/sqrt(1+sigma_u**2);
ESTIMATE c1mar = c1/sqrt(1+sigma_u**2);
ESTIMATE d1mar = d1/sqrt(1+sigma_u**2);
ESTIMATE e1mar = e1/sqrt(1+sigma_u**2);
RUN;
```

6.7 Discussion

In this chapter, we contrasted two joint modeling (i.e., multivariate) approaches with the univariate models for a mixed bivariate continuous and binary outcome. Two main points are noteworthy. First, if the two outcomes share the same covariates, the results of a multivariate approach are identical to those of a univariate approach that ignores the correlation between the outcomes. Although counterintuitive, this result is consistent with other multivariate models. In the setting of seemingly unrelated regressions with normally distributed outcomes, and for the particular case of a common set of covariates associated with the outcomes, the ordinary least squares estimate is still the best linear unbiased estimate, despite the correlation between the outcomes (Zellner, 1963; Zellner, 1962). However, for binary outcomes jointly modeled with the same covariates, there is a small gain in efficiency by taking into account the correlation. This only occurs if the outcomes are strongly associated. The two aforementioned properties are combined for the non-commensurate outcomes. The estimates of the parameters associated with the continuous outcome have the same SEs as in the univariate approach. The estimates of the parameters associated with the binary outcome show a small gain in efficiency when compared with the univariate approach, but only for cases of high correlation between the outcomes.

Finally, the efficiency gain is higher when the outcomes are associated with a different set of covariates, and there is stronger correlation between the outcomes. This suggests that if one anticipates that different covariates may be associated with the outcomes, the multivariate approach offers serious advantages. Fitzmaurice and Laird (1995) have previously shown higher gains in efficiency when compared with the univariate approach; however, the efficiency gains observed by Fitzmaurice and Laird (1995) were inflated as a consequence of heteroscedasticity in the data. If data are generated under the factorization model, the variance depends on the covariates. In this case, the univariate approach, assuming homoscedasticity, will lead to less efficient estimates due to misspecification of the variance.

One important aspect not discussed in this chapter is the situation of missing data in one of the outcomes. In fact, the joint model for the outcomes may be more advantageous, in comparison to separate analyses of the outcomes, when some observations are missing for one of the outcomes. Teixeira–Pinto and Normand (2011) and Teixeira–Pinto and Mauri (2011) present a number settings with missing data, where multivariate models clearly outperform the univariate approach.

A final remark should be made regarding extensions to more than two outcomes and clustered data within each outcome, such as repeated and longitudinal data (see Chapter 8). In this case, the latent variable model provides a more tractable model structure than the factorization approach, as it may be cumbersome to choose the most appropriate factorization to accommodate multiple outcomes. With k outcomes, there are $k!$ possible factorizations; hence, even with a small number of outcomes, say five, there are 120 possible ways of factorizing the likelihood with no clear guidance on which one to choose. As described in Chapter 9, for the latent variable model, on the other hand, it is sufficient to consider a different latent variable scale factor for each outcome. Although this imposes some constraints on the covariance between the outcomes, the resulting covariance structure is flexible enough to accommodate most cases.

Chapter 7

Regression models for analyzing clustered binary and continuous outcomes under the assumption of exchangeability

E. Olusegun George, Dale Bowman, and Qi An

7.1 Introduction

One of the major concerns of regulatory agencies such as the FDA and EPA, is the evaluation or assessment of risk to unborn fetuses as a result of maternal exposure to hazardous compounds. Since the evaluation of such risks in humans by clinical trials is unethical, preclinical teratology experiments, usually with rodents, are often performed. In such studies, pregnant mice are treated with the compound under investigation, and several measurements are recorded on the fetuses or pups in each litter. The endpoints recorded usually include binary indicators of fetal death or malformation, and continuous measurements, such as fetal weight or length. The common feature in such studies is that measurements and responses within each cluster (i.e., litter) are dependent.

Previous methods for analyzing mixed discrete and continuous outcomes include the general location model (GLOM) of Olkin and Tate (1961), in which the discrete outcome is assumed to have a multinomial distribution and the continuous outcome is modeled by a multivariate normal (i.e., Gaussian) distribution; and a model by Cox (1972), in which the marginal distribution of the continuous outcome is Gaussian, while the conditional distribution of the binary response is modeled by a logistic regression. Extensions of GLOM have been proposed by Little and Rubin (2002), Laird (1995), and Little and Schluchter (1985), among others. Laird's (1995) model, which relates the marginal means to covariates through a link function, may be considered as a special case of the class of partly exponential models of Zhao *et al.* (1992).

In the context of analyzing developmental toxicity data, Catalano and Ryan (1992) and Fitzmaurice and Laird (1995) propose different models for the joint distribution of fetal malformation and weight. Catalano and Ryan (1992) assume that corresponding to the binary malformation variable, there exists an unobservable continuous (Gaussian) latent variable, and that within each litter, the latent malformation variables and fetal weights have a multivariate normal distribution. By conditioning on the fetal weight, they obtain a log-likelihood function as the sum of the marginal log-likelihood for fetal weight plus the conditional log-likelihood for malformation, given fetal weight. The model of Fitzmaurice and Laird (1995) is based on GLOM and the extension of this model by Laird (1995), and is implemented by the use of general estimating equations (GEE) of Liang and Zeger (1986) and Zeger and Liang (1986), to model the marginal discrete outcomes and the conditional continuous outcomes, given the discrete outcomes. In related work, Geys *et al.* (1999) introduce a pseudo-likelihood model based on a multivariate exponential family of Molenberghs and Ryan (1999), to estimate benchmark dose levels. Diggle *et al.* (1994) note that models for multivariate correlated binary data can be grouped into conditionally specified, marginal, and cluster-specific models. In problems related to dose response in developmental toxicity studies, where fetuses are clustered within litters, and the exchangeability of fetal response is a reasonable assumption, it would seem appropriate to focus on cluster-specific models in order to assess risks of maternal exposure to potentially toxic agents. In follow-up studies, Molenberghs and Geys (2001), Regan and Catalano (1999a), and Geys et al. (1999) explore the use of multivariate latent variables in specifying cluster-specific models.

The distinguishing feature of the model that we propose in this chapter is that data from the same cluster are assumed to be exchangeable. Specifically, for a single cluster, let $\{(X_1, W_1), \cdots, (X_n, W_n)\}$ be a set of observations from the cluster, where X_i stands for the binary outcome, i. e., $X_i = 1$ if the ith fetus is malformed, and $X_i = 0$ if it is not malformed. Also, let W_i denote the weight of the ith fetus. We assume that the vector of bivariate random variables $\{(X_1, W_1), \cdots, (X_n, W_n)\}$ is exchangeable in the sense of de Finetti (1974). That is, for any permutation $(\pi(1), \cdots, \pi(n))$ of $(1, \cdots, n)$, we have

$$\{(X_{\pi(1)}, W_{\pi(1)}), \cdots, (X_{\pi(n)}, W_{\pi(n)})\} \overset{\mathscr{L}}{=} \{(X_1, W_1), \cdots, (X_n, W_n)\},$$

where "$\overset{\mathscr{L}}{=}$" denotes equality in distribution. Note that the exchangeability of $\{(X_1, W_1), \cdots, (X_n, W_n)\}$ implies that each of $(X_1, \cdots, X_n)$ and $(W_1 \cdots, W_n)$ is an exchangeable set of variables. However, $\{(X_1, W_1), \cdots, (X_n, W_n) | X_1 = x_1, \cdots, X_n = x_n\}$ is not necessarily exchangeable. Such a model, in which the marginal distributions of a set of bivariate random variables are exchangeable, but the conditional distribution is not, belongs to a class of distributions called quasi-symmetric (Darroch, 1981). As shown in the model described in Section 7.3, and in the example of Section 7.4, for such a class of distributions the subset of random variables associated with a fixed value are exchangeable. An advantage of the exchangeable model proposed here over the Bahadur (Bahadur, 1961) and multivariate exponential model of Molenberghs and Ryan (1999), is that it is reproducible in the sense of Liang et al. (1992). That is, the form of the model used is preserved and does not become increasingly complex as the dimension of the cluster increases. This is a significant advantage in developmental toxicity applications, where cluster sizes vary and are not fixed. Moreover, efficient estimates of all moments of the distribution of the cluster sum can be calculated readily without extra effort using the results in Bowman and George (1995) and George and Bowman (1995).

In this chapter, we express the joint distribution of $\{(X_1, W_1), \cdots, (X_n, W_n)\}$ as a product of the marginal distribution of $(X_1, \cdots, X_n)$ and the conditional distribution of $(W_1, \cdots, W_n)$, given $X_1, \cdots, X_n$. In applications involving developmental data, this has the interpretation of modeling fetal weights in a litter conditional on the malformation status of the fetuses in the same litter. In addition to the difference in mathematical models, our approach is different from Catalano and Ryan (1992), in which the malformation variables are modeled conditional on the fetal weights, and from Fitzmaurice and Laird (1995), in which quasi-likelihood methods are used to estimate the marginal mean parameters for the binary outcome. Fitzmaurice and Laird (1995) first consider the problem with a single (i.e., non-clustered) bivariate observation (X_i, W_i) on each of N experimental units. After linking $E(X_i)$ and $E(W_i | X_i)$ to specified covariates, they obtain joint estimating equations for the parameters of the binary and continuous responses. By analogy, they use this procedure to motivate moment structures for GEEs for the case when X_i is replaced by a cluster of correlated binary data, and W_i by a vector of multivariate continuous random variables.

Other proposed methods include those of Gueorguieva and Agresti (2001), Dunson (2000), and Regan and Catalano (1999a). Gueorguieva and Agresti (2001) and Regan and Catalano (1999a) both use continuous latent variables with correlated probit models to model the discrete malformation response. Regan and Catalano (1999**a**) estimate model parameters by maximum likelihood estimation (MLE) method, and make a number of simplifying assumptions to make their computations feasible. Gueorguieva and Agresti (2001) use an approach similar to the Bayesian method investigated by Dunson (2000). However, they use a Monte Carlo EM algorithm for parameter estimation. Each of the mentioned procedures implicitly assumes exchangeability, but with restriction to a probit, quasi-likelihood or GEE approach, none uses a parametric model that fully exploits the exchangeable structure. For example, Catalano and Ryan (1992) specify an exchangeable correlation matrix for the continuous outcome, while Fitzmaurice and Laird (1995) specify it for both continuous and discrete outcomes. However, the property of exchangeability is not fully exploited by these authors. None of these procedures could estimate third or higher order correlations among litter mates, whereas expression for MLEs of such higher level dependence parameters can be obtained with the exchangeable model, as shown in George and Bowman (1995).

In Sections 7.2 and 7.3, we show that this difficulty is overcome under an exchangeable model. Furthermore, as shown in Section 7.3, the explicit use of exchangeability also overcomes another difficulty concerning intractability of the full likelihood-based approach, when cluster sizes are large. Parsimonious regression models are then constructed for efficient estimation of marginal mean parameters and correlations.

In Section 7.2, we review the distribution theory results of Bowman and George (1995) and George and Bowman (1995), and Kuk (2004), on the joint distribution of exchangeable binary random variables. In Section 7.3, the models for the binary and continuous endpoints are presented, and procedures for obtaining MLEs of parameters are described. In Section 7.4, we use the National Toxicology Program data on exposure of pregnant CD-1 mice to diethylhexylphthalate (DEHP) (Tyl *et al.*, 1983), to estimate fetal malformation rates and distribution of number of malformations and weight reduction. In addition, we use the q-power family model of Kuk (2004) and the folded logistic model, jointly with a Gaussian model for continuous measurements, for assessing the joint malformation-reduced weight risks. In Section 7.5, we develop a cluster-specific risk assessment based on joint adverse effects of malformation and reduced weights of fetuses within a litter. We compare estimates of safe dose obtained by the folded logistic and q-power family. Finally, we conclude with possible extensions and work in progress in Section 7.6.

7.2 Distribution theory and likelihood representation

Corresponding to a cluster of size n, let $\{(X_1, W_1), \cdots, (X_n, W_n)\}$ be a sequence of exchangeable bivariate random variables, with X_i binary and W_i continuous, $i = 1, \cdots, n$. In this chapter, for the joint distribution of $(X_1, W_1), \cdots, (X_n, W_n)$, we use the factorization

$$f_{\mathbf{X},\mathbf{W}}\{(x_1, w_1), \cdots, (x_n, w_n)\} = f_{\mathbf{W}|\mathbf{X}}(w_1, \cdots, w_n | x_1, \cdots, x_n) f_{\mathbf{X}}(x_1, \cdots, x_n), \quad (7.1)$$

where $\mathbf{X} = (X_1, \cdots, X_n)^\top$ and $\mathbf{W} = (W_1, \cdots, W_n)^\top$. Let $\boldsymbol{\lambda} = (\lambda_0, \lambda_1, \cdots, \lambda_n)^\top$, where

$$\lambda_k = P(X_1 = 1, \cdots, X_k = 1) \quad (7.2)$$

where, $1 \leq k \leq n$, and $\lambda_0 = 1$.

Let $R = \sum_{k=1}^{n} X_k$, and $r = \sum_{k=1}^{n} x_k$. Using a standard inclusion-exclusion argument, it can be shown that

$$P(X_1 = x_1, \cdots, X_n = x_n) = \sum_{j=0}^{n-r} (-1)^j \binom{n-r}{j} \lambda_{r+j} \quad (7.3)$$

$$P(R = r) = \binom{n}{r} \sum_{j=0}^{n-r} (-1)^j \binom{n-r}{j} \lambda_{r+j}. \quad (7.4)$$

Note that the λ_ks satisfy the property

$$1 = \lambda_0 \geq \lambda_1 \geq \cdots \geq \lambda_n \geq 0, \quad (7.5)$$

and, as pointed out by Bowman and George (1995) and Kuk (2004), they form a completely monotone sequence; that is,

$$\triangle^{n-r}(\lambda_r) \geq 0, \quad (7.6)$$

for all r, where $\triangle$ denotes the finite difference operator $\triangle(\lambda_r) = \lambda_r - \lambda_{r+1}$, and $\triangle^k(\lambda_r)$ denotes the kth forward difference. Thus, in a quest to model the λ_ks with a continuous function $H(\cdot)$, we require that $H(\cdot)$ must be completely monotone, i.e., for any positive integer k, $(-1)^k d^k H(x)/dx^k \geq 0$ (Feller, 1971). Consequently, in the application to developmental toxicity data discussed in this chapter, we model λ by a completely monotonic parametric function $H(\mathbf{z}^\top \boldsymbol{\beta})$, where $\mathbf{z}$ is a vector of covariates that includes the dose level of the compound. Using the notation $(\mathbf{X}, \mathbf{W}) = \{(X_1, W_1), \cdots, (X_n, W_n)\}$, assume that

$$\mathbf{W}|\mathbf{X} = \mathbf{x} \quad = \quad (W_1, \cdots, W_n)^\top | X_1 = x_1, \cdots, X_n = x_n \quad \sim \quad N_n(\boldsymbol{\mu}, \boldsymbol{\Sigma}), \tag{7.7}$$

where the mean vector $\boldsymbol{\mu}$ and covariance matrix $\boldsymbol{\Sigma}$ are given by

$$\mu_k = \begin{cases} \mu_1 & , \text{if } x_k = 1, \text{ i.e., } k\text{th fetus is malformed} \\ \mu_2 & , \text{if } x_k = 0, \text{ i.e., } k\text{th fetus is normal} \end{cases},$$

and

$$\sigma_{k\ell} = cov(W_k, W_\ell | X_k = x_k, X_\ell = x_\ell) = \begin{cases} \rho_1 \sigma_1^2 & , \text{if } k\text{th and } \ell\text{th fetuses are malformed} \\ \rho_{12} \sigma_1 \sigma_2 & , \text{if only one of } k\text{th and } \ell\text{th fetuses is} \\ & \quad \text{malformed} \\ \rho_2 \sigma_2^2 & , \text{if both } k\text{th and } \ell\text{th fetuses are normal} \end{cases},$$

for $1 \leq k, \ell \leq n$. Let $\boldsymbol{\theta}^\top = (\mu_1, \mu_2, \sigma_1^2, \sigma_2^2, \rho_1, \rho_2, \rho_{12})$, $(\mathbf{x}, \mathbf{w}) = \{(x_1, w_1), \cdots, (x_n, w_n)\}$, and $(\mathbf{x}^*, \mathbf{w}^*)$ be a permutation of the components of $\{(x_1, w_1), \cdots, (x_n, w_n)\}$ such that $x_{k_1}^* = \cdots = x_{k_r}^* = 1$, and $x_{k_{(r+1)}}^* = \cdots = x_{k_n}^* = 0$. Then, using the fact that $\{(X_1, W_1), \cdots, (X_n, W_n)\}$ is exchangeable, the likelihood function of $(\boldsymbol{\beta}, \boldsymbol{\theta})$ can be expressed as

$$L(\boldsymbol{\beta}, \boldsymbol{\theta}; \mathbf{x}, \mathbf{w}) \quad = \quad L(\boldsymbol{\beta}, \boldsymbol{\theta}; \mathbf{x}^*, \mathbf{w}^*) \quad = \quad L_{\mathbf{x}^*}(\boldsymbol{\beta}) L_{\mathbf{w}^*|\mathbf{x}^*}(\boldsymbol{\beta}, \boldsymbol{\theta}), \tag{7.8}$$

where $L_{\mathbf{x}^*}(\boldsymbol{\beta})$ is the marginal likelihood based on the marginal density of $\mathbf{X}^*$, and $L_{\mathbf{w}^*|\mathbf{x}^*}(\boldsymbol{\beta}, \boldsymbol{\theta})$ is the conditional likelihood based on the conditional density of $\mathbf{W}^*$, given $\mathbf{X}^*$. It is easy to see that if $0 < r < n$, then

$$\mathbf{W}^*|\mathbf{X}^* = \mathbf{x}^* \quad \sim \quad N_n \left(\begin{pmatrix} \mu_1 \mathbf{1}_r \\ \mu_2 \mathbf{1}_{n-r} \end{pmatrix}, \begin{pmatrix} \boldsymbol{\Sigma}_{11r} & \boldsymbol{\Sigma}_{12r} \\ \boldsymbol{\Sigma}_{12r}^\top & \boldsymbol{\Sigma}_{22r} \end{pmatrix} \right), \tag{7.9}$$

where

$$\begin{aligned} \boldsymbol{\Sigma}_{11r} &= \sigma_1^2 \{(1 - \rho_1)\mathbf{I}_r + \rho_1 \mathbf{J}_r\}, \\ \boldsymbol{\Sigma}_{22r} &= \sigma_2^2 \{(1 - \rho_2)\mathbf{I}_{n-r} + \rho_2 \mathbf{J}_{n-r}\}, \\ \boldsymbol{\Sigma}_{12r} &= \rho_{12} \sigma_1 \sigma_2 \mathbf{1}_r \mathbf{1}_{n-r}^\top, \end{aligned}$$

and $\mathbf{I}_p$ is a $p \times p$ identity matrix, $\mathbf{J}_p$ is a $p \times p$ matrix of 1s, and $\mathbf{1}_p$ is a $p \times 1$ vector of 1s. Also, if $r = 0$, then $\mathbf{W}^*|\mathbf{X}^* = \mathbf{x}^* \sim N_n(\mu_2 \mathbf{1}_n, \boldsymbol{\Sigma}_{22n})$, and if $r = n$, $\mathbf{W}^*|\mathbf{X}^* = \mathbf{x}^* \sim N_n(\mu_1 \mathbf{1}_n, \boldsymbol{\Sigma}_{11n})$. In what follows, we assume that $\mu_i = \mu_i(\boldsymbol{\lambda})$, $i = 1, 2$.

7.3 Parametric models

7.3.1 Model descriptions

Let $\{(X_{ijk}, W_{ijk}), i = 1, \cdots, g; j = 1, \cdots, m_i; k = 1, \cdots, n_{ij}\}$, where X_{ijk} represents the binary response and W_{ijk} represents the continuous measurement from the kth member of the jth cluster of the ith treatment group. In developmental toxicity studies, the binary responses X_{ijk} correspond to adverse responses, such as fetal malformation or death, and the continuous measurements W_{ijk} may be fetal weights. Let $\lambda_r^{(i)} = P(X_{ij1} = 1, \cdots, X_{ijr} = 1)$ and $\lambda_r^{'(i)} = P(X_{ij1} = 0, \cdots, X_{ijr} = 0)$, $0 \le r \le n_{ij}$. Motivated by a loglog link function, Kuk (2004) proposes the q-power family of models for $\lambda_r^{'(i)}$. The q-power family is defined by

$$\lambda_r^{'(i)}(\boldsymbol{\beta}) = \exp\{-r^{\xi_i} \exp(\beta_1 + \beta_2 d_i)\}, \tag{7.10}$$

where ξ_i is defined in terms of the intra-litter correlation ϕ_i of the binary response. We model ϕ_i as

$$\phi_i(\boldsymbol{\beta}) = \frac{1 - \exp(\beta_3 + \beta_4 d_i)}{1 + \exp(\beta_3 + \beta_4 d_i)}.$$

The value of ξ_i can be calculated from the inversion equation given by Kuk (2004), and may be expressed as

$$\xi_i = \frac{1}{\log 2} \log \left(\frac{\log\left\{ (1 - \lambda_1^{(i)})^2 + \phi_i \lambda_1^{(i)}(1 - \lambda_1^{(i)}) \right\}}{\log(1 - \lambda_1^{(i)})} \right).$$

For the purpose of comparison, we also use the folded logistic model for $\lambda_r^{(i)} = P(X_{ij1} = 1, \cdots, X_{ijr} = 1)$, defined by

$$\lambda_r^{(i)}(\boldsymbol{\beta}) = \frac{2}{1 + \exp\{(\beta_1 + \beta_2 d_i)\log(r+1)\}}. \tag{7.11}$$

The rationale given by the q-power family is that the folded logistic model is underparameterized, in the sense that correlations and higher moments are all explained by the same parameters β_1 and β_2. Consequently, the q-power family introduces extra parameters to account for second order correlation. We compare the two models in an application to the DEHP data in Section 7.4.

Based on the factorization (7.1) of the joint likelihood in Section 7.2, we express the joint distribution of the continuous outcomes, given the binary outcomes, by the multivariate normal distribution. Specifically, let $(\mathbf{X}_{ij}, \mathbf{W}_{ij}) = \{(X_{ijk}, W_{ijk}), k = 1, \cdots, n_{ij}\}$. We assume that, conditional on the binary outcomes,

$$\mathbf{W}_{ij}|\mathbf{X}_{ij} = \mathbf{x}_{ij} \sim N_{n_{ij}}\left(\mathbf{A}_{ij}\boldsymbol{\mu}_i(\boldsymbol{\beta}), \boldsymbol{\Sigma}_{ij}\right), \tag{7.12}$$

where $\boldsymbol{\Sigma}_{ij} = \sigma^2\{(1-\rho)\mathbf{I}_{n_{ij}} + \rho\mathbf{J}_{n_{ij}}\}$, for $j = 1, \cdots, m_i$, $i = 1, \cdots, g$, $\boldsymbol{\mu}_i = (\mu_{1i}, \mu_{2i})^\top$, $\mathbf{I}_{n_{ij}}$ and $\mathbf{J}_{n_{ij}}$ are the identity matrix and matrix of ones, respectively, each of dimension $n_{ij} \times n_{ij}$. Following George et al. (2007), we assume a common exchangeable correlation structure for the weights. The common correlation parameter for the weight within a litter is $\rho = \rho_1 = \rho_2 = \rho_{12}$, and the common variance for the weights is $\sigma^2 = \sigma_1^2 = \sigma_2^2$. We model the mean weight parameters, as in George et al. (2007), by

$$E(\mathbf{W}_{ij}|\mathbf{X}_{ij} = \mathbf{x}_{ij}) = \mathbf{A}_{ij}\boldsymbol{\mu}_i(\boldsymbol{\beta}) = \mathbf{A}_{ij}\mathbf{D}_{ij}(\boldsymbol{\beta})\boldsymbol{\eta}, \tag{7.13}$$

where,

$$\mathbf{A}_{ij} = \begin{pmatrix} \mathbf{1}_{r_{ij}} & \mathbf{0} \\ \mathbf{0} & \mathbf{1}_{s_{ij}} \end{pmatrix}, \quad \mathbf{D}_{ij}(\boldsymbol{\beta}) = \begin{pmatrix} 1 & d_i & 1 - \lambda_1^{(i)}(\boldsymbol{\beta}) & \{1 - \lambda_1^{(i)}(\boldsymbol{\beta})\}d_i \\ 1 & d_i & -\lambda_1^{(i)}(\boldsymbol{\beta}) & -\lambda_1^{(i)}(\boldsymbol{\beta})d_i \end{pmatrix}, \tag{7.14}$$

and $\boldsymbol{\eta}^\top = (\alpha_1, \alpha_2, \gamma_1, \gamma_2)$, with $\boldsymbol{\theta}^\top = (\boldsymbol{\eta}^\top, \sigma^2, \rho)$. Note here that $\lambda_1^{(i)}(\boldsymbol{\beta})$ is the marginal malformation rate and is equal to $1 - \lambda_r^{'(i)}(\boldsymbol{\beta})$, when the q-power family of models is used.

7.3.2   Estimation of parameters

Let $K < \infty$ denote the maximum possible cluster size, and let $A_{r,n}^{(i)}$ be the number of clusters of size n with r positive binary responses, in the ith treatment group. Then the full log-likelihood function is given by

$$
\begin{aligned}
\ell(\boldsymbol{\beta}, \boldsymbol{\theta}; \mathbf{x}^*, \mathbf{w}^*) &= \ell_{\mathbf{x}^*}(\boldsymbol{\beta}) + \ell_{\mathbf{w}^*|\mathbf{x}^*}(\boldsymbol{\beta}, \boldsymbol{\theta}) \\
&= \sum_{i=1}^{g} \sum_{n=1}^{K} \sum_{r=0}^{n} A_{r,n}^{(i)} \log P(R = r|n, \boldsymbol{\beta}) - \frac{1}{2} \sum_{i=1}^{g} \sum_{j=1}^{m_i} \Big\{ n_{ij} \log \sigma^2 + (n_{ij} - 1) \log(1 - \rho) \\
&\quad + \log \varphi_{ij} + (\mathbf{W}_{ij} - \mathbf{A}_{ij}\mathbf{D}_{ij}\boldsymbol{\eta})^{\top} \boldsymbol{\Sigma}_{ij}^{-1} (\mathbf{W}_{ij} - \mathbf{A}_{ij}\mathbf{D}_{ij}\boldsymbol{\eta}) \Big\},
\end{aligned} \tag{7.15}
$$

where $\ell_{\mathbf{x}^*}(\boldsymbol{\beta}) = \log L_{\mathbf{x}^*}(\boldsymbol{\beta})$, $\ell_{\mathbf{w}^*|\mathbf{x}^*}(\boldsymbol{\beta}, \boldsymbol{\theta}) = \log L_{\mathbf{w}^*|\mathbf{x}^*}(\boldsymbol{\beta}, \boldsymbol{\theta})$, $\varphi_{ij} = 1 + (n_{ij} - 1)\rho$, and

$$
P(R = r; n, \boldsymbol{\beta}) = \begin{cases} \binom{n}{r} \sum_{k=0}^{n-r} (-1)^k \binom{n-r}{k} \lambda_{r+k}^{(i)}(\boldsymbol{\beta}) & \text{, for the folded logistic model} \\ \binom{n}{r} \sum_{k=0}^{r} (-1)^k \binom{r}{k} \lambda_{n-r+k}^{\prime(i)}(\boldsymbol{\beta}) & \text{, for the } q\text{-power family model} \end{cases},
$$

and $\lambda_k^{(i)}(\boldsymbol{\beta})$ and $\lambda_k(\boldsymbol{\beta})^{\prime(i)}$ are modeled with the completely monotonic functions in (7.10) and (7.11).

The maximization of the full likelihood is implemented with the aid of the optimization function `optim` in R. The above iterative procedure is used at each iteration in which a $\boldsymbol{\beta}$ value is calculated. The value of $\lambda_1^{(i)}(\boldsymbol{\beta})$ obtained is then used in the above iteration. The optimization is conducted iteratively based on the BFGS method. The BFGS is a quasi-Newton method (also known as a variable metric algorithm) by Broyden (1970), Fletcher (1987, 1970), Goldfarb (1970), and Shanno (1970). In jointly maximizing the likelihood for $(\boldsymbol{\beta}, \boldsymbol{\theta})$, the component involving the conditional likelihood of $\boldsymbol{\theta}(\boldsymbol{\beta})$ is expedited with the help of relationships from George *et al.* (2007):

$$
\widehat{\boldsymbol{\eta}(\widehat{\boldsymbol{\beta}})} = \widehat{\boldsymbol{\eta}} = \left(\sum_{i=1}^{g} \sum_{j=1}^{m_i} \mathbf{D}_{ij}^{\top} \mathbf{A}_{ij}^{\top} \widehat{\boldsymbol{\Sigma}}_{ij}^{-1} \mathbf{A}_{ij} \mathbf{D}_{ij} \right)^{-1} \sum_{i=1}^{g} \sum_{j=1}^{m_i} \mathbf{D}_{ij}^{\top} \mathbf{A}_{ij}^{\top} \widehat{\boldsymbol{\Sigma}}_{ij}^{-1} \mathbf{W}_{ij} \tag{7.16}
$$

$$
\widehat{\sigma}^2 = \frac{1}{N(1-\widehat{\rho})} \sum_{i=1}^{g} \sum_{j=1}^{m_i} (\mathbf{W}_{ij} - \mathbf{A}_{ij}\mathbf{D}_{ij}\widehat{\boldsymbol{\eta}})^{\top} \left(\mathbf{I}_{ij} - \frac{\widehat{\rho}}{\widehat{\varphi}_{ij}} \mathbf{J}_{ij} \right) (\mathbf{W}_{ij} - \mathbf{A}_{ij}\mathbf{D}_{ij}\widehat{\boldsymbol{\eta}}), \tag{7.17}
$$

and

$$
\sum_{i=1}^{g} \sum_{j=1}^{m_i} \left(\widehat{B}_{ij} - n_{ij}\widehat{b}_{ij}\{1 + (n_{ij}-1)\widehat{\rho}^2\} \right) = \frac{\widehat{\rho}}{N} \sum_{i=1}^{g} \sum_{j=1}^{m_i} \frac{n_{ij}(n_{ij}-1)}{\widehat{\varphi}_{ij}} \sum_{i=1}^{g} \sum_{j=1}^{m_i} \left(\widehat{B}_{ij} - \frac{\widehat{\rho} n_{ij}\widehat{b}_{ij}}{\widehat{\varphi}_{ij}} \right), \tag{7.18}
$$

where $N = \sum_{i=1}^{g} \sum_j n_{ij}$, and

$$
\widehat{B}_{ij} = (\mathbf{W}_{ij} - \mathbf{A}_{ij}\mathbf{D}_{ij}\widehat{\boldsymbol{\eta}})^{\top} (\mathbf{W}_{ij} - \mathbf{A}_{ij}\mathbf{D}_{ij}\widehat{\boldsymbol{\eta}}), \tag{7.19}
$$

$$
\widehat{b}_{ij} = \frac{1}{n_{ij}} (\mathbf{W}_{ij} - \mathbf{A}_{ij}\mathbf{D}_{ij}\widehat{\boldsymbol{\eta}})^{\top} \mathbf{J}_{n_{ij}} (\mathbf{W}_{ij} - \mathbf{A}_{ij}\mathbf{D}_{ij}\widehat{\boldsymbol{\eta}}). \tag{7.20}
$$

7.4   Application to DEHP data

7.4.1   Data summary

The data set used in this application comes from a National Toxicology Program experiment in which several dose levels (i.e., 0, 0.025, 0.05, 0.1, and 0.15g/kg body weight) of diethylhexyl phthalate (DEHP) were administered to timed-pregnant CD-1 mice on gestational days 0 through 17;

Table 7.1 *Frequency distribution of female mice according to numbers of live (n) and malformed (r) fetuses following exposure to DEHP.*

Dose	r	n														
		2	3	4	5	6	7	8	9	10	11	12	13	14	15	16
0g/kg	0	0	1	0	0	1	3	2	4	0	1	3	3	1	2	1
	1	0	0	0	0	0	0	1	0	0	0	2	1	0	0	1
	2	0	0	0	0	0	0	0	0	0	0	1	0	0	0	0
0.025g/kg	0	0	1	0	1	0	1	0	2	1	4	5	4	2	0	0
	1	0	0	0	0	0	0	0	0	1	2	0	0	1	1	0
0.05g/kg	0	0	1	0	0	1	0	1	0	1	0	4	1	2	0	0
	1	0	0	0	0	0	1	0	0	2	2	1	1	0	0	0
	2	0	0	0	0	0	0	0	0	0	3	1	0	0	0	0
	3	0	0	0	0	0	0	1	0	0	0	0	0	0	0	0
	4	0	0	0	0	0	0	0	1	0	0	0	0	1	0	0
	6	0	0	0	0	0	0	0	0	0	1	0	0	0	0	0
0.1g/kg	0	0	0	0	0	0	0	0	0	1	0	2	0	0	0	0
	1	2	0	0	0	0	0	1	0	0	0	0	0	0	0	0
	2	0	0	0	1	0	0	0	1	0	0	1	0	0	0	0
	3	0	0	1	0	0	0	0	1	0	0	0	0	0	0	0
	4	0	0	0	0	0	0	1	0	0	0	0	1	0	0	0
	5	0	0	0	0	0	0	1	0	0	0	0	0	0	0	0
	6	0	0	0	0	1	0	1	1	0	0	0	0	0	0	0
0.15g/kg	1	0	0	1	0	0	0	0	0	0	0	0	0	0	0	0
	2	1	0	0	0	0	0	0	0	0	0	0	0	0	0	0
	3	0	0	1	0	0	0	0	0	0	0	0	0	0	0	0
	4	0	0	0	1	0	0	0	1	0	0	0	0	0	0	0
	5	0	0	0	1	1	0	0	0	0	0	0	0	0	0	0
	6	0	0	0	0	1	0	0	0	0	0	0	0	0	0	0
	8	0	0	0	0	0	0	0	1	0	0	0	0	0	0	0

see http://ntp.niehs.nih.gov/go/7887 (Tyl *et al.*, 1983). After sacrifice, the dams were evaluated by their weights and the status of the uterine implantation sites (e.g., resorptions, dead fetuses and live fetuses).

The live fetuses were examined for several developmental endpoints. These include fetal weight and visceral malformations. For our application, we focus on the analysis of malformation (discrete endpoint) and fetal weight (continuous endpoint), collected on each live fetus. Using notations introduced in Section 7.2, we let $(\mathbf{X}_{ij}, \mathbf{W}_{ij}) = \{(X_{ij1}, W_{ij1}), \cdots, (X_{ijn_{ij}}, W_{ijn_{ij}})\}$ represent the binary malformation and continuous weight data from the jth litter of the ith dose group. The numbers in the body of Table 7.1 are the values of $A_{r,n}^{(i)}$, the number of litters of size n with r malformations in the ith dose group. Here, n ranges from litters of size 2 to 16 in each dose group, and the number of malformations range from 0 to 8 over the range of doses (0, 0.025, 0.05, 0.1, and 0.15g/kg body weight). This summary in fact, represents sufficient statistics, since, for each i, and for all litters of size n, $\{A_{r,n}^{(i)}, 0 \leq r \leq n\}$ has a multinomial distribution. The universal use of the cluster summaries $A_{r,n}$ as sufficient statistics, is a further indication that exchangeability has been implicity assumed in applications in developmental toxicity studies.

The data summary is given in Table 7.2. It shows preliminary evidence of increase in malformation rate as dose increases, and a corresponding decrease in average fetal weight. There does not appear to be an effect of dose on the variance of the weights for either malformed or normal fetuses, and the variance does not appear to differ for the two binary outcomes, thus leading to the

Table 7.2 *Summary of DEHP malformation and weight data.*

Dose	Dams	Live	Litter size		Malformation		Weight			
							Malformed		Normal	
			Mean	SD	No.	%	Mean	SD	Mean	SD
0	28	301	10.8	3.3	7	2.1	0.976	0.145	0.94	0.108
0.025	26	288	11.1	2.8	5	1.6	1.006	0.093	0.972	0.107
0.05	26	277	10.7	2.6	32	11.7	0.867	0.118	0.927	0.105
0.1	17	137	8.1	3.4	46	40.3	0.842	0.118	0.957	0.102
0.15	9	50	5.6	2.3	38	77.4	0.795	0.124	0.845	0.157

assumption previously discussed, of a common exchangeable correlation structure for the weights. Note that there is a decrease in malformation rate from control dose (i.e., 0g/kg) to dose group 2 (i.e., 0.025g/kg). This phenomenon hints at hormesis. In fact, there have been documented evidence of possible hormetic effect of DEHP, and some other phthalates in other studies; see, for example, Diamanti−Kandarakis *et al.* (2009). In the absence of further information, such as location in the womb, and considering maternal influence on litter mates, it is reasonable to assume that these bivariate data within litters, have an exchangeable dependence structure. That is, for any permutation $(\pi(1), \cdots, \pi(n_{ij}))$ of $(1, \cdots, n_{ij})$,

$$\{(X_{\pi(1)}, W_{\pi(1)}), \cdots, (X_{\pi(n_{ij})}, W_{\pi(n_{ij})})\} \stackrel{\mathscr{L}}{=} \{(X_1, W_1), \cdots, (X_{n_{ij}}, W_{n_{ij}})\}. \qquad (7.21)$$

Thus, the corresponding components of $\mathbf{X}_{ij}$ and $\mathbf{W}_{ij}$ are also exchangeable, while the elements of $\mathbf{W}_{ij}$ are partially exchangeable, given $\mathbf{X}_{ij} = \mathbf{x}_{ij}$. Under this condition, we fit the joint distribution of $\{X_{ij1}, \cdots, X_{ijn_{ij}}\}$ or equivalently, the distribution of $R_{ij} = \sum_{k=1}^{n_{ij}} X_{ijk}$ by the q-power family model of Kuk (2004), and for comparison, by the folded logistic model of George and Bowman (1995). We also fit the conditional distribution of $\mathbf{W}_{ij}|\mathbf{X}_{ij} = \mathbf{x}_{ij}$ by a multivariate normal distribution with a partially exchangeable covariance matrix, as described in Sections 7.2 and 7.3.

7.4.2   Estimates and model comparisons

In Table 7.3, we give MLEs of all the parameters, and use these to compute estimates of malformation rates and expected fetal weights for normal and malformed fetuses at each dose level using the q-power family and folded logistic models. We also summarize the estimates of inter-litter correlations in Table 7.4 and Figures 7.1 and 7.2. Overall, the estimates produced by the the two models are quite comparable, although the q-power family could be judged to give estimates that are slightly closer to the observed malformation rates and average malformation counts. This is to be expected, since with additional two parameters, the q-power family specifically models the second order correlation. Indeed, estimates of the second order inter-litter correlation with the q-power family are significantly bigger than those of the folded logistic at low to moderate dose levels. However, overall the difference between the two models in this application is not as striking as what is reported in Kuk (2004), especially when the estimates of fetal mean weights are included in the comparison. Note that both models fail to pick up any hint of hormesis at low dose (as suggested by the summary data), since our model did not allow for such an effect. In order to model such an effect, perhaps a quadratic dose coefficient or a more flexible dose-response model should be included in the regression model for fetal malformation. In addition, the extra two parameters β_3 and β_4 were found not to be significant. However, An *et al.* (2012) have developed an iterative weighted Fisher scoring algorithm for MLE in nonlinear models, such as the folded logistic and the q-power family. In a just completed application of this algorithm to the DEHP data, the estimated SEs of β_3 and β_4, were found to be considerably smaller than those obtained with `optim`, although the estimates of β_3 and β_4 were essentially the same as those obtained from `optim`. With the new algorithm, we found

β_3 and β_4 to be significant. Finally, we note that we have not added a litter size term to the λs or λ's, to account for an effect of litter size on malformation; although the addition of a litter size term is straightforward, the coefficient of $n_{ij} - \bar{n}_i$, where $\bar{n}_i = \sum_{j=1}^{m_i} n_{ij}/m_i$, is found to be statistically not significant in George *et al.* (2007).

Table 7.3 *Estimates and their standard errors (SEs) for the folded logistic and q-power family models for DEHP data.*

Parameter	Folded logistic model			q-power family model		
	Est	SE	p-value	Est	SE	p-value
β_1	6.348	0.344	< 0.001	−3.635	0.316	< 0.001
β_2	−37.754	3.016	< 0.001	27.01	2.907	< 0.001
β_3	—	—	—	−0.366	0.308	0.235
β_4	—	—	—	−1.062	3.327	0.750
σ^2	0.013	0.001	< 0.001	0.013	0.001	< 0.001
ρ	0.445	0.041	< 0.001	0.450	0.041	< 0.001
α_1	0.965	0.011	< 0.001	0.965	0.011	< 0.001
α_2	−0.823	0.179	< 0.001	−0.831	0.181	< 0.001
γ_1	0.011	0.021	0.595	0.011	0.021	0.590
γ_2	−0.330	0.255	0.195	−0.329	0.255	0.198

Table 7.4 *Estimates, with their SEs, of fetal malformation rates (MR), counts (MC), intra-litter correlation (ILC), normal (NFW) and malformed fetal (MFW) weights.*

Dose	Folded logistic model									
	MR		MC		ILC		NFW		MFW	
	Est	SE	Est	SE	Est	SE	Est	SE	Est	SE
0	0.024	0.006	7.3	1.7	0.054	0.006	0.965	0.011	0.976	0.024
0.025	0.046	0.009	13.1	2.5	0.071	0.005	0.944	0.009	0.947	0.018
0.05	0.087	0.013	24.1	3.5	0.091	0.005	0.924	0.008	0.919	0.013
0.1	0.288	0.027	39.4	3.7	0.141	0.006	0.889	0.013	0.867	0.015
0.15	0.767	0.070	38.4	3.5	0.293	0.032	0.871	0.027	0.833	0.021
Dose	q-power family model									
	MR		MC		ILC		NFW		MFW	
	Est	SE	Est	SE	Est	SE	Est	SE	Est	SE
0	0.026	0.008	7.8	2.4	0.181	0.149	0.965	0.011	0.976	0.024
0.025	0.051	0.012	14.6	3.6	0.194	0.114	0.944	0.009	0.947	0.018
0.05	0.097	0.018	26.8	5.0	0.207	0.085	0.924	0.008	0.919	0.014
0.1	0.325	0.038	44.5	5.2	0.232	0.072	0.889	0.013	0.867	0.015
0.15	0.781	0.070	39	3.5	0.257	0.124	0.870	0.027	0.832	0.021

We explore the advantages of the exchangeable models over marginal models by actually estimating the probability distribution of the number of fetal malformations per litter, without the introduction of a latent variable, as in beta-binomial or random effects models. We illustrate this with estimated plots of the distribution of the number of responses for litter sizes 12 and 16 in Figure 7.3. Examination of these graphs shows considerable similarity between the estimated distributions computed under the q-power family and folded logistic models. Both methods produce expected profiles of these distributions as the dose level increases: they have modes at or close to zero, at 0g/kg and 0.025g/kg dose levels, suggesting that most fetuses are not adversely affected at the control and low dose 0.025g/kg. However, as dose increases, the number of fetuses malformed in the litter increases. This is consistent with the modal values at dose level 0.1g/kg in Figure 7.3. At the higher dose level 0.15g/kg, most fetuses within the litter are affected, as reflected in the shift of the modal values to the right tail of the distributions.

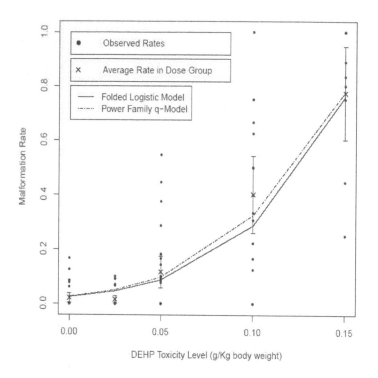

Figure 7.1 *Comparisons of estimated malformation rates from the q-power family and folded logistic models.*

7.5   Litter-specific joint quantitative risk assessment

One of the major goals in developing dose-response models for adverse effects in a developmental toxicity study is to derive a safe dose of exposure for humans. When more than one endpoint is considered, the typical approach has been to examine each endpoint separately, and then to base the derivation of the safe exposure level on the endpoint that is most sensitive to the exposure. This assumes that the safe dose for the most sensitive endpoint will be safe for other adverse outcomes. It has been found, however, that the correlation between endpoints, such as weight and malformation, can significantly affect the estimation of a safe dose (Faes *et al.*, 2004). Estimation of a safe dose level based on the joint model in this case would lead to more appropriate estimates of safe exposure levels.

To derive an estimate of a safe dose, define $P(d)$ as the probability of an adverse effect. For the joint model under consideration here, this is the probability that a fetus has low birth weight or exhibits a malformation at dose level d. In order to determine the probability of a low birth weight, a birth weight must be specified that is sufficiently low so as to characterize an adverse event. The cut off weight is typically specified as a certain number of standard deviations below the average fetal weight in the control group. Assume the cut off weight W_c is specified. Then, following the approach of Geys *et al.* (2001), the expression for $P(d)$, the probability of an adverse effect at dose level d, is given by

$$P(d) \quad = \quad \Phi\left(\frac{W_c - \mu_2(d)}{\sigma}\right)\{1 - \lambda_1(d)\} + \lambda_1(d), \qquad (7.22)$$

Folded Logistic Model

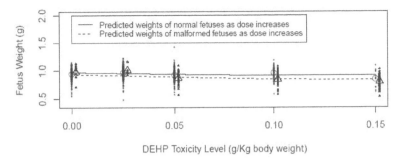

Power Family q-Model

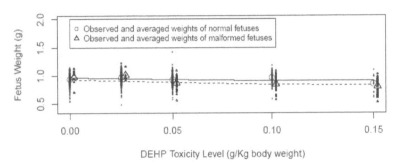

Figure 7.2 *Comparisons of estimated fetal weights from the q-power family and folded logistic models.*

where $\lambda_1(d)$ is the probability of malformation at dose level d, $\mu_2(d)$ is the mean of the weights for fetuses that are not malformed, and σ is the common variance of fetal weights.

Based on $P(d)$, the risk function is defined as

$$r(d) = \frac{P(d) - P(0)}{1 - P(0)}. \tag{7.23}$$

The risk function $r(d)$, called the excess risk, measures the increase in response at dose level d over background risk in the control group with $d = 0$. The risk function is a function of unknown parameters relating adverse outcomes to dose levels and as such, can be used to determine a safe level of exposure. One such measure is the benchmark dose BMD_q, and is defined as the dose at which the excess risk over background is equal to some small quantity q, typically equal to $0.0001, 0.01$, or 0.05. If we write the excess risk as $r(d, \boldsymbol{\theta})$, the benchmark dose may be found as the dose level d, satisfying the equation

$$r(d, \boldsymbol{\theta}) = q, \tag{7.24}$$

and can be estimated as a solution to $\widehat{r(d, \widehat{\boldsymbol{\theta}})} = q$, where $\widehat{\boldsymbol{\theta}}$ is our estimate of the parameters in the model. The resulting estimate of the benchmark dose is an estimate of the true dose that results in the excess risk. To reflect this, the BMD_q is often replaced by a lower confidence limit. The confidence limit used here is that suggested by Kimmel and Gaylor (1988). This estimate is termed the lower effective dose LED_q, and is the value of d that solves

$$q = \widehat{r(d, \widehat{\boldsymbol{\theta}})} + 1.645 \left\{ \widehat{var\{r(d, \widehat{\boldsymbol{\theta}})\}} \right\}^{1/2}, \tag{7.25}$$

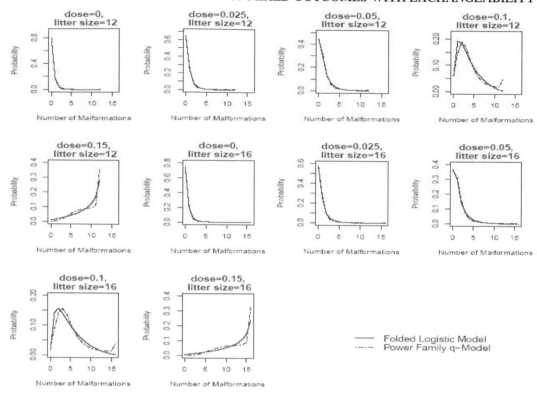

Figure 7.3 *Comparisons of estimated distributions of number of malformed fetuses in a litter, using the q-power family and folded logistic models.*

Table 7.5 *BMD_q and LED_q corresponding to $q = 10\%$ excess risk for DEHP data.*

Dose	Folded logistic	q-power family
BMD_{10}	0.0551	0.0520
LED_{10}	0.0436	0.0419
Malformation BMD_{10}	0.0636	0.0596
Weight BMD_{10}	0.1251	0.1235

where

$$\widehat{var\{r(d,\widehat{\boldsymbol{\theta}})\}} = \left\{\frac{\partial}{\partial\boldsymbol{\theta}}r(d)\right\}^{\top} \widehat{var(\widehat{\boldsymbol{\theta}})}\left\{\frac{\partial}{\partial\boldsymbol{\theta}}r(d)\right\}, \tag{7.26}$$

evaluated at $\boldsymbol{\theta} = \widehat{\boldsymbol{\theta}}$.

Estimated BMDs and LEDs for the DEHP data are shown in Tables 7.5 and 7.6. Table 7.5 shows estimates for 10% excess risk over control, and Table 7.6 shows estimates for 1% excess risk. Following Geys *et al.* (2001), the cut-off value W_c used for low birth weight is the value two standard deviations (SDs) below the average birth weight in the control group (i.e., dose = 0). The cut off value W_c for the DEHP data is 0.7243g/kg.

Included in these tables for comparison are estimated benchmark doses, computed separately for each endpoint. As discussed, one approach to estimating the safe dose for human exposure is to obtain estimated safe doses for each endpoint separately, and then use the most sensitive endpoint to set the safe exposure levels. For the malformation endpoint, the corresponding excess risk function

Table 7.6 BMD_q and LED_q corresponding to $q = 1\%$ excess risk for DEHP data.

Dose	Folded logistic	q-power family
BMD_1	0.0093	0.0087
LED_1	0.0064	0.0057
Malformation BMD_1	0.0131	0.0120
Weight BMD_1	0.0259	0.0254

is given by

$$r_m(d,\boldsymbol{\theta}) \quad = \quad \frac{\lambda_1(d,\boldsymbol{\theta}) - \lambda_1(0,\boldsymbol{\theta})}{1 - \lambda_1(0,\boldsymbol{\theta})}. \tag{7.27}$$

The risk function is estimated using estimates of $\boldsymbol{\theta}$ derived from the joint estimation procedure described in previous sections. The benchmark dose for a given value of q for the malformation endpoint is found by solving $\widehat{r_m(d,\boldsymbol{\theta})} = q$. The excess risk function for the weight endpoint is given by

$$r_w(d,\boldsymbol{\theta}) \quad = \quad \frac{P(d,\boldsymbol{\theta}) - P(0,\boldsymbol{\theta})}{1 - P(0,\boldsymbol{\theta})} \tag{7.28}$$

for

$$P(d,\boldsymbol{\theta}) \quad = \quad \Phi\left(\frac{W_c - \mu_1(d,\boldsymbol{\theta})}{\sigma}\right)\lambda_1(d,\boldsymbol{\theta}) + \Phi\left(\frac{W_c - \mu_2(d,\boldsymbol{\theta})}{\sigma}\right)\{1 - \lambda_1(d,\boldsymbol{\theta})\}, \tag{7.29}$$

the probability of low birth weight at dose level d.

For the DEHP data, the malformation endpoint is the more sensitive endpoint for all models examined. This is clear from the graph of the estimated excessive risk functions $r(d,\boldsymbol{\theta})$, $r_m(d,\boldsymbol{\theta})$, and $r_w(d,\boldsymbol{\theta})$, shown in Figure 7.5. It is also clear from both Tables 7.5 and 7.6 that for each model, the joint benchmark doses are more conservative than those calculated for malformation alone. Thus, using benchmark doses from the malformation endpoint alone could overestimate the safe levels of exposure. In this case, estimates of benchmark doses for the separate endpoints were computed using estimates derived from the joint estimation procedure described here. If each endpoint were modeled separately, and these dose-response curves were used to obtain these benchmark doses, they would likely give estimates even less conservative than those given here (Geys et al., 2001). The estimated lower effective dose LED in (7.25) are included in the tables as well. For both 10% and 1% excess risk, the q-power family and folded logistic models appear to yield very similar estimates of BMDs and LEDs.

It has been noted by several authors (Hunt and Bowman, 2006; Schwartz et al., 1995) that the DEHP data exhibit a non-monotonic dose response curve. Doull et al. (1999) demonstrate that carcinogenic studies of DEHP indicate strong evidence of a threshold level, that is, a dose level below which there is no add-on risk over the background level. In developmental studies, Narotsky (1995) finds evidence of a U-shaped dose response relationship for endpoints, such as maternal weight gain and fetal death, for a mixture of chemicals including DEHP. Such a U-shaped dose response curve may be evidence of a hormetic effect. Hormesis is the appearance of beneficial or stimulatory effects at low exposure levels of a material that has detrimental effects at higher exposure levels.

For the malformation endpoint alone, Hunt and Bowman (2006) assume a U-shaped threshold dose response model, where the probability of malformation at dose level d is given by

$$\mu(d) \quad = \quad \left\{ \begin{array}{ll} \gamma_1 d^2 + \gamma_2 d & , \text{ for } 0 \le d \le \tau \\ \{1 + \exp(-\beta_0 - \beta_1(d - \tau))\}^{-1} & , \text{ for } d > \tau \end{array} \right. , \tag{7.30}$$

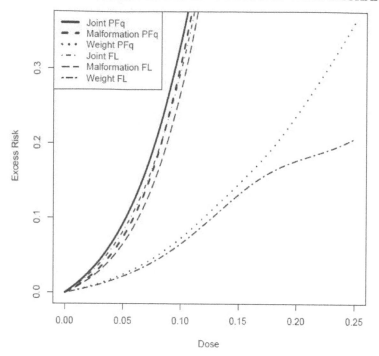

Figure 7.4 *Comparisons of excess risks of q-power family and folded logistic models.*

where τ is the assumed threshold level. The curve is logistic above threshold, and U-shaped below threshold. Hunt and Bowman (2006) use a beta-binomial model to obtain parameter estimates of the U-shaped dose response model. All parameter estimates were highly significant, and the U-shaped threshold model provided better fit to the DEHP data than either a monotonic dose response curve (Chen and Kodell, 1989), or a constant below-threshold model (Schwartz *et al.*, 1995). This is not however, enough evidence to support a hormetic effect (Gaylor, 1994), as in general, binary data from a developmental toxicity study cannot adequately distinguish between several plausible dose response curves. However, if DEHP does exhibit a hormetic effect at low doses, the estimate of benchmark doses can be affected.

For example, using the U-shaped dose response curve of (7.30) for the malformation endpoint alone, the BMD_{10} is found to be 0.0349 and BMD_1 is 0.0068. These estimates are much more conservative than those of the q-power family and folded logistic models; see Figure 7.5. Further complicating the situation, unpublished results of Hunt and Bowman for a joint threshold model for weight and malformation for the DEHP data, find a significant threshold level for malformation response rate, but no significant threshold level for fetal weight. This result may help to explain the difference in benchmark doses shown in Tables 7.5 and 7.6, when each endpoint is modeled separately.

7.6   Discussion

In this chapter, we described regression models for clustered bivariate binary and continuous responses under the assumption that data within clusters are exchangeable. We used the folded logistic (George and Bowman, 1995) and q-power family (Kuk, 2004) models to model the binary data, and a Gaussian regression model for the continuous measurements, given the binary outcome. The assumption of exchangeability seems particularly suitable to data from developmental toxicity studies, where data on each fetus may include continuous measurements, such as weight, and binary indicators of malformation. The joint regression models are based on modeling the moments λ_ks

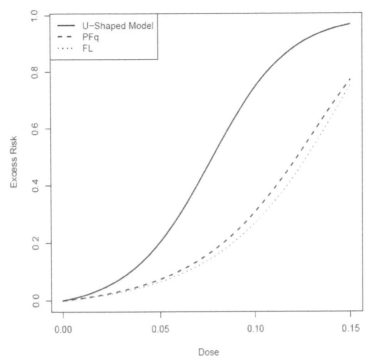

Figure 7.5 *Comparisons of excess risk curves of malformation endpoint for the q-power family, folded logistic, and U-shaped models.*

or λ'_ks of the binary responses by completely monotonic functions, and a 'partially' exchangeable Gaussian model for continuous responses, conditional on binary responses. Although the q-power family represents an effort to correct for the under-parameterization of the folded logistic model, by adding extra parameters to account for the inter-litter correlation, we find that the folded logistic and q-power family models perform quite similarly, and the q-power family provides only a slight improvement. Work is in progress extending these results to the analysis of clustered multinomial discrete and continuous data. In addition to consideration of the use of GEE for the continuous data, given the discrete outcomes, with the mean and covariance structures specified as in Section 7.3, we are developing a Bayesian approach that borrows from some of the work of Dunson (2000). Moreover, the definition of exchangeability can be broadened beyond de Finetti's (1974) that is used in this chapter. For example, Draper *et al.* (1993) give several generalizations of the concept of exchangeability, and highlight some of the advantages of this concept as a tool for data analysis.

We note here that the use of exchangeability allows for accommodating the full cluster response, even when interest is in estimating the marginal response; it also overcomes one of the problems mentioned in Fitzmaurice and Laird (1995), namely, that of finding a representation of the joint distribution of the discrete and continuous variables that yields tractable estimating equations for MLE.

Chapter 8

Random effects models for joint analysis of repeatedly measured discrete and continuous outcomes

Ralitza Gueorguieva

8.1 Introduction

In many applications, data are collected repeatedly over time (i.e., longitudinal) or on groups of related individuals (i.e., clustered). For example, one or more variables may be measured on each patient at a number of hospital visits; a number of questions can be asked at a series of interviews; the same measures can be collected on members of the same family, or different measurements can be made on different parts of a subject's body. Correlations between observations on the same subject, or within the same cluster of subjects, complicate data analysis. Statistical methods and estimation techniques are well developed for repeated measures on a univariate normally distributed variable. Furthermore, much research is dedicated to repeated observations on a binary variable and more generally, on variables with distributions in the exponential family. Zeger and Liang (1992) provide an overview of methods for longitudinal data. Fahrmeir and Tutz (2001), Diggle *et al.* (1994), and Lindsey (1993) cover many details.

However, in the case of multiple outcome variables, two types of correlations must be taken into account: correlations between measurements on different variables, and correlations between measurements on the same variable within a cluster or subject. This complicates data analysis even further. When the variables are of the same type (e.g., normal, Poisson, binary), models for a single repeatedly measured variable can be used almost directly. But when the variables are of different types (e.g., discrete and continuous), model formulation is difficult because of the lack of appropriate multivariate distributions. Thus, there are multiple approaches for joint modeling of repeatedly measured discrete and continuous outcomes in recent years without a consensus about the best approach. In this chapter, we focus on random effects models that seamlessly take into account the between-subject heterogeneity and handle the various correlations. We consider extensions of the generalized linear mixed model (GLMM) and correlated probit models, and focus on maximum likelihood estimation (MLE) (Gueorguieva, 2001; Gueorguieva and Agresti, 2001). Both models consider the multivariate normal distribution for the random effects. Extensions to discrete random effects are possible (McCulloch, 2008).

Alternative methods include approaches where the joint distribution of the responses is factored into the marginal distribution of one outcome variable and then the conditional distribution of the other outcomes given that first outcome are used to estimate effects (Catalano and Ryan, 1992, Fitzmaurice and Laird, 1997, 1995); marginal models, where the regression of the responses on the predictors are modeled separately of the association among repeated observations for each individual (e.g., Geys *et al.*, 2001; Rochon, 1996; Spiess and Hamerle, 1996); Bayesian approaches with underlying random variables (e.g., Dunson *et al.*, 2003; Dunson, 2000), and approaches based on copulas (de Leon and Wu, 2011; Song *et al.*, 2009; Zimmer and Trivedi, 2006). Overviews of approaches for the analysis of mixed outcomes are provided by de Leon and Carrière Chough (2010), Faes *et al.* (2009), and Regan and Catalano (2002).

Many models trace their origins to the general location model (GLOM) for discrete and continuous outcomes of Olkin and Tate (1961). It is based on a multinomial model for the discrete outcomes, and a multivariate normal model for the continuous outcomes, conditional on the discrete outcomes. Fitzmaurice and Laird (1995) discuss a generalization of this model for clustered binary and continuous variables, whereas other authors reverse the conditioning order. Catalano and Ryan (1992) consider standard random effects models for the continuous variable and a correlated probit model for the binary variables, conditional on the continuous outcome. Catalano (1997) extends their procedure to ordinal data. Faes *et al.* (2004) also consider clustered continuous and ordinal outcomes and propose an extension of the Plackett–Dale approach (Geys *et al.*, 2001). Najita *et al.* (2009) incorporate the sampling scheme in the correlated probit formulation. Due to the complexity of dealing with the full likelihood, most authors consider pseudo-likelihood or generalized estimation equations (GEE) for estimation. A potential drawback of this conditioning approach is that the outcomes are not treated symmetrically.

A number of the proposed models are appropriate only for equicorrelated data (e.g., George *et al.*, 2007; Regan and Catalano, 1999*a*), and some models assume shared rather than correlated random effects or latent variables (Gueorguieva and Sanacora, 2006; Dunson, 2003; Molenberghs *et al.*, 2001; Sammel *et al.*, 1997). Both assumptions may be restrictive in some scenarios. Random effects models, conditional models, and some marginal models face difficulties when the number of outcome variables increases. Extensions to high-dimensional data have been proposed by Faes *et al.* (2008).

Marginal models are usually applied when the effects of predictors on the marginal distribution of the outcomes is of primary interest and the correlation structure of the data is treated as a nuisance. In one of the first extensions of GEE for mixed discrete and continuous data, Rochon (1996) combines a pair of GEE models (one for each outcome) into an overall analysis framework via seemingly unrelated regression. The model of Fitzmaurice and Laird (1995) also uses GEE for estimation and can also be regarded as a special case of the partly exponential model introduced by Zhao *et al.* (1992). Partly exponential models for the regression analysis of multivariate response data are parametrized in terms of the response mean and a general shape parameter. A fully parametric approach to estimation leads to asymptotically independent MLEs of the mean and shape parameters. The score equations for the mean parameters are essentially the same as in GEE. The MLEs are fully efficient if the covariance matrix is correctly specified but lose efficiency under incorrect specification. The partial exponential models are useful for multivariate outcome data of the same type, but are difficult to generalize to mixtures of discrete and continuous outcomes.

The Bayesian approach places prior distributions on the unknown parameters and in cases when prior knowledge about plausible ranges of the parameters exists, can stabilize estimation (Dunson *et al.*, 2003; Chib and Greenberg, 1998). However, selection of appropriate priors and ascertaining identifiability are not trivial. Furthermore, usually such an approach cannot be directly applied in existing software, whereas MLE of random effects models is easily programmable using PROC NLMIXED in SAS.

Since full MLE methods may be very computationally intensive, especially when the number of random effects is high, some authors have considered tractable analytical approximations to the marginal likelihood (Breslow and Clayton, 1993; Wolfinger and O'Connell, 1993). Although these methods lead to inconsistent estimates in some cases, they have considerable advantage in computational speed over 'exact' methods and can be fitted with standard software.

Joint analysis of multiple outcome variables has several important advantages over separate analyses of the variables that ignore the correlation across variables. In particular, joint analysis can help answer multivariate questions (e.g., estimation of the effect of a treatment on disease severity and side effects simultaneously), keeps actual level of tests closer to the nominal level by limiting the number of tests that need to be performed, may improve efficiency in estimating the effects

Table 8.1 *Summary statistics of EG mice data.*

Dose (g/kg)	Number of dams	Number of live fetuses	Malformations		Weight (g)	
			Number	Percent	Mean	SD
0.00	25	297	1	0.337	0.972	0.098
0.75	24	276	26	9.42	0.877	0.104
1.50	22	229	89	38.865	0.764	0.107
3.00	23	226	129	57.08	0.704	0.124

of covariates on each outcome variable by borrowing information from other related outcome variables, and may provide additional information about the relationship between variables, such as how the correlations and variances change over time. However, these advantages come at the expense of greater modeling and computational complexity, the need for more model assumptions, and often the need for the development of custom-made software.

Two data examples are used for illustration of the models we describe. The first is from a developmental toxicity study of ethylene glycol (EG) in mice (Price *et al.*, 1985). The experiment involved randomizing pregnant mice to one of four possible levels of the teratogen EG during major organogenesis. After sacrifice, measurements of weight and malformation status (yes/no) were taken on each fetus in the uterus of each pregnant dam. The goal of the experiment was to assess the joint effects of increasing dose on fetal weight and on malformation status. Table 8.1 shows some descriptive statistics by dose. A number of authors have considered this example for illustration of their methods (Catalano and Ryan, 1992; Fitzmaurice and Laird, 1995; Gueorguieva, 2001, 2005; Gueorguieva and Agresti, 2001; Dunson *et al.*, 2003; Lin *et al.*, 2010).

The second example is from a depression clinical trial. Patients with major depression were randomized either to the combination of fluoxetine and yohimbine (experimental treatment) or to fluoxetine alone (standard treatment) (Sanacora *et al.*, 2004). Compared to the standard treatment group, the experimental treatment group was expected to show faster improvement in depression severity as measured by the Hamilton Depression Rating score (HAMD) and the Clinical Global Improvement score (CGI). Both measures were collected on each subject at each visit during the 7 week treatment period. The HAMD is a 25-item scale that is scored from 0 to 90 and could be assumed to be normally distributed. The CGI is an ordinal variable measuring severity of illness on a scale from 1 to 7, with 1 indicating "normal, not at all ill" and 7 indicating "among the most extremely ill patients." Descriptive statistics for these data are provided in Table 8.2. Models for repeatedly measured continuous and ordinal outcomes have been proposed and applied to these data by Gueorguieva and Sanacora (2006).

The chapter is organized as follows. In Section 8.2, we define multivariate GLMMs that can handle any combination of outcomes with distributions in the exponential family, and the correlated probit models that provide a very general correlation structure for mixed binary and continuous outcomes. In Section 8.3, we detail procedures for MLE and inference, while in Section 8.4, we apply the described models and estimation procedures to the two data examples. A brief description of available software is also provided. The chapter concludes with a discussion in Section 8.5. For alternative models and estimation techniques for mixed longitudinal or clustered outcomes, see Chapters 5, 7, 9–10, and 12–13.

8.2 Models

GLMMs are random effects models that are well suited for subject-specific inference for repeated measures data. They are designed for correlated outcomes in the exponential family (e.g., normal, Poisson, gamma, Bernoulli). Herein, we define an extension of this class of models to repeatedly

Table 8.2 *Descriptive statistics for depression data.*

Group	Time	HAMD Mean	HAMD SD	CGI Mean	CGI SD
	0	29.3	6.0	4.1	0.5
	1	23.3	5.9	3.6	0.6
	2	18.4	7.9	3.1	0.8
Active	3	15.6	8.8	2.9	1.0
	4	14.4	8.5	2.7	1.0
	5	10.6	6.0	2.6	0.9
	6	8.9	6.6	2.0	1.0
	0	31.1	6.3	4.3	0.7
	1	26.1	6.9	3.9	0.7
	2	20.9	7.6	3.5	0.8
Control	3	19.8	7.9	3.5	1.0
	4	18.6	8.6	3.4	1.0
	5	14.3	8.4	2.9	0.9
	6	11.7	6.5	2.5	1.1

measured response variables of different types. This extension relies on the assumption of conditional independence of the observations on the different outcome variables, given the random effects. This assumption can be restrictive in some applications, hence alternative models (i.e., correlated probit models) are also described for combinations of binary, ordinal, and continuous outcomes.

8.2.1 Multivariate GLMMs

Let Y_{ijv} be response variable $v = 1, \cdots, V$, on occasion $j = 1, \cdots, n_i$, for individual $i = 1, \cdots, n$. At each occasion, at least one of the response variables is measured. We assume that for each v, Y_{ijv}, $j = 1, \cdots, n_i$, are conditionally independent given a random effect $\mathbf{B}_{iv}$ and whose density $f_{Y_{ijv}}(\cdot)$ belongs to the exponential family. Note that the densities $f_{Y_{ijv}}(\cdot)$ need not be the same but each one is of the form

$$f_{Y_{ijv}|\mathbf{B}_{iv}}(y_{ijv}|\mathbf{b}_{iv}) = \exp\left\{\frac{y_{ijv}\theta_{ijv} - b(\theta_{ijv})}{\phi}\omega_{ijv} + c(y_{ijv}, \phi, \omega_{ijv})\right\}, \tag{8.1}$$

where $\mu_{ijv} = \partial b(\theta_{ijv})/\partial \theta_{ijv} = b'(\theta_{ijv})$ is the conditional mean, ϕ is the dispersion parameter, $b(\cdot)$ and $c(\cdot)$ are specific functions corresponding to the type of exponential family, and ω_{ijv} are known weights.

The responses $\mathbf{Y}_{iv} = (Y_{i1v}, \cdots, Y_{in_iv})^\top$, $v = 1, \cdots, V$, are conditionally independent given $\mathbf{B}_{iv}$, and the responses on different subjects are independent. Let $g_v(\cdot)$ be appropriate link functions for $f_{Y_{ijv}}(\cdot)$. Denote the conditional means of Y_{ijv} by μ_{ijv}, respectively, with $\boldsymbol{\mu}_{iv} = (\mu_{iv1}, \cdots, \mu_{in_iv})^\top$. At stage one of the mixed model specification, we assume

$$g_v(\boldsymbol{\mu}_{iv}) = \mathbf{X}_{iv}\boldsymbol{\beta}_v + \mathbf{Z}_{iv}\mathbf{B}_{iv}, \tag{8.2}$$

where $\boldsymbol{\beta}_v$ are unknown parameter vectors, $\mathbf{X}_{iv}$ are design matrices for the fixed effects, $\mathbf{Z}_{iv}$ are design matrices for the random effects, and $g_v(\cdot)$ are applied componentwise to $\boldsymbol{\mu}_{iv}$. At stage two, a joint distribution for all $\mathbf{B}_{iv}$s is specified. The normal distribution is a very good candidate, as it provides a rich covariance structure. Hence, we assume

$$\mathbf{B}_i = \begin{pmatrix} \mathbf{B}_{i1} \\ \vdots \\ \mathbf{B}_{iV} \end{pmatrix} \overset{iid}{\sim} N_P\left(\begin{pmatrix} \mathbf{0} \\ \vdots \\ \mathbf{0} \end{pmatrix}, \boldsymbol{\Sigma} = \begin{pmatrix} \boldsymbol{\Sigma}_{11} & \cdots & \boldsymbol{\Sigma}_{1V} \\ \vdots & \ddots & \vdots \\ \boldsymbol{\Sigma}_{1V}^\top & \cdots & \boldsymbol{\Sigma}_{VV} \end{pmatrix}\right), \tag{8.3}$$

where $\boldsymbol{\Sigma}, \boldsymbol{\Sigma}_{11}, \boldsymbol{\Sigma}_{22}, \cdots, \boldsymbol{\Sigma}_{VV}$, are in general, unknown positive-definite matrices, "$\overset{iid}{\sim}$" means "independent identically distributed", and $P = n_i V$.

When $\boldsymbol{\Sigma}_{kk'} = \mathbf{0}$, for all $k \neq k'$, then the above model is equivalent to separate GLMMs for the outcome variables. If the vectors of random effects are perfectly correlated, then this is equivalent to assuming a common latent variable with a multivariate normal distribution. It is possible to allow the random effects in these models to depend on covariates, but this extension is not considered here. See Sammel $et\ al.$ (1997) for details about potential extensions.

The regression parameters $\boldsymbol{\beta}_v$ in the multivariate GLMM (MGLMM) have subject-specific interpretations, i.e., they describe the individual's, rather than the average population, response to changes in the covariates. For example, consider the case of a single predictor variable for a particular outcome, say, dose. Then the regression coefficients are interpreted as the expected changes in the conditional mean response for a particular subject should the dose for this subject be increased by one unit.

The marginal mean of each outcome can be obtained as

$$\mu_{ijv}^* = E(Y_{ijv}) = E\left\{E(Y_{ijv}|\mathbf{B}_i)\right\} = \int g_v^{-1}(\mathbf{x}_{ijv}^\top\boldsymbol{\beta}_v + \mathbf{z}_{ijv}^\top\mathbf{b}_{iv})f_{\mathbf{B}_i}(\mathbf{b}_i; \boldsymbol{\Sigma})d\mathbf{b}_i.$$

In general, $g_v(\mu_{ijv}) \neq \mathbf{x}_{ijv}^\top\boldsymbol{\beta}_v$, if $g_v(\cdot)$ is a nonlinear function. However, this equation holds approximately if the standard deviations of the random effects distributions are small. Closed-form expressions for the marginal means exist for certain link functions (Zeger $et\ al.$, 1988).

As mentioned above, the formulation of GLMMs relies on the potentially restrictive assumption of independence of outcome variables conditional on random effects. The correlated probit model does not make this assumption but is applicable only for combinations of binary, ordinal, and continuous outcomes.

8.2.2 *Correlated probit model*

Let Y_{ijv} be response variable $v = 1, \cdots, V$, on occasion $j = 1, \cdots, n_i$, for individual $i = 1, \cdots, n$. Here Y_{ijv} is continuous, binary, or ordinal. Assume an underlying latent continuous variable Y_{ijv}^* for each ordinal or binary response with categories $1, \cdots, R_v$, for ordinal variables. The observed and the latent responses are related by $Y_{ijv} = \mathrm{I}\{Y_{ijv}^* > 0\}$ for binary outcomes, and $Y_{ijv} = r$, if $\tau_{r-1,v} \leq Y_{ijv}^* < \tau_{r,v}, r = 1, \cdots, R_v$, for ordinal outcomes, where $\tau_{0,v} = -\infty < \tau_{1v} < \tau_{2v} < \cdots < \tau_{R_v-1,v} < \tau_{R_v,v} = +\infty,$.

The correlated probit model is defined as

$$Y_{ijv}^* = \mathbf{x}_{ijv}^\top\boldsymbol{\beta} + \mathbf{z}_{ijv}^\top\mathbf{B}_i + \varepsilon_{ijv}, \tag{8.4}$$

$$\mathbf{B}_i \overset{iid}{\sim} N_P(\mathbf{0}, \boldsymbol{\Sigma}_b), \tag{8.5}$$

$$\boldsymbol{\varepsilon}_i = \begin{pmatrix} \varepsilon_{ij1} \\ \vdots \\ \varepsilon_{ijV} \end{pmatrix} \overset{iid}{\sim} N_V(\mathbf{0}, \boldsymbol{\Sigma}_e). \tag{8.6}$$

The parameter vectors could be common or separate for the different variables. Identifiability constraints need to be imposed on the thresholds, intercepts, and/or the variances of the underlying latent variables. One possible way to impose constraints is by setting all diagonal elements of $\boldsymbol{\Sigma}_e$ corresponding to binary and ordinal outcomes, to be equal to one, and to set all intercepts in the linear predictors for ordinal responses equal to zero. The covariate vectors are in general different for the different variables.

This model is a generalization of the models of Catalano and Ryan (1992), Catalano (1997), and Gueorguieva and Agresti (2001), and includes several other models as special cases. For example, considering only binary outcomes, the model becomes the multivariate probit model proposed by Lesaffre and Molenberghs (1991). Considering only normally distributed continuous outcomes, the model is a GLMM. If there is no correlation among the error terms for a binary and a continuous variable, the model is a bivariate GLMM with a probit link for the Bernoulli response and an identity link for the normal response. If there are no correlations among both the random effects and the errors corresponding to different variables, the model is equivalent to specifying separate GLMMs for the response variables.

As in GLMMs, the regression parameters in the correlated probit model have subject-specific interpretations. In the case of one continuous predictor, the regression coefficients are interpreted as the expected changes in the continuous responses (or latent continuous responses) for a particular subject should dose for this subject be increased by one unit.

The correlated probit model is very flexible in accounting for the correlation structure of repeatedly measured multiple outcome variables because of the underlying multivariate linear mixed model for normal data. Multiple variables, spatially or temporally correlated data and from hierarchical sampling designs, can be accommodated. Correlations among binary variables and among ordinal variables are the corresponding correlations among the latent continuous variables and are called tetrachoric and polychoric correlations, respectively. Correlations between continuous and binary variables, and between continuous and ordinal variables, are also the corresponding latent correlations, and are called biserial and polyserial correlations, respectively.

8.3 Estimation and inference

Let Ψ denote the vector of all unknown parameters in either model formulation from the previous section. Let $\mathbf{Y}$ be the vector of all observations and $\mathbf{Y}_i$ be the vector of responses on the ith subject. For both the MGLMM and the correlated probit model, the marginal log-likelihood is equal to the sum of the log-likelihoods of the individual subjects

$$\log L(\Psi; \mathbf{y}) \;=\; \sum_{i=1}^{n} \log L_i(\Psi; \mathbf{y}_i). \tag{8.7}$$

Each individual likelihood involves integration over the random effects distribution that is not analytically tractable, as in

$$\log L_i(\Psi; \mathbf{y}_i) \;=\; \int_{\mathbb{R}^P} \prod_{j=1}^{n_i} f_{Y_{ij1}, \cdots, Y_{ijV} | \mathbf{B}_i}(y_{ij1}, \cdots, y_{ijV} | \mathbf{b}_i; \Psi) |2\pi \Sigma_b|^{-1/2}$$
$$\times \exp\left(-\frac{1}{2} \mathbf{b}_i \Sigma_b^{-1} \mathbf{b}_i\right) d\mathbf{b}_i, \tag{8.8}$$

where $\mathbb{R}^P$ is the P-dimensional real plane. Direct approaches to MLE are based on numerical or stochastic approximations of the integrals and on numerical maximizations of those approximations. The most widely used numerical approximation procedure for this type of integral is Gauss–Hermite quadrature, which is appropriate when the dimension of the random effects is small. It involves evaluating the integrands at m prespecified quadrature points and substituting weighted sums in place of the intractable integrals for the n subjects. If the number m of quadrature points is large, the approximation can be made very accurate; however, to keep the numerical effort low, m should be kept as small as possible. Adaptive Gaussian quadrature allows a reduction in the required number of quadrature points by centering and scaling them around the mode of the integrand function.

An alternative for high-dimensional random effects is to approximate the integrals by Monte Carlo sums. This involves generating m random values from the random effects distribution for each subject, evaluating the conditional densities $f_{\mathbf{Y}_i | \mathbf{B}_i}(\mathbf{y}_i | \mathbf{b}_i; \Psi)$ at those values, and then taking averages.

Details about Gauss–Hermite quadrature and Monte Carlo approximation for the GLMM can be found in Fahrmeir and Tutz (2001), for the MGLMM in Gueorguieva (2001), and for the correlated probit model in Gueorguieva and Agresti (2001). The general algorithms are presented below for the technically minded reader. Sections 6.3.1 and 6.3.2 can be skipped by readers not interested in technical details.

8.3.1   *Estimation via Gauss–Hermite quadrature*

To perform Gauss–Hermite quadrature, $L_i(\boldsymbol{\Psi}; \mathbf{y}_i)$ is approximated by

$$L_i^{GQ}(\boldsymbol{\Psi}; \mathbf{y}_i) = \sum_{k_1=1}^{m} v_{k_1}^{(1)} \cdots \sum_{k_q=1}^{m} v_{k_q}^{(q)} h(\mathbf{z}^{(k)}),$$

where $\mathbf{z}^{(k)} = \sqrt{2} \mathbf{L} \mathbf{d}^{(k)}$ (with $\boldsymbol{\Sigma} = \mathbf{L}\mathbf{L}^{\top}$) for the multiple index $k = (k_1, \cdots, k_q)$, $\mathbf{d}^{(k)}$ denotes the tabled nodes of univariate Gauss–Hermite integration of order m (Abramowitz and Stegun, 1972), and

$$h(\mathbf{z}^{(k)}) = \prod_{j=1}^{n_{i1}} f_{Y_{ij1}, \cdots, Y_{ijV}|\mathbf{B}_i}(y_{ij1}, \cdots, y_{ijV}|\mathbf{z}^{(k)}; \boldsymbol{\Psi}).$$

The corresponding weights are given by $v_{k_\ell}^{(\ell)} = \pi^{-1/2} w_{k_\ell}^{(\ell)}$, where $w_{k_\ell}^{(\ell)}$ are the tabled univariate weights, $\ell = 1, \cdots, m$. The maximization algorithm then proceeds as follows:

1. Choose initial estimate for the parameter vector $\widehat{\boldsymbol{\Psi}}^{(0)}$. Set the iteration counter $c = 0$.

2. Increase c by 1. Approximate each of the integrals $L_i(\widehat{\boldsymbol{\Psi}}^{(c-1)}; \mathbf{y}_i)$ by $L_i^{GQ}(\widehat{\boldsymbol{\Psi}}^{(c-1)}; \mathbf{y}_i)$ using m quadrature points in each dimension.

3. Update the estimate of $\boldsymbol{\Psi}$ using a step of a numerical maximization routine. For example, the Newton–Raphson method would involve calculation of

$$\widehat{\boldsymbol{\Psi}}^{(c)} = \widehat{\boldsymbol{\Psi}}^{(c-1)} + \frac{\partial}{\partial \boldsymbol{\Psi}} \log L^{GQ}(\boldsymbol{\Psi}; \mathbf{y}_i) \left\{ \frac{\partial^2}{\partial \boldsymbol{\Psi} \partial \boldsymbol{\Psi}^{\top}} \log L^{GQ}(\boldsymbol{\Psi}; \mathbf{y}_i) \right\}^{-1} \Bigg|_{\boldsymbol{\Psi} = \boldsymbol{\Psi}^{(c-1)}}.$$

4. Iterate between steps 2 and 3 until convergence.

After the algorithm has converged, standard errors (SEs) of the estimates can be obtained from observed information matrix, which is a by-product of the Newton–Raphson algorithm.

In the MGLMM, the integrand function $f_{Y_{ij1}, \cdots, Y_{ijV}|\mathbf{B}_i}(y_{ij1}, \cdots, y_{ijV}|\mathbf{b}_i; \boldsymbol{\Psi})$ factors into V conditionally independent components because all the responses on a particular subject are conditionally independent given the random effects. However, in the correlated probit formulation, because of the correlation of the errors of the continuous and latent continuous variables, we can not simply factor the integrand function into separate components for the outcome variables. To deal with this issue, we use the properties of the multivariate normal distribution.

Consider a correlated probit model with one continuous response variable and one ordinal (binary) outcome. In this case we can rewrite the joint conditional density of the bivariate response given the random effects as the product of the univariate normal density of the continuous outcome and the conditional multinomial density of the ordinal outcome. That is,

$$f_{Y_{ij1}, Y_{ij2}|\mathbf{B}_i}(y_{ij1}, y_{ij2}|\mathbf{b}_i; \boldsymbol{\Psi}) = f_{Y_{ij1}|\mathbf{B}_i}(y_{ij1}|\mathbf{b}_i; \boldsymbol{\Psi}) f_{Y_{ij2}|Y_{ij1}, \mathbf{B}_i}(y_{ij2}|y_{ij1}, \mathbf{b}_i; \boldsymbol{\Psi}), \tag{8.9}$$

$$f_{Y_{ij1}|\mathbf{B}_i}(y_{ij1}|\mathbf{b}_i; \boldsymbol{\Psi}) = (2\pi\sigma_{e1}^2)^{-1/2} \exp\left\{ -\frac{1}{2\sigma_{e1}^2}(y_{ij1} - \eta_{ij1})^2 \right\}, \tag{8.10}$$

$$f_{Y_{ij2}|Y_{ij1}, \mathbf{B}_i}(y_{ij2}|y_{ij1}, \mathbf{b}_i; \boldsymbol{\Psi}) = \Phi\left(\frac{\tau_1 - \eta_{ij2}}{\sigma^*} \right)^{\mathrm{I}\{y_{ij2}=1\}} \left\{ \Phi\left(\frac{\tau_2 - \eta_{ij2}}{\sigma^*} \right) - \Phi\left(\frac{\tau_1 - \eta_{ij2}}{\sigma^*} \right) \right\}^{\mathrm{I}\{y_{ij2}=2\}}$$

$$\times \cdots \times \left\{ 1 - \Phi\left(\frac{\tau_{1,R-1} - \eta_{ij2}}{\sigma^*} \right) \right\}^{\mathrm{I}\{y_{ij2}=R\}} \tag{8.11}$$

where $\eta_{ij1} = \mathbf{x}_{ij1}^\top \boldsymbol{\beta} + \mathbf{z}_{ij1}^\top \mathbf{b}_i$, $\eta_{ij2} = \mathbf{x}_{ij2}^\top \boldsymbol{\beta} + \mathbf{z}_{ij2}^\top \mathbf{b}_i + (\sigma_{e12}/\sigma_{e1}^2)(Y_{ij1} - \eta_{ij1})$, and $\sigma^* = (1 - \sigma_{e12}^2/\sigma_{e1}^2)^{1/2}$. This approach is extended to the case of multiple binary, ordinal, and continuous outcomes by specifying the multivariate conditional density of the observed continuous outcomes and then, conditional on the continuous outcomes, specifying a multinomial likelihood for all binary and ordinal outcomes, with category probabilities obtained using the multivariate normal cumulative density function. Gaussian quadrature then proceeds as described for the MGLMM.

8.3.2   Estimation via the EM algorithm

The EM algorithm is an iterative technique for finding MLEs when direct maximization of the observed likelihood $L(\boldsymbol{\Psi}; \mathbf{y})$ is either not feasible or not practical. It involves augmenting the observed data by unobserved data (i.e., random effects in MGLMMs and correlated probit models), so that maximization at each step of the algorithm is considerably simplified. The EM algorithm can be summarized as follows:

1. Select starting values $\widehat{\boldsymbol{\Psi}}^{(0)}$ for the parameters. Set the iteration counter $c = 0$.

2. **E-step:** Increase c by 1 and calculate $E\{\log f_{\mathbf{Y}|\mathbf{B}}(\mathbf{y}|\mathbf{b}; \widehat{\boldsymbol{\Psi}}^{(c-1)})\}$.

3. **M-step:** Find a value $\widehat{\boldsymbol{\Psi}}^{(c)}$ of $\boldsymbol{\Psi}$ that maximizes this conditional expectation.

4. Iterate between (2) and (3) until convergence is achieved.

In the MGLMM, the complete data are $\mathbf{u} = (\mathbf{y}^\top, \mathbf{b}^\top)^\top$, and the complete data log-likelihood is

$$\log L(\boldsymbol{\Psi}; \mathbf{u}) = \sum_{i=1}^n \log f_{\mathbf{Y}_i|\mathbf{B}_i}(\mathbf{Y}_i|\mathbf{b}_i; \boldsymbol{\Psi}) + \sum_{i=1}^n \log f_{\mathbf{B}_i}(\mathbf{b}_i; \boldsymbol{\Sigma}).$$

The cth E-step of the EM algorithm involves computing $E\{\log L(\mathbf{u}, \boldsymbol{\Psi})|\mathbf{y}, \widehat{\boldsymbol{\Psi}}^{(c-1)}\}$ and the cth M-step maximizes this quantity with respect to $\boldsymbol{\Psi}$ and updates the parameter estimates. Under the assumption of conditional independence and different random effects for the different outcomes, the log-likelihood of the complete data can be separated in $V + 1$ different terms, as in

$$\log L(\mathbf{u}, \boldsymbol{\Psi}) = \sum_{i=1}^n \sum_{j=1}^{n_{i1}} \log f_{Y_{ij1}|\mathbf{B}_{i1}}(y_{ij1}|\mathbf{b}_{i1}; \boldsymbol{\Psi}) + \cdots + \sum_{i=1}^n \sum_{j=1}^{n_{iV}} \log f_{Y_{ijV}|\mathbf{B}_{iV}}(y_{ijV}|\mathbf{b}_{iV}; \boldsymbol{\Psi})$$

$$+ \sum_{i=1}^n \log f_{\mathbf{B}_i}(\mathbf{b}_i; \boldsymbol{\Sigma}).$$

and each M-step of the algorithm consists of $V + 1$ separate maximizations, evaluated at the current parameter estimate of $\boldsymbol{\Psi}$. The conditional expectations for each of the first V terms $E\{\sum_{i=1}^n \sum_{j=1}^{n_{iv}} \log f_{Y_{ijv}|\mathbf{B}_i}(y_{ijv}|\mathbf{b}_i; \boldsymbol{\Psi})|\mathbf{y}\}$ do not have closed-form expressions but can be approximated by Gaussian quadrature or Monte Carlo estimates. Gaussian quadrature may be preferrable when there is a large number of within-cluster observations, but Monte Carlo sums are better when the dimension of the random effects is high. The Monte Carlo sums are of the form

$$\frac{1}{m} \sum_{k=1}^m \sum_{i=1}^n \sum_{j=1}^{n_{iv}} \log f_{Y_{ijv}|\mathbf{B}_{iv}}(y_{ijv}|\mathbf{b}_{iv}^{(k)}; \boldsymbol{\Psi}) \tag{8.12}$$

or

$$\frac{1}{m} \sum_{k=1}^m \sum_{i=1}^n \log f_{\mathbf{B}_i}(\mathbf{b}_i^{(k)}; \boldsymbol{\Sigma}_b), \tag{8.13}$$

where $\mathbf{b}_i^{(k)}$, $k = 1, \cdots, m$, are simulated values from the conditional distribution of $\mathbf{B}_i$, given $\mathbf{Y}$, evaluated at the current parameter estimate of $\boldsymbol{\Psi}$. To obtain simulated values for $\mathbf{B}_i$, several different techniques can be used, among which are importance sampling, multivariate rejection sampling (Gueorguieva, 2001; Booth and Hobert, 1999), and Markov Chain Monte Carlo methods (Dunson et al., 2003).

Once m simulated values for the random effects are available, the Monte Carlo sums of the conditional expectations can be maximized using numerical maximization procedures such as the Newton–Raphson algorithm. After convergence, SEs can be calculated using Louis' method of approximation of the observed information matrix (Tanner, 1991), as detailed in Gueorguieva (2001).

In the correlated probit model, the complete data are $\mathbf{u} = (\mathbf{y}^\top, \mathbf{y}^{*\top}, \mathbf{b}^\top)^\top$, where $\mathbf{y}$ is a vector of all observed normal response variables, $\mathbf{y}^*$ is a vector of all observed latent normal variables, and $\mathbf{b}$ are the random effects. Then, the complete data log-likelihood is

$$
\begin{aligned}
\log L(\mathbf{u}, \mathbf{\Psi}) = {} & -\frac{1}{2}\sum_{i=1}^{n}\sum_{j=1}^{n_i}\log|\mathbf{\Sigma}_e| - \frac{1}{2}\sum_{i=1}^{n}\sum_{j=1}^{n_i}(\mathbf{y}_{ij}^* - \mathbf{X}_{ij}\boldsymbol{\beta} - \mathbf{Z}_{ij}\mathbf{b}_i)^\top \mathbf{\Sigma}_e^{-1}(\mathbf{y}_{ij}^* - \mathbf{X}_{ij}\boldsymbol{\beta} - \mathbf{Z}_{ij}\mathbf{b}_i) \\
& -\frac{1}{2}\sum_{i=1}^{n}\log|\mathbf{\Sigma}_b| - \frac{1}{2}\sum_{i=1}^{n}\mathbf{b}_i^\top \mathbf{\Sigma}_b^{-1}\mathbf{b}_i,
\end{aligned}
$$

where $\mathbf{y}_{ij}^*$ is a vector containing all observed continuous and latent continuous variables for individual i at occasion j, $\mathbf{X}_{ij}$ and $\mathbf{Z}_{ij}$ are the corresponding design matrices for the fixed and random effects, respectively. Because of the underlying multivariate normality, the following closed-form expressions are available for calculating the full data MLEs:

$$
\widehat{\mathbf{\Sigma}}_b = \frac{1}{n}\sum_{i=1}^{n}\widehat{\mathbf{b}}_i\widehat{\mathbf{b}}_i^\top; \tag{8.14}
$$

for a fixed $\mathbf{\Sigma}_e$,

$$
\widehat{\boldsymbol{\beta}} = \left(\sum_{i=1}^{n}\sum_{j=1}^{n_i}\mathbf{X}_{ij}^\top\mathbf{\Sigma}_e^{-1}\mathbf{X}_{ij}\right)^{-1}\left(\sum_{i=1}^{n}\sum_{j=1}^{n_i}\mathbf{X}_{ij}^\top\mathbf{\Sigma}_e^{-1}(\mathbf{y}_{ij}^* - \mathbf{Z}_{ij}\widehat{\mathbf{b}}_i)\right), \tag{8.15}
$$

and for a fixed $\boldsymbol{\beta}$,

$$
\widehat{\mathbf{\Sigma}}_e = \frac{1}{N}\sum_{i=1}^{n}\sum_{j=1}^{n_i}(\mathbf{y}_{ij}^* - \mathbf{X}_{ij}\boldsymbol{\beta} - \mathbf{Z}_{ij}\widehat{\mathbf{b}}_i)(\mathbf{y}_{ij}^* - \mathbf{X}_{ij}\boldsymbol{\beta} - \mathbf{Z}_{ij}\widehat{\mathbf{b}}_i)^\top. \tag{8.16}
$$

Iterating between the last two equations in the EM algorithm provides convergence to the true ML estimates. At each step, the new estimate of $\mathbf{\Sigma}_e$ uses the previous value of $\widehat{\boldsymbol{\beta}}$, and then the new value of $\widehat{\mathbf{\Sigma}}_e$ is used to update $\widehat{\boldsymbol{\beta}}$.

Conditional versions of these equations that depend only on the mean and covariance matrices of the conditional distributions of the latent continuous responses replace (8.14), (8.15), and (8.16) in the E-step. The Gibbs sampler is used to simulate values from the truncated multivariate normal distributions (Gueorguieva and Agresti, 2001) in this case.

Note that since the variances of the underlying latent continuous variables are not estimable from the observed data, they need to be restricted (e.g., fixed equal to 1). Alternatively, they can be estimated within the EM algorithm via parameter expanded EM (PX-EM) algorithm (Liu et al., 1998), so that the advantage of having closed-form expressions for the complete data is not lost. The PX-EM expands the complete data model while preserving the observed data model. It is shown to improve efficiency in the EM-algorithm while preserving the simplicity and stability of ordinary EM.

At the final step of the EM algorithm, estimable functions of the parameters are calculated (e.g., the ratios of regression components and the variance of the corresponding underlying continuous variables). Since the EM algorithm may be quite slow, an alternative fitting procedure based on stochastic approximation EM algorithm (Delyon et al., 1999) is available. For details of the EM algorithm and extensions for correlated probit models, see Gueorguieva and Agresti (2001).

8.3.3 Inference

The estimates considered herein for both MGLMMs and the correlated probit models are approximate MLEs, and hence confidence intervals and hypothesis tests can be constructed according to asymptotic maximum likelihood theory. For example, significance of regression coefficients can be tested using Wald or likelihood ratio tests; nested models can be compared using likelihood-ratio tests; score tests can be used in scenarios where fitting a more general model is too computationally intensive. The conditional independence assumption in the MGLMM model can be tested using the score test proposed by Gueorguieva (2001), without fitting a model with conditional dependence of the response variables. In general, when testing the significance of variance components, caution needs to be applied as the parameters may be on the boundary of the parameter space (Lin, 1997; Verbeke and Molenberghs, 2003). Empirical Bayes estimates can be used for estimating the random effects or individual response for a particular subject. Comparison of models with different covariance structures can be based on information criteria, such as Akaike information criterion and Schwartz Bayesian criterion.

Since there are no closed-form expressions for the marginal log-likelihood (8.7), the score, and the information matrix, approximations (numerical, stochastic, or analytical) must be used in all inference. This aggravates the problem of determining coverage probabilities and actual error rates and simulations are needed to study the behavior of tests and confidence intervals (Gueorguieva, 2001).

8.4 Applications

8.4.1 Developmental toxicity study in mice

In this example, the primary interest lies in simultaneously estimating the effect of dose on fetal weight and the probability of malformation for fetus within litter while properly taking into account all correlations between the outcomes. For the jth fetus within the ith litter, let fetal weight be denoted by Y_{ij1}, and observed malformation status by Y_{ij2}. We assume only a linear effect of dose, since we found no evidence that the quadratic effect was significant. However, other authors have considered quadratic effects of dose (Lin $et\ al.$, 2010; Regan and Catalano, 1999a). In the MGLMM formulation for this example, we consider both logit and probit links for the binary response. The logit link is the more popular choice and has the advantage of better interpretation of the estimated regression coefficients as log odds ratios. The probit link allows comparison of the MGLMM with the more general correlated probit model for this example. Thus, for the continuous response (fetal weight) we have

$$Y_{ij1} = \beta_{10} + x_i\beta_{11} + B_{i1} + \varepsilon_{ij1}.$$

For the binary response (malformation), we have

$$\text{logit}\{P(Y_{ij2} = 1)\} = \beta_{20} + x_i\beta_{21} + B_{i2} + \varepsilon_{ij2},$$

in the logit formulation, where

$$\mathbf{B}_i = \begin{pmatrix} B_{i1} \\ B_{i2} \end{pmatrix} \overset{iid}{\sim} N_2(\mathbf{0}, \mathbf{\Sigma}),$$

and $\varepsilon_{ij1}, \varepsilon_{ij2} \overset{iid}{\sim} N(0, \sigma^2)$, with the random effects and errors independent. Alternatively, in the probit formulation, we assume a continuous latent variable Y_{ij2}^*, such that $Y_{ij2} = \text{I}\{Y_{ij2}^* > 0\}$, and

$$Y_{ij2}^* = \beta_{20} + x_i\beta_{21} + B_{i2} + \varepsilon_{ij2},$$

where in addition to the assumption of normality of the random effects, we have

$$\boldsymbol{\varepsilon}_{ij} = \begin{pmatrix} \varepsilon_{ij1} \\ \varepsilon_{ij2} \end{pmatrix} \overset{iid}{\sim} N_2(\mathbf{0}, \mathbf{\Sigma}_e).$$

In the correlated probit formulation, we set $\sigma_{e2}^2 = var(\varepsilon_{ij2}) = 1$, for identifiability.

Table 8.3 contains the approximate MLEs obtained for the MLGMM-logit, the MGLMM-probit (i.e., $\boldsymbol{\Sigma}_e$ is diagonal), and the correlated probit model (CPM); note that $\sigma_{e1}^2 = \sigma^2$, for MGLMM-logit and MGLMM-probit. Dose significantly affects both fetal weight and the probability of malformation. An increase of 1g/kg in dose is associated with an average decrease in fetal weight of -0.087 (SE = 0.009) with 95% confidence interval $(-0.105, -0.069)$. This estimate of the dose effect on fetal weight is the same, with precision up to three places after the decimal point, as the estimate from a separate linear mixed model when fetal weight is considered separately from malformation. This estimate is also very similar to the estimates that other authors have observed in the same data set for the effect of dose on fetal weight using different approaches. Fitzmaurice and Laird (1995) considered a GEE logit model for the binary outcome and a multivariate normal model for the continuous outcome, conditional on the binary outcome, and estimated the effect of dose on fetal weight to be -0.089 (SE = 0.008). Dunson et al. (2003) used a Bayesian approach with one common random effect for both outcomes at the cluster level and one common random effect for both outcomes at the subject level. Their estimate was -0.088 (SE = 0.009).

The estimate of Catalano and Ryan (1992) in the multivariate normal model for the continuous outcome was slightly larger than our estimate (i.e., -0.095, SE = 0.008), which is most likely due to the fact that they used litter size as a covariate in the model for fetal weight. On the other hand, the estimate of Lin et al. (2010) was slightly smaller (i.e., -0.084) than ours with smaller SE (i.e., 0.006), probably due to the conditioning on the effect of the binary outcome in the regression model for the continuous outcome. Lin et al. (2010) and Regan and Catalano (1999a) considered models with quadratic dose effects; however, we cannot compare the estimates from our models with linear dose effects to their estimates from models with quadratic dose effects. Our estimate of the within-litter correlation in fetal weight (i.e., 0.48) is consistent with the estimates obtained by Catalano and Ryan (1992) (i.e., 0.48) and by Fitzmaurice and Laird (1995) (i.e., 0.45).

Note that because of the equivalence of the marginal and subject-specific effects for the continuous outcome, we can directly compare the regression parameter estimates across different models; however, this is not the case for the binary malformation outcome. In the logit formulation for the malformation outcome (see Table 8.3), an increase of 1g/kg in dose for a particular fetus within a litter is associated with an increase of 1.653 (SE = 0.194) in the log-odds for malformation. This translates into an odds ratio of 5.223 and a 95% confidence interval $(3.571, 7.639)$. The dose estimate from the joint model of fetal weight and malformation is slightly smaller than the corresponding estimate from the model when malformation is considered separately (i.e., 1.690) and has a smaller SE (i.e., 0.194 compared to 0.206).

Similar to the logit model, the dose effect estimates in the joint models with probit link for malformation in MGLMM (i.e., 0.918, SE = 0.103) and in the correlated probit model (i.e., 0.917, SE = 0.103) were slightly smaller with smaller SEs than the corresponding estimate from the separate model for malformation (i.e., 0.942, SE = 0.110). The better precision of the joint estimates is expected as information from the continuous outcome helps improve estimates for the binary outcome in the joint analysis.

Both dose estimates from the joint and the separate models have subject-specific interpretations, and hence, are expected to be larger than the corresponding marginal estimates from GEE models. Indeed, the dose estimate for malformation in the GEE model of Fitzmaurice and Laird (1995) was 1.198 (SE = 0.146), and the dose estimate in the GEE model of Lin et al. (2010) was 1.163 (SE = 0.118). Both estimates are smaller than our estimate from the logit MGLMM. Our dose effect estimates in the joint models with probit link are slightly larger than the corresponding estimate of Catalano and Ryan (1992) (i.e., 0.831, SE = 0.109) and the corrected estimate of Dunson et al. (2003) (i.e., 0.856).

Note that the estimates of the regression parameters in the logit and probit formulations are different in magnitude, but the Wald test statistics for testing the significance of the parameters are very similar (i.e., $z = 1.653/0.194 = 8.521$ in the logit model versus $z = 0.918/0.103 = 8.913$ in the probit model). This is because of the different scales of the underlying logit and probit latent continuous

Table 8.3 *MLEs (approximate) and their large-sample SEs for EG mice data.*

Parameter	MGLMM-logit		MGLMM-probit		CPM	
	Est	SE	Est	SE	Est	SE
β_{10}	0.944	0.015	0.944	0.015	0.944	0.015
β_{11}	−0.087	0.009	−0.087	0.009	−0.087	0.009
β_{20}	−4.200	0.389	−2.332	0.201	−2.331	0.201
β_{21}	1.653	0.194	0.918	0.103	0.917	0.103
σ_{b1}	0.088	0.007	0.088	0.007	0.089	0.007
σ_{b2}	1.424	0.183	0.789	0.100	0.789	0.100
ρ_b	−0.664	0.095	−0.649	0.096	−0.644	0.097
σ_{e1}	0.095	0.002	0.095	0.002	0.095	0.002
ρ_e	—	—	—	—	−0.036	0.055
BIC	−983.6		−983.3		−979.2	

variables. With respect to the variance components, both random intercepts seem to be necessary for describing the data accurately while the correlation between the errors (i.e., $\rho_e = \sigma_{e12}/\sigma_{e1}$) is not significantly different from zero. Furthermore, according to the Bayesian information criterion (BIC), the MGLMM with logit link fits the data the best among the models we considered. Note that when comparing models with different variance components, we assume that the mean response is specified correctly.

In the correlated probit formulation, the direct interpretation of the regression coefficient for malformation is not as straightforward as in the logit model. However, in the correlated probit formulation, correlations between measurements on different variables are more easily assessed. For example, correlation between weight and latent malformation measured on different fetuses within a litter is −0.271 (SE = 0.052), correlations between weight and latent malformation within the same fetus is −0.292 (SE = 0.058), correlation between weight measurements within litter is 0.464 (SE = 0.043), and between latent malformation within litter is 0.383 (SE = 0.060). These are all statistically significant.

8.4.2 Depression clinical trial

In the depression clinical trial, we consider the correlated probit model, since it allows to seamlessly model ordinal and continuous variables, and since both HAMD and CGI are evaluating depression severity, and are thus likely to be conditionally dependent rather than independent. We denote by Y_{ij1} the measured depression severity at time $j = 0, \cdots, 6$, for subject $i = 1, \cdots, 50$, and assume an underlying continuous outcome Y_{ij2}^* for CGI. We also assume that $I_i = 1$, if the ith subject is assigned to the experimental group, and $I_i = 0$, if this subject is assigned to the control group. We consider fixed effects of treatment, time, treatment-by-time, quadratic time effect, and treatment-by-quadratic time effect; however, since the treatment-by-quadratic time effects were not statistically significant, they were dropped from the final models. The final joint model was defined as follows:

$$Y_{ij1} = \beta_{10} + \beta_{11}I_i + \beta_{12}j + \beta_{13}I_ij + \beta_{14}j^2 + B_{i11} + B_{i12}j + \varepsilon_{ij1},$$
$$Y_{ij2}^* = \beta_{20}I_i + \beta_{21}j + \beta_{23}I_ij + \beta_{24}j^2 + B_{i21} + B_{i22}j + \varepsilon_{ij2},$$

where

$$\begin{pmatrix} B_{i11} \\ B_{i12} \\ B_{i21} \\ B_{i22} \end{pmatrix} \overset{iid}{\sim} N_4(\mathbf{0},\Sigma_b) \quad \text{and} \quad \begin{pmatrix} \varepsilon_{ij1} \\ \varepsilon_{ij2} \end{pmatrix} \overset{iid}{\sim} N_2(\mathbf{0},\Sigma_e).$$

The MLEs of regression parameters and error variance components are provided in Table 8.4. The treatment-by-time effects were not significant (i.e., $t = -1.24$, p-value $= 0.22$ for HAMD, and $t = -1.434$, p-value $= 0.16$ for CGI), suggesting that there were no sizeable differences in treatment effectiveness over time. Note that the error degrees of freedom $df = 46$ are equal to the number of subjects minus the number of random effects, as calculated by default in PROC NLMIXED. The formal test for overall treatment effect in the joint model is a test of whether β_{13} and β_{23} are simultaneously significantly different from zero. This is a general linear hypothesis F-test. The test statistic and p-value for this test are $F = 0.24$, p-value $= 0.79$, on $(2, 46)$ degrees of freedom.

Since the two outcomes in this example are highly correlated, it is of interest to compare parameter estimates for treatment effects in the joint model to those in models fitted separately to HAMD and CGI. Table 8.4 shows that the regression parameter estimates for continuous outcome HAMD and their SEs are almost exactly the same across models, while those for ordinal outcome CGI differ. For example, the estimate of the treatment-by-time effect for CGI in the separate model was larger in absolute value (i.e., $\beta_{23} = -0.28, \text{SE} = 0.18$) than the estimate from the joint model (i.e., $\beta_{23} = -0.22, \text{SE} = 0.15$). This suggests that borrowing information from the "information-richer" HAMD in the inference concerning CGI changes the estimates for CGI. However, borrowing information from the "information-poorer" CGI does not affect inference concerning HAMD. This is useful information for pharmaceutical researchers, as they decide which outcome variable to specify as a primary measure of depression severity in future clinical trials.

The estimated correlations from the joint model between the intercepts for HAMD and CGI, and the slopes for HAMD and CGI were quite high ($r_{13} = 0.82, \text{SE} = 0.09$, and $r_{24} = 0.91, \text{SE} = 0.05$), suggesting that a joint model with common random intercepts and common slopes may adequately describe the data. Such models were considered by Gueorguieva and Sanacora (2006). The estimated polyserial correlations between HAMD and CGI at each time point were all greater than 0.75, thus suggesting that both outcomes indeed capture similar aspects of underlying depression severity. Since in this particular application, HAMD is missing less often than CGI, from a subject-matter perspective, it may be better to consider only HAMD for interpretability.

Software

When the repeatedly measured outcomes are of the same type, several programs can be used for inference. For example, PROC GLIMMIX in SAS uses analytical approximations for inference on GLMMs (and hence, on MGLMMs) with outcomes in the exponential family; MIXOR (Gibbons, 2000) can be used for ordinal outcomes; and several functions in R can be used for outcomes in the exponential family (e.g., glmer from library lme4, glmmML from library glmmML, glmmPQL from library MASS). MPlus allows consideration of models for continuous, count, ordinal, and binary outcomes, and performs Monte Carlo EM estimation (Muthèn and Shedden, 1999). For outcomes of different types that are conditionally independent, the gllamm function in Stata can be used (Rabe–Hesketh *et al.*, 2001). An advantage of this procedure is that it can handle multilevel models. The popular SAS PROC NLMIXED can be used for non-nested Gaussian integration for any conditional likelihood that can be explicitly specified. The NLMIXED codes for the two data examples described in the previous section can be obtained from the author upon request.

8.5   Discussion

In summary, in the current chapter, we presented random effects models for the joint analysis of repeatedly measured discrete and continuous outcomes. The MGLMM can be used for any combinations of outcomes in the exponential family, but it makes a potentially restrictive assumption about the correlation structure of the data. The correlated probit model allows for a very general correlation structure but is restricted to binary, ordinal, and continuous outcomes. Joint models can answer multivariate questions, preserve actual level of tests at nominal level, may improve efficiency in parameter estimates, and provide additional information about the relationship among outcome

Table 8.4 *MLEs (approximate) of regression parameters, error variance, and covariance, and their large-sample SEs for depression data.*

Effect		Separate analyses				Joint model	
		HAMD		CGI			
		Est	SE	Est	SE	Est	SE
HAMD							
Intercept	β_{10}	3.08	0.13	—	—	3.08	0.13
Treatment	β_{11}	−0.21	0.17	—	—	−0.21	0.17
Time	β_{12}	−0.50	0.05	—	—	−0.49	0.05
Treatment×Time	β_{13}	−0.06	0.05	—	—	−0.06	0.07
Time^2	β_{14}	0.04	0.01	—	—	0.03	0.01
Error variance	σ_{e1}^2	0.15	0.01	—	—	0.15	0.01
Error covariance	σ_{e12}	—	—	—	—	0.30	0.03
CGI							
Threshold 1	τ_1	—	—	−6.60	0.61	−6.07	0.55
Threshold 2	τ_2	—	—	−3.79	0.42	−4.03	0.43
Threshold 3	τ_3	—	—	−1.81	0.34	−2.01	0.35
Threshold 4	τ_4	—	—	0.87	0.32	0.90	0.31
Treatment	β_{21}	—	—	−0.41	0.37	−0.52	0.40
Time	β_{22}	—	—	−0.65	0.19	−0.80	0.18
Treatment×Time	β_{23}	—	—	−0.28	0.18	−0.22	0.15
Time^2	β_{24}	—	—	0.02	0.02	0.05	0.02

variables. However, these advantages come at the expense of greater modeling and computational complexity than considering the outcome variables separately.

Random effects models are appropriate when the emphasis is on the subject-specific effects of covariates on the outcomes, i.e., when one is interested primarily in the effect on a subject's response of a change in a covariate. For non-normal response variables, this effect is usually larger than the marginal effect of a change in the covariate in the population of subjects. When interest lies primarily in the marginal effects, a GEE approach or an approach based on copulas may be more appropriate.

Compared to estimates in GEE models, MLEs from random effects models are asymptotically efficient, and thus, parameters are estimated with greater precision. Random effects models are more flexible than GEE in handling missing data, since they provide valid results when data are missing at random (MAR), while GEE models provide valid results when data are missing completely at random (MCAR, Little and Rubin, 2002). If data are informatively missing, then both approaches may be biased. Full MLE as used in random effects models, allows for model comparison and evaluation of model fit. In GEE models, model comparison and evaluation of fit are more difficult because of the lack of a likelihood function. One potential advantage of GEE over random effects models occurs when marginal models are correctly specified but the full likelihood model is incorrectly specified. In this case, estimates from GEE models will be unbiased while estimates from mixed effects models can be biased. However, simulation studies show that the magnitude of this bias is small (Lin *et al.*, 2010). Random effects models also allow derivation of the distribution of one outcome given the other outcomes while this is not possible in GEE models.

The advantages of random effects models, however, come at the expense of greater computational complexity than GEE, as the number of random effects increases. The computational burden of Gaussian quadrature, in particular, increases exponentially with the number of random effects. Practical considerations limit the Gaussian quadrature approach to a 4- or 5-dimensional random effects vector. The Monte Carlo EM algorithm is more flexible and can be used more readily than

Gaussian quadrature, when the dimension of the random effects is high. However, it converges slowly and although there are modifications for speeding it up (e.g., ECM, PX-EM, stochastic approximation EM), they require substantial computational resources. Methods based on analytical approximations are faster but may be biased, especially when data are binary.

Similar to GEE models, copula models focus on estimation of marginal effects. However, while GEE are semiparametric models, the full joint distribution is completely specified in copula models. Therefore, they are fully parametric models and share many of the advantages of random effects models over GEE. Copulas are a very recent development for mixed discrete and continuous responses (de Leon and Wu, 2011; de Leon and Carrière, 2010; Song et al., 2009; Zimmer and Trivedi, 2006) and have not been applied to situations with mixed outcomes that are also repeatedly measured yet. The challenge in extending copula approaches to longitudinal data is in model specification that guarantees proper joint distributions while allowing for different levels of association among outcomes.

One of the reasons for joint modeling rather than modeling the outcomes separately, is the potential efficiency gains for the parameter estimates. However, as can be seen in the data examples that we considered, efficiency gains can be small. This is consistent with previous results (McCulloch, 2008; Gueorguieva and Agresti, 2001; Fitzmaurice and Laird, 1997; Matsuyama and Ohashi, 1997; Lesaffre and Molenberghs, 1991). Indeed, when the mean structure of all outcomes depends on the same set of covariates, efficiency gains by adopting a multivariate approach are negligible. In contrast, when the mean outcomes depend on different covariate sets, large efficiency gains can be realized (Teixeira–Pinto and Normand, 2009).

Efficiency gains can also be observed when there is a large proportion of missing data (Fitzmaurice and Laird, 1997) on some outcomes, and in the case of binary data, when the probability of "success" is close to 0 or 1 (Gueorguieva and Agresti, 2001). McCulloch (2008) demonstrates that when the missing mechanism for one of the outcomes depends on the other outcome and the outcomes are analyzed separately, inconsistent estimates can result. Joint modeling assures the consistency of the estimates when the model is correctly specified.

In some applications, different outcome variables can be observed on a different set of subjects, and this sampling scheme can be taken into account for proper inference. Najita et al. (2009) proposed such an approach for clustered binary and continuous outcomes, when the continuous response is missing by design and missingness depends on the binary trait.

Since MLE ensures consistency and asymptotic normality of the estimates when the model is correctly specified, care must be taken to ensure that model assumptions are reasonable and supported by the data. One particular aspect of model specification that has been demonstrated to significantly affect the parameter estimates is the random effects specification (McCulloch, 2008; Gueorguieva, 2005). A common simplifying assumption is that one or more random effects are shared across outcomes. When reasonable, this assumption may improve efficiency. However, when not supported by the data, it can lead to bias in parameter estimates (Gueorguieva, 2006). Further work is necessary to develop methods for checking all aspects of model assumptions.

In conclusion, random effects models for repeatedly measured discrete and continuous outcomes are intuitively appealing, and flexible both in assessing the effect of covariates on the means and in modeling the correlation structure of the data. They allow simultaneous assessment of treatment or other covariate effects on multiple outcomes of different types and better control of Type I error rates. MLEs for such models can be obtained using existing software; however, computational problems may become severe when there are a large number of random effects. Missing at random data are handled seamlessly, and efficiency can be gained by analyzing outcomes together rather than separately. Such models are useful in many application areas, such as risk assessment with mixed outcomes, and longitudinal studies with outcomes of different types.

Chapter 9

Hierarchical modeling of endpoints of different types with generalized linear mixed models

Christel Faes

9.1 Introduction

In many health surveys, a large collection of measurements are recorded, often containing measurements of different types. Perhaps the most common non-commensurate situation is that of a continuous, often assumed normally distributed, outcome, and a binary or ordinal outcome. For example, in dose-toxicology experiments, both the body temperature (typically recorded on a continuous scale) and specific reflexes (recorded on a binary or ordinal scale) are measured as indicators for toxicity (Faes *et al.*, 2008). In addition, it is not uncommon that these measurements are recorded repeatedly in time, for example, in a repeated-dose toxicity experiment; or that sampling units are clustered, for example, in surveys where individuals within the same households are interviewed. In this chapter, our general interest is in the analysis of multivariate hierarchical, also called multivariate multi-level, data. The strength of association between the different endpoints as well as the association of measurements within the same sampling unit might be important questions of interest in this case.

When the different endpoints follow different distributions, there is not a standard methodology to use, and several approaches to construct a joint model exist (for an overview, see Chapter 1). A first approach is to directly specify a multivariate distribution for the endpoints. While the multivariate normal distribution is well-known for continuous outcomes (Johnson and Wichern, 2002), multivariate distributions for endpoints of different types are not widely available. In recent years, the derivation of a joint distribution from the marginal distributions and a copula function, which describes the association between the components, has gained a lot of popularity, and has been used also for the joint analysis of clustered or longitudinal endpoints of different types (de Leon and Wu, 2011; Faes *et al.*, 2004; Burzykowski *et al.*, 2001; Geys *et al.*, 2001), see Chapter 10. A second approach is the use of a factorization model, where one endpoint is conditioned on the other endpoint, as presented in Chapters 5 and 6. In the situation of continuous and discrete outcomes, the joint density is factorized in a marginal component and a conditional component, where the conditioning can be done either on the discrete or on the continuous outcome. Conditional models are discussed by Cox and Wermuth (1994, 1992), Little and Schluchter (1985), Krzanowski (1988), Olkin and Tate (1961), and Tate (1954). Extensions of these models in the context of clustered data are discussed by Fitzmaurice and Laird (1995) and Catalano and Ryan (1992). A drawback of these factorization models is that they are difficult to generalize to more than two endpoints, and that the intra-subject correlation cannot be estimated directly. A third approach, as presented in Chapter 8, is the use of generalized linear mixed models (GLMMs) for the analysis of multivariate multi-level data, and as discussed in Section 9.2. Shared parameter models and correlated random effects models are special cases in this general framework, and can be used for the analysis of multivariate multi-level data of different types. The GLMM approach is a general paradigm and has been used in a wide variety of ways to model correlated endpoints of mixed types, see e.g., Lin *et al.* (2010), de Leon and Carrière Chough (2010), Najita *et al.* (2009), Teixeira–Pinto and Normand (2009), McCulloch (2008), George *et al.* (2007), Faes *et al.* (2006), Gueorguieva and Sanacora (2006), Gueorguieva and Agresti (2001), and Regan and Catalano (2002; 1999*a*). In this chapter, our focus is on the GLMM approach to model both the correlation between the different endpoints and the correlation as a result of clustering (or repeated measurements), via the specification of shared and correlated random effects and correlated residual errors.

Extensions to more than two outcomes in the GLMM context is straightforward, but computational problems arise as the number of outcomes increases. To avoid maximization of the full likelihood expression, inference can be based on non-likelihood methods. The pseudo-likelihood methodology as proposed by Arnold and Strauss (1991), also called composite likelihood, is a general and flexible framework. The idea is to replace the full likelihood by a function which is easier to evaluate to reduce the computational costs. A convenient pseudo-likelihood function in this setting is one that replaces the joint density by the product of all pairwise densities, the so-called pairwise pseudo-likelihood. It is in this spirit that Fieuws and Verbeke (2006) propose a pairwise modeling strategy in which all possible pairs are modeled separately. The estimates of the separate bivariate model fits are then combined into a single set of estimates for the joint model. In Faes *et al.* (2008), a similar approach is applied based on the pairwise pseudo-likelihood, but the pseudo-likelihood expression is maximized at once, leading directly to the parameter estimates of interest. These pairwise modeling strategies are described in Section 9.4.

9.2 Multivariate multi-level models

Random effects models are probably the most frequently used models to analyze hierarchical data. But random effects can also be used to analyze multivariate data. A general form of the GLMM allows the joint modeling of outcomes of different types, and accounts for the association due to the hierarchy in the data.

Assume we have K sequences of n_i outcomes for sampling unit $i = 1, \cdots, N$. We denote the kth sequence for unit i by $\mathbf{Y}_{ki} = (Y_{ki1}, Y_{ki2}, \cdots, Y_{kin_i})^\top$, with Y_{kij} referring to the jth measurement of outcome k, measured on unit (subject) i. The sequences can refer to measurements recorded repeatedly in time on the same subject or measurements of individuals within a cluster. Possibly, the K sequences of outcomes are of different types, e.g., continuous, categorical, time-to-event; and have different distributions, e.g., normal, Bernoulli, Poisson, Weibull.

For the multivariate response vector $\mathbf{Y}_i = (\mathbf{Y}_{1i}^\top, \mathbf{Y}_{2i}^\top, \cdots, \mathbf{Y}_{Ki}^\top)^\top$, we assume a GLMM of the form (Fitzmaurice *et al.*, 2009; Molenberghs and Verbeke, 2005)

$$\mathbf{Y}_i = \boldsymbol{\mu}_i(\boldsymbol{\eta}_i) + \boldsymbol{\varepsilon}_i, \qquad (9.1)$$

where the data is decomposed into the mean $\boldsymbol{\mu}_i$ and an appropriate error term $\boldsymbol{\varepsilon}_i$. The mean $\boldsymbol{\mu}_i$ is specified in terms of fixed and random effects

$$\boldsymbol{\mu}_i = \mathbf{h}(\boldsymbol{\eta}_i) = \mathbf{h}(\mathbf{X}_i \boldsymbol{\beta} + \mathbf{Z}_i \mathbf{B}_i),$$

where $\mathbf{X}_i$ and $\mathbf{Z}_i$ are $(Kn_i \times p)$- and $(Kn_i \times q)$-dimensional matrices of known covariate values, respectively, $\boldsymbol{\beta}$ is a vector of unknown fixed regression coefficients, and $\mathbf{B}_i \sim N_s(\mathbf{0}, \mathbf{D})$ are the unit-specific random effects, with $s = Kn_i q$. The model is written in its most general form where both the mean and error term are allowed to change with the nature of the endpoints, and can account for various dependencies in the data. The components of the inverse link function $\mathbf{h}(\cdot)$ depend on the nature of the outcomes in $\mathbf{Y}_i$. The components of the residual error $\boldsymbol{\varepsilon}_i$ have the appropriate distributions, depending on the type of the endpoint, with $(Kn_i \times Kn_i)$-dimensional covariance matrix

$$var(\boldsymbol{\varepsilon}_i) = \boldsymbol{\Sigma}_i \simeq \boldsymbol{\Xi}_i^{1/2} \mathbf{A}_i^{1/2} \mathbf{R}_i(\boldsymbol{\alpha}) \mathbf{A}_i^{1/2} \boldsymbol{\Xi}_i^{1/2},$$

where $\mathbf{A}_i$ is a diagonal matrix containing the variances from the generalized linear model specification of Y_{kij}, given the random effects $\mathbf{B}_i = \mathbf{0}$, i.e., with diagonal elements $v(\mu_{kij}|\mathbf{B}_i = \mathbf{0})$, and

$v(\cdot)$ is the variance function corresponding to the exponential family distribution. Likewise, $\boldsymbol{\Xi}_i$ is a diagonal matrix with the overdispersion parameters along the diagonal, and $\mathbf{R}_i(\boldsymbol{\alpha})$ is a correlation matrix. In a standard GLMM, $\boldsymbol{\varepsilon}_i$ is assumed to be uncorrelated and hence, does not lead to additional parameters in $\mathbf{R}_i(\boldsymbol{\alpha})$. For continuous endpoints, the decomposition is straightforward with the inverse link function $\mathbf{h}(\cdot)$ and variance function $v(\cdot)$ corresponding to the identity function, and overdispersion parameter equal to the residual variance σ^2. For binary endpoints with a logistic link function, the inverse link function $\mathbf{h}(\cdot)$ is given by the expit-function (i.e., inverse-logit function) and the error terms ε_{ij} equal to $1 - \pi_{ij}$ with probability π_{ij}, and $-\pi_{ij}$ with probability $1 - \pi_{ij}$. Based on this general formulation, one can derive a general first-order approximate expression for the covariance matrix of $\mathbf{Y}_i$ (Molenberghs and Verbeke, 2005), which is useful to investigate the association between the endpoints:

$$\mathbf{V}_i \;=\; var(\mathbf{Y}_i) \;\simeq\; \boldsymbol{\Delta}_i \mathbf{Z}_i \mathbf{D} \mathbf{Z}_i^\top \boldsymbol{\Delta}_i^\top + \boldsymbol{\Sigma}_i, \qquad (9.2)$$

with

$$\boldsymbol{\Delta}_i \;=\; \left. \frac{\partial \boldsymbol{\mu}_i}{\partial \boldsymbol{\eta}_i} \right|_{\mathbf{B}_i = \mathbf{0}}.$$

Dependence between the bivariate outcomes and dependence within the outcomes can be introduced in a variety of ways, leading to different models. In a first possible model, all dependencies amongst the outcomes can be introduced by means of the residual covariance matrix $\boldsymbol{\Sigma}_i$, or via the specification of a residual correlation $\mathbf{R}_i(\boldsymbol{\alpha})$. This model is a marginal model, because no random effects are incorporated in it. A second example is a model where the dependencies are modeled via a shared random effect. A third possible model assumes different, but correlated, random effects for every outcome, leading to a correlated random effects model. A fourth model uses random effects to account for the hierarchical structure in the data, but uses correlated error terms to account for the associations between the different outcomes.

9.3 Special cases

As an illustration, bivariate hierarchical models are explained for the special case of two endpoints, of which one is continuous and the other is binary.

9.3.1 Marginal model

For the special case of continuous and binary endpoints, a model for $\mathbf{Y}_i = (\mathbf{Y}_{1i}^\top, \mathbf{Y}_{2i}^\top)^\top$ with no random effects, has building blocks of the form

$$\begin{pmatrix} Y_{1ij} \\ Y_{2ij} \end{pmatrix} \;=\; \begin{pmatrix} \alpha_0 + \alpha_1 x_{1ij} \\ \dfrac{\exp(\beta_0 + \beta_1 x_{2ij})}{1 + \exp(\beta_0 + \beta_1 x_{2ij})} \end{pmatrix} + \begin{pmatrix} \varepsilon_{1ij} \\ \varepsilon_{2ij} \end{pmatrix}, \qquad (9.3)$$

in which the first component of the inverse link function $\mathbf{h}(\cdot)$ equals the identity link, and the second component corresponds to the logit-link. The correlations between the endpoints $\mathbf{Y}_{1i}$ and $\mathbf{Y}_{2i}$ are specified via the residual error variance

$$var\left\{ \begin{pmatrix} \boldsymbol{\varepsilon}_{1i} \\ \boldsymbol{\varepsilon}_{2i} \end{pmatrix} \right\},$$

with $\boldsymbol{\varepsilon}_{ki} = (\varepsilon_{ki1}, \cdots, \varepsilon_{kin_i})^\top$, by the specification of a correlation matrix $\mathbf{R}_i(\boldsymbol{\alpha})$, allowing each pair of outcomes to have their own correlation coefficient. The correlations in $\mathbf{R}_i(\boldsymbol{\alpha})$ are exactly the correlations between any two endpoints Y_{1ij} and Y_{2ik}.

While such a marginal model, with a fully unstructured $2n_i \times 2n_i$ covariance matrix, is very appealing because of ease of interpretation, it can become computationally very demanding, especially when the number n_i of measurements per subject gets large. Another specification of the correlation structure of interest could be as a product of an unstructured 2×2 correlation matrix and a structured $n_i \times n_i$ correlation matrix (e.g., AR(1), compound symmetry), where the first correlation matrix corresponds to the multivariate structure in the data, and the second corresponds to the multi-level structure. However, these covariance structures are not readily available in current software packages.

9.3.2   Shared random effects model

Another approach is to use a shared random parameter model. More specifically, for a mixed continuous-binary setting, the GLMM (9.1) can be written in the form

$$
\begin{pmatrix} Y_{1ij} \\ Y_{2ij} \end{pmatrix} = \begin{pmatrix} \alpha_0 + \alpha_1 x_{1ij} + \lambda B_i \\ \frac{\exp(\beta_0 + \beta_1 x_{2ij} + B_i)}{1 + \exp(\beta_0 + \beta_1 x_{2ij} + B_i)} \end{pmatrix} + \begin{pmatrix} \varepsilon_{1ij} \\ \varepsilon_{2ij} \end{pmatrix}, \tag{9.4}
$$

where B_i denotes a shared random effect for the two outcomes. The residual errors are assumed to be uncorrelated, and hence do not lead to additional parameters. Note that a scale parameter λ in the continuous component of the model is included, given that the continuous and binary outcomes are measured on different scales. For this model, the variance of $\mathbf{Y}_i$ can be derived from (9.2), from which the approximate correlation between the two outcomes Y_{1i} and Y_{2i} can be calculated. The correlation between a continuous and discrete endpoint measured on the same unit i is

$$
\rho_{12} \approx \frac{v_{2ij} \lambda \tau^2}{\sqrt{\lambda^2 \tau^2 + \sigma^2} \sqrt{v_{2ij}^2 \tau^2 + v_{2ij}}}, \tag{9.5}
$$

where $v_{2ij} = \pi_{2ij}(\mathbf{B}_i = 0)\left\{1 - \pi_{2ij}(\mathbf{B}_i = 0)\right\}$. It can be observed that this correlation depends on the fixed effects through the variance function v_{i2}. The correlation between two continuous measurements on the same unit is

$$
\rho_{11} \approx \frac{\lambda \tau^2}{\lambda^2 \tau^2 + \sigma^2}, \tag{9.6}
$$

and that between two binary measurements on the same unit is

$$
\rho_{22} \approx \frac{v_{2ij}^2 \tau^2}{v_{i2}^2 \tau^2 + v_{i2}}. \tag{9.7}
$$

9.3.3   Correlated random effects model

Another approach is to specify a correlated random intercepts model, with a general covariance matrix $\mathbf{D}$ for the random effects. This model can be written in the following form:

$$
\begin{pmatrix} Y_{1ij} \\ Y_{2ij} \end{pmatrix} = \begin{pmatrix} \alpha_0 + \alpha_1 x_{1ij} + B_{1i} \\ \frac{\exp(\beta_0 + \beta_1 x_{2ij} + B_{2i})}{1 + \exp(\beta_0 + \beta_1 x_{2ij} + B_{2i})} \end{pmatrix} + \begin{pmatrix} \varepsilon_{1ij} \\ \varepsilon_{2ij} \end{pmatrix}, \tag{9.8}
$$

where the random effects B_{1i} and B_{2i} are normally distributed as

$$
\begin{pmatrix} B_{1i} \\ B_{2i} \end{pmatrix} \sim N_2\left(\mathbf{0}, \begin{pmatrix} \tau_1^2 & \rho \tau_1 \tau_2 \\ \rho \tau_1 \tau_2 & \tau_2^2 \end{pmatrix}\right) \tag{9.9}
$$

and where ε_{1ij} and ε_{2ij} are independent. The random effects B_{1i} and B_{2i} are used to accommodate the longitudinal or clustered structure of the data. The correlation ρ_{12} between the continuous and binary endpoints is induced by the incorporation of a correlation ρ between the two random effects. The covariance matrix of two observations Y_{1ij} and Y_{2ij} can be derived from (9.2), resulting in

$$\mathbf{V}_{ij} = \begin{pmatrix} \tau_1^2 + \sigma^2 & \rho\,\tau_1\tau_2 v_{2ij} \\ \rho\,\tau_1\tau_2 v_{2ij} & v_{2ij}^2\tau_2^2 + v_{2ij} \end{pmatrix}.$$

Thus, the correlation ρ_{12} between the binary and continuous outcomes of the same unit follows from the fixed effects and variance components, and is approximately equal to

$$\rho_{12} \approx \frac{\rho\,\tau_1\tau_2 v_{2ij}}{\sqrt{\tau_1^2 + \sigma^2}\sqrt{v_{2ij}^2\tau_2^2 + v_{2ij}}}.$$

In the case of conditional independence ($\rho \equiv 0$), the approximate marginal correlation function ρ_{12} also equals zero. In case $\rho \equiv 1$, this model reduces to the shared parameter model. The correlation between the two endpoints Y_{1ij} and Y_{2ij} is then given by (10.21), in which the scale factor λ is equal to τ_1/τ_2. The correlation between the continuous endpoints of the same unit can be derived in a similar way, and is given by

$$\rho_{11} \approx \frac{\tau_1^2}{\tau_1^2 + \sigma^2},$$

the typical expression for the intra-cluster correlation (ICC). For two binary endpoints, we have an intra-class correlation given by

$$\rho_{22} \approx \frac{v_{2ij}^2\tau_2^2}{v_{2ij}^2\tau_2^2 + v_{2ij}}.$$

9.3.4   *Independent random effects and correlated errors model*

Another approach is to use a random intercepts model for each endpoint to account for the hierarchy in the data, and in addition, assume a correlated residual error structure to account for the association between the endpoints. In this model, random effects are introduced at the level of the linear predictor after application of the link function to account for the multi-level structure in the data, while the residual correlation is introduced at the level of $\boldsymbol{\varepsilon}_i$ to accommodate the association among the binary and continuous measurements. The model is also of the form (9.8), but where B_{1i} and B_{2i} are independent zero-mean normally distributed random effects with variances τ_1^2 and τ_2^2, respectively. The residual error for $(Y_{1ij}, Y_{2ij})^\top$ is assumed to have covariance matrix

$$\begin{pmatrix} \sigma^2 & \rho\sigma\sqrt{v_{2ij}} \\ \rho\sigma\sqrt{v_{2ij}} & v_{2ij} \end{pmatrix}. \tag{9.10}$$

Here, ρ now denotes the correlation between ε_{1ij} and ε_{2ij}. The variances of the jth measurements of the continuous and binary endpoints for subject i, can be derived from the covariance matrix (9.2), and is approximately equal to

$$\mathbf{V}_{ij} \approx \begin{pmatrix} \tau_1^2 + \sigma^2 & \rho\sigma\sqrt{v_{2ij}} \\ \rho\sigma\sqrt{v_{2ij}} & v_{2ij}^2\tau_2^2 + v_{2ij} \end{pmatrix}.$$

As a result, the approximate correlation between the two endpoints is equal to

$$\rho_{12} \approx \frac{\rho\sigma\sqrt{v_{2ij}}}{\sqrt{\tau_1^2 + \sigma^2}\sqrt{v_{2ij}^2\tau_2^2 + v_{2ij}}}.$$

In the special case of no random effects, the model is a marginal model and the correlation reduces to ρ. Further, ρ_{12} equals zero in the case of conditional independence. The intra-class correlation between the continuous endpoints within the same unit is given by

$$\rho_{11} \approx \frac{\tau_1^2}{\tau_1^2 + \sigma^2},$$

and between the binary endpoints within the same unit is

$$\rho_{22} \approx \frac{v_{2ij}^2 \tau_2^2}{v_{2ij}^2 \tau_2^2 + v_{2ij}}.$$

9.4   Likelihood inference

In the general case of $K > 2$, the sequences can be simultaneously modeled by specifying a joint distribution for the random effects, in analogy with the correlated random effects model as presented in the previous section, but with a $(K \times q)$-dimensional random effects vector $\mathbf{B}_i$. However, computational problems often arise when K is large, owing to the $(K \times q)$-dimensional integral involved, especially when outcomes are of different types. In this case, rather than considering the full likelihood contribution of each subject i, one can avoid the computational complexity by using a pseudo-likelihood approach, similar to the pairwise modeling approach proposed by Arnold and Strauss (1991). Varin *et al.* (2011) provide a state-of-the-art overview of the methodology. The methodology has been used for the analysis of multivariate observations by, e.g., Faes *et al.* (2004), Geys *et al.* (2001), and Poon and Lee (1987).

9.4.1   Pseudo-likelihood estimation

Faes *et al.* (2008) propose to replace the full likelihood contribution of subject i by the pairwise pseudo-likelihood function

$$PL_i = \prod_{k=1}^{K-1} \prod_{k'=k+1}^{K} L_{kk'i}(\mathbf{\Theta}; \mathbf{y}_{ki}, \mathbf{y}_{k'i}),$$

where each contribution $L_{kk'i}(\mathbf{\Theta}; \mathbf{y}_{ki}, \mathbf{y}_{k'i})$ is equal to the bivariate likelihood function for outcomes k and k', and $\mathbf{\Theta}$ is the vector of all model parameters. This methodology is similar to the method proposed by de Leon (2005). In this way, we have simplified the $(K \times q)$-dimensional integration problem to a product of $(2 \times q)$-dimensional integrations. In practice, this is achieved by restructuring the data in all possible pairs of outcomes, and making a working assumption that, conditional on the random effects, all combinations of pair (k,k') and subject i are independent. Inference for $\mathbf{\Theta}$ follows from pseudo-likelihood theory, and is based on a sandwich-type robust variance estimate (Arnold and Strauss, 1991). The asymptotic covariance matrix of the estimate $\widehat{\mathbf{\Theta}}$ is given by

$$\mathbf{J}(\mathbf{\Theta})^{-1}\mathbf{K}(\mathbf{\Theta})\mathbf{J}(\mathbf{\Theta})^{-1}, \tag{9.11}$$

where $\mathbf{J}(\mathbf{\Theta})$ is a matrix with elements defined by

$$J_{pq} = -\sum_{k=1}^{K-1} \sum_{k'=k+1}^{K} E\left(\frac{\partial^2}{\partial \mathbf{\Theta}_p \partial \mathbf{\Theta}_q} \log L_{ikk'}(\mathbf{\Theta}; \mathbf{y}_{ik}, \mathbf{y}_{ik'}) \right),$$

and $\mathbf{K}(\mathbf{\Theta})$ is a symmetric matrix with elements

$$K_{pq} = -\sum_{k=1}^{K-1} \sum_{k'=k+1}^{K} E\left(\frac{\partial}{\partial \mathbf{\Theta}_p} \log L_{ikk'}(\mathbf{\Theta}; \mathbf{y}_{ik}, \mathbf{y}_{ik'}) \frac{\partial}{\partial \mathbf{\Theta}_q} \log L_{ikk'}(\mathbf{\Theta}; \mathbf{y}_{ik}, \mathbf{y}_{ik'}) \right).$$

This method simplifies the computations by reducing the dimensions of integration, and results in asymptotically unbiased estimates.

As an alternative, computations can further be reduced by using a pairwise modeling methodology as proposed by Fieuws and Verbeke (2006). They use the pseudo-likelihood function

$$PL_i \; = \; \prod_{k=1}^{K-1} \prod_{k'=k+1}^{K} L_{ikk'}(\boldsymbol{\Theta}_{k,k'}; \mathbf{y}_{ik}, \mathbf{y}_{ik'}), \tag{9.12}$$

where $\boldsymbol{\Theta}_{k,k'}$ represents the vector of parameters in the bivariate model for the pair of outcomes (k,k'). With this pseudo-likelihood function, the bivariate likelihood functions can be maximized separately, resulting in different parameter estimates of $\boldsymbol{\Theta}^* = (\boldsymbol{\Theta}_{1,2}, \boldsymbol{\Theta}_{1,3}, \cdots, \boldsymbol{\Theta}_{K-1,K})$ for different pairs of outcomes. Estimates of parameters $\boldsymbol{\Theta}$ of the joint model are then derived by taking averages over estimates obtained from the different bivariate models, or $\widehat{\boldsymbol{\Theta}} = \mathbf{A}\widehat{\boldsymbol{\Theta}}^*$ (for an appropriate weight matrix $\mathbf{A}$). The covariance matrix of $\widehat{\boldsymbol{\Theta}}$ is given by $\mathbf{A}\Sigma(\widehat{\boldsymbol{\Theta}}^*)\mathbf{A}^\top$, where $\Sigma(\boldsymbol{\Theta}^*)$ is the covariance matrix of $\widehat{\boldsymbol{\Theta}}^*$. This method further simplifies a computationally very challenging problem, making the approach quite attractive for high-dimensional multivariate modeling.

Both the pseudo-likelihood method proposed by Faes *et al.* (2008) and the pairwise method proposed by Fieuws and Verbeke (2006) have important advantages. While the first one is slightly more principled, avoids post-hoc combination of various estimates of the same parameter, and ensures that resulting covariance matrices are positive definite, it is also computationally considerably more complex, both in terms of computation time requirements, as well as with regard to convergence. For a typical data set with, for example, 4 endpoints of different types, the full likelihood will be computationally (almost) impossible to evaluate, and the first method would take about 5 times longer than the second pairwise approach.

9.5   Applications

In this section, two applications are used to illustrate the described methodology. The first application is a developmental toxicity study (similarly analyzed in Chapters 8 and 10), in which fetuses of a dam are investigated for adverse developmental effects of ethylene glycol. The second application is a health interview survey, where individuals within households are questioned about their social contact network. For both analyses, the procedures PROC NLMIXED or PROC GLIMMIX in SAS are used.

9.5.1   Toxicology of ethylene glycol

The U. S. National Toxicology Program develops scientific information about potentially toxic chemicals that can be used for protection of public health and prevention of chemically induced diseases. One of the goals is to determine the developmental toxicity of chemicals to which a wide segment of the population is exposed. A developmental toxicity study of ethylene glycol conducted by the Research Triangle Institute, under contract with the National Toxicology Program of the U.S., is investigated (Price *et al.*, 1985). Ethylene glycol (EG) is a high-volume industrial chemical with diverse applications. It is used to make antifreeze and de-icing solutions for cars, airplanes and boats, as a solvent in the paint and plastics industries, and to make polyester compounds.

Price *et al.* (1985) describe a study in which timed-pregnant CD-1 mice were dosed by gavage with EG in distilled water. Dosing occurred during the period of major organogenesis and structural development of fetuses. The doses selected for the study were 0, 750, 1500, and 3000 mg/kg/day, with 25, 24, 23 and 23 timed-pregnant mice randomly assigned to each of these dose groups. Figure 9.1 presents the probability of an affected fetus per cluster and the average fetal birth weight per cluster; a summary of the data is also presented in Table 8.1 in Chapter 8. The observed overall malformation rate in the dose groups increases quickly from 0.3% in the control group to 57.1%

in the highest dose group. The average fetal weight decreases monotonically with dose, declining from 0.972 g on average in the control group to 0.7404 g on average in the highest dose group.

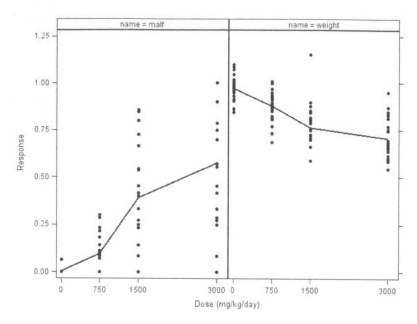

Figure 9.1 *Observed malformation rate and average birth weight per litter (dots) and per dose group (line) for EG data.*

9.5.1.1   Separate analysis of endpoints

In this section, we investigate the effects of the chemical substance for both endpoints separately. For each endpoint, two separate models are used. First, a marginal model is used, where the ICC is specified via the residual error variance. The mean structure for birth weight Y_{1ij} for fetus j in dam i, is specified as

$$E(Y_{1ij}) \quad = \quad \alpha_0 + \alpha_1 d_i,$$

with d_i the dosage given to dam i. The mean structure for malformation indicator Y_{2ij} is specified using a logit-link, as

$$E(Y_{2ij}) \quad = \quad \frac{\exp(\beta_0 + \beta_1 d_i)}{1 + \exp(\beta_0 + \beta_1 d_i)}.$$

An exchangeable correlation structure is assumed for the outcomes of fetuses within a litter. Second, a random intercept model is assumed to account for effects in common to all fetuses within a litter, with conditional means

$$E(Y_{1ij}|B_{1i}) \quad = \quad \alpha_0 + \alpha_1 d_i + B_{1i} \quad \text{and} \quad E(Y_{2ij}|B_{2i}) \quad = \quad \frac{\exp(\beta_0 + \beta_1 d_i + B_{2i})}{1 + \exp(\beta_0 + \beta_1 d_i + B_{2i})},$$

for birth weight and malformation status, respectively. The random effects B_{1i} and B_{2i} are normally distributed with mean zero and variances τ_1^2 and τ_2^2, respectively. Parameter estimates and their standard errors (SEs) are displayed in Table 9.1.

The exposure seems to result in a significant increase of the malformation probability and in a

Table 9.1 *Estimates and their SEs from separate analyses of malformation and fetal birth weight from EG data, based on a marginal model and a random intercepts (RI) GLMM.*

Effect		Marginal Model		RI Model	
		Est	SE	Est	SE
Birth weight					
Intercept	α_0	0.952	0.011	0.952	0.014
Dose	α_1	−0.087	0.008	−0.087	0.008
Error variance	σ^2	0.006	0.000	0.006	0.000
Variance component	τ_1^2			0.008	0.001
ICC		0.578	—	0.572	0.041
Malformation					
Intercept	β_0	−2.886	0.239	−4.251	0.437
Dose	β_1	1.147	0.136	1.780	0.219
Overdispersion	ϕ	0.600	0.028		
Variance component	τ_2^2			2.471	0.663
ICC		0.273	—	0.429	0.066

significant decrease of mean fetal body weight. While the estimated fixed effect parameters based on the two models are very similar for the continuous endpoint, we note large differences between parameter estimates for the binary endpoint. This follows from the fact that the parameters in both models have different interpretations. The parameters in the marginal model represent the population-averaged dose effect, while the parameters in the random intercepts model represent the individual-specific effect. The intra-class correlation of the malformation outcome for fetuses from the same litter, is also significant. Also, the body weight seems to be correlated for fetuses within a cluster.

9.5.1.2 *Bivariate clustered data model*

We now focus on the bivariate hierarchical model, modeling both the malformation rate and fetal body weight. We consider models with a shared random intercept, as in (9.4), with correlated random intercepts, as in (9.8) and (9.9), and with uncorrelated random intercepts, as in (9.10). In the uncorrelated random intercepts model, correlation is added in the residual structure. The marginal model with an unstructured residual error covariance is not feasible in this setting because the litter sizes are different per dam. The marginal model could be extended by allowing a block-structured covariance matrix, with blocks referring to an exchangeable correlation structure within each endpoint and an association between the two endpoints. However, since there is no readily available software for this model, we do not consider this model. Table 9.4 shows estimates and SEs corresponding to the two fitted models.

The first column in Table 9.2 shows parameter estimates for the shared parameter model. This model is a special case of the correlated random intercepts model, where the correlation is set equal to 1. We see a significant dose-effect for both malformation and birth weight. The ICC is estimated as 0.2504 (SE = 0.0494) and 0.5648 (SE = 0.0418), for malformation and birth weight, respectively. The correlation between malformation and birth weight is negative, estimated as 0.0307 (SE = 0.0032).

The second column in Table 9.2 shows parameter estimates of the uncorrelated random intercepts model. The correlation parameter represents the correlation in the residual error structure, and thus, the correlation between malformation and birth weight. A negative correlation between the binary and continuous endpoints is estimated.

The third column presents results for the correlated random intercepts model. The fixed effects parameter estimates and their SEs for the continuous endpoints are approximately the same for all

Table 9.2 *Estimates and SEs from joint analysis of malformation and fetal birth weight from EG data, based on uncorrelated and correlated random intercepts model (RIM).*

Effect		Shared RIM		Uncorrelated RIM		Correlated RIM	
		Est	SE	Est	SE	Est	SE
Birth Weight							
Intercept	α_0	0.952	0.014	0.952	0.014	0.952	0.014
Dose	α_1	−0.087	0.008	−0.087	0.008	−0.087	0.008
Error variance	σ^2	0.006	0.000	0.006	0.000	0.006	0.000
Variance component	τ_1^2	1.099	0.289	0.008	0.001	0.007	0.001
Scale parameter	λ	−0.082	0.009	—	—	—	—
Malformation							
Intercept	β_0	−3.768	0.292	−3.992	0.337	−4.331	0.408
Dose	β_1	1.485	0.146	1.638	0.014	1.749	0.206
Overdispersion	ϕ	—	—	0.591	0.027	—	—
Variance component	τ_2^2	1.099	0.289	2.291	0.478	2.259	0.586
Association							
Correlation	ρ	—	—	−0.128	0.032	−0.687	0.088

considered models, while the fixed effects estimates for the binary endpoint differ somewhat more. The correlation parameter ρ in this model is larger in magnitude, but represents the correlation between the random effects. The ICC is estimated as 0.4071 (SE = 0.0626) for malformation and 0.5713 (SE = 0.0414) for birth weight. The correlation between the continuous and binary outcomes is estimated to be −0.3313 (SE = 0.0560).

The models are consistent in the estimation of the effect of dose on the endpoints. In all models, the dose significantly reduces birth weight and increases risk of a malformation. A negative association between birth weight and presence/absence of malformation is observed.

9.5.2 Health interview survey

In 2001, a health interview survey (HIS) was conducted in Belgium. The Belgian HIS uses a complex multistage probability sampling scheme to select individuals for interviewing. In order to reach respondents, one first selects towns (municipalities), then a number of households within towns, and finally, a number of household members within a household. A consequence of such a sampling scheme is that a number of respondents stem from the same household and the same town. One then cannot ignore the possibility of individuals within families being more alike than between families, with the same, to a lesser extent, holding for towns.

Information on the social health of individuals is obtained in this survey. Social health variables were measured via a self-administered questionnaire and are limited to the 15-years-and-older population in Belgium. Via a set of questions, the following indicators of social health are investigated: quality of social support (binary), number of close relationships (continuous), and satisfaction about social relationships (binary). Satisfaction is recorded as a binary indicator, equal to 1, if satisfied, and 0, otherwise. Quality of social support is reported as 1, if of high quality, and 0, otherwise. The average profiles as function of age are given in Figure 9.2. Household sizes vary from 1 up to 12, with frequencies given in Table 9.3.

It is investigated whether age and gender have an impact on the social health. Further, it is of interest to know if there is an association between individuals within households. We also want to know if satisfaction and quality of support are related with the number of social contacts.

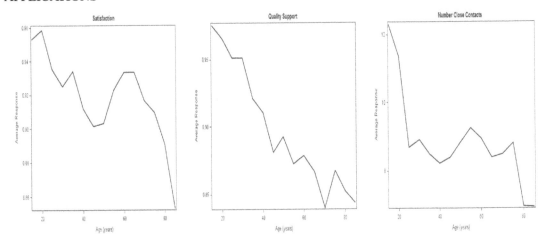

Figure 9.2 *Plots of average satisfaction, quality of support, and number of social contacts, as functions of age for HIS.*

Table 9.3 *Distribution of household sizes in sampled population for Belgian HIS.*

Household size	Frequency	Percent
1	1,538	18.46
2	2,673	32.08
3	1,641	19.70
4	1,830	21.96
5	440	5.28
6	140	1.68
7	50	0.60
> 8	20	0.44

9.5.2.1 Bivariate hierarchical model

We now focus on bivariate analysis of quality of social support and number of close relationships. The association of the social health of individuals within the same household is taken into account in the following analysis.

We consider a model for number of close relationships Y_{1ij} for indiviual j in household i, and the quality of social support Y_{2ij}, with a shared random effect, specified as

$$\begin{pmatrix} Y_{1ij} \\ Y_{2ij} \end{pmatrix} = \begin{pmatrix} \alpha_0 + \alpha_1 x_{1ij} + \alpha_2 x_{2ij} + \lambda B_i \\ \dfrac{\exp(\beta_0 + \beta_1 x_{1ij} + \beta_2 x_{2ij} + B_i)}{1 + \exp(\beta_0 + \beta_1 x_{1ij} + \beta_2 x_{2ij} + B_i)} \end{pmatrix} + \begin{pmatrix} \varepsilon_{1ij} \\ \varepsilon_{2ij} \end{pmatrix}, \qquad (9.13)$$

and a similar model, but with correlated random intercepts model, as in (9.8) and (9.9), with x_{1ij} and x_{2ij} representing the age and gender of the individual. Table 9.4 shows the estimates and their SEs corresponding to each model. The first column shows results for the shared parameter model. We can see a significant decrease of quality of social support with age. Gender does not have a significant effect on the quality of support. The ICC for support of individuals within a household is as high as 0.442 (SE = 0.043). The number of close relationships decreases also as a function of age, and males have significantly more close relationships than females do. The ICC of the number of relationships is 0.138 (SE = 0.011), which is somewhat smaller than that for the binary response.

Table 9.4 *Estimates and their SEs from joint analysis of number of close relationships (continuous endpoint) and satisfaction about social contact (binary endpoint) from HIS. Asterisk indicates significance at 5% level.*

Effect		Shared RIM		Correlated RIM	
		Est	SE	Est	SE
Number close relationships					
Intercept	α_0	9.458	0.269*	9.502	0.269*
Age	α_1	−0.432	0.066*	−0.454	0.068*
Gender	α_2	1.528	0.198*	1.597	0.197*
Error variance	σ^2	69.295	1.614*	68.928	1.606*
Scale parameter	λ	4.264	0.380*	—	—
Variance component	τ_2^2	2.610	0.456*	47.838	2.059*
ICC	ρ_{11}	0.138	0.011*	0.410	0.014*
Quality of support					
Intercept	β_0	3.443	0.152*	5.176	0.270*
Age	β_1	−0.102	0.027*	−0.392	0.367*
Gender	β_2	0.147	0.092	−0.100	0.101
Variance component	τ_2^2	2.610	0.456*	6.354	0.936*
ICC	ρ_{22}	0.442	0.043*	0.659	0.033*
Association					
RI correlation	ρ	—	—	0.783	0.049*
Correlation	ρ_{12}	0.511	0.026*	0.407	0.028*

The correlation between the two endpoints is estimated as 0.511 (SE $= 0.026$), indicating a strong positive association between quality of support and size of social network.

The second column shows parameter estimates for the correlated random intercepts model. The fixed effects parameter estimates and their SEs for the continuous endpoint are approximately the same as in the previous model, while the fixed effects estimates for the binary endpoint differ somewhat more. This follows from the fact that interpretation of the parameters in the binary case is conditional on the random effects. Note that the correlation ρ in this model represents the correlation between the random intercepts. This estimate can be used to calculate the correlation ρ_{12} between the continuous and discrete outcomes. The ICC estimate is higher than that for the shared parameter model, but the bivariate correlation is slightly smaller.

The models are consistent in the estimation of the effect of age and gender on the endpoints. In all models, age significantly reduces social health, while gender has only an effect on size of the social network. There is a strong association of social health within a household, and also a strong association between size of social network and quality of social support.

9.5.2.2   *Trivariate hierarchical model*

In this section, the three endpoints are considered together, namely, number of close relationships Y_{1ij}, quality of social support Y_{2ij}, and satisfaction about social network Y_{3ij}, and a trivariate model is constructed. While this is an easy extension of the previously described models, the computational challenges increase enormously. Already with only three endpoints, we are faced with some computational problems. Here, we focus on the correlated random intercepts model

$$
\begin{pmatrix} Y_{1ij} \\ Y_{2ij} \\ Y_{3ij} \end{pmatrix} = \begin{pmatrix} \alpha_0 + \alpha_1 x_{1ij} + \alpha_2 x_{2ij} + \lambda B_{1i} \\ \dfrac{\exp(\beta_0 + \beta_1 x_{1ij} + \beta_2 x_{2ij} + B_{2i})}{1 + \exp(\beta_0 + \beta_1 x_{1ij} + \beta_2 x_{2ij} + B_{2i})} \\ \dfrac{\exp(\gamma_0 + \gamma_1 x_{1ij} + \gamma_2 x_{2ij} + B_{3i})}{1 + \exp(\gamma_0 + \gamma_1 x_{1ij} + \gamma_2 x_{2ij} + B_{3i})} \end{pmatrix} + \begin{pmatrix} \varepsilon_{1ij} \\ \varepsilon_{2ij} \\ \varepsilon_{3ij} \end{pmatrix}, \tag{9.14}
$$

Table 9.5 *Estimates and their uncorrected (SE$_u$) and corrected (SE$_c$) SEs from joint analysis of number of close relationships (continuous endpoint), satisfaction about social contact (binary endpoint), and quality of support (binary endpoint) from HIS.*

Effect	Close relations			Satisfaction			Support		
	Est	SE$_u$	SE$_c$	Est	SE$_u$	SE$_c$	Est	SE$_u$	SE$_c$
Intercept	9.516	0.189	0.255	3.650	0.129	0.186	5.125	0.178	0.251
Age	−0.443	0.046	0.068	−0.104	0.020	0.029	−0.383	0.025	0.036
Gender	1.562	0.140	0.188	0.142	0.068	0.094	−0.126	0.071	0.098
Error variance	69.065	1.139	6.007	—	—	—	—	—	—
Variance component	47.491	1.446	6.723	2.900	0.348	0.487	5.735	0.558	0.778

with x_{1ij} and x_{2ij} representing age and gender of the individual, respectively, and

$$\begin{pmatrix} B_{1i} \\ B_{2i} \\ B_{2i} \end{pmatrix} \sim N_3\left(\mathbf{0}, \begin{pmatrix} \tau_1^2 & \rho_{12}\tau_1\tau_2 & \rho_{13}\tau_1\tau_3 \\ \rho_{12}\tau_1\tau_2 & \tau_2^2 & \rho_{23}\tau_2\tau_3 \\ \rho_{13}\tau_1\tau_2 & \rho_{23}\tau_2\tau_3 & \tau_3^2 \end{pmatrix}\right), \tag{9.15}$$

and use the pairwise pseudo-likelihood method to gain computational efficiency. Starting values from the univariate models are used to aid in convergence. Table 9.5 shows parameter estimates for the trivariate model. Both uncorrected and corrected SEs are presented.

We see a decrease with age in the number of close relationships, the satisfaction about the social network, and in the quality of social support. For gender, only a significant effect is seen for the number of close relationships, with males having more relations as compared to females. Correlations between random effects corresponding to different endpoints are given by the following correlation matrix:

$$corr\left\{ \begin{pmatrix} B_{1i} \\ B_{2i} \\ B_{3i} \end{pmatrix} \right\} = \begin{pmatrix} 1 & 0.772 & 0.839 \\ 0.772 & 1 & 0.785 \\ 0.839 & 0.785 & 1 \end{pmatrix}. \tag{9.16}$$

We note that all the random effects are highly correlated, indicating a positive association between the endpoints. Based on this correlation matrix, the correlation matrix between the endpoints are calculated, resulting in

$$corr\left\{ \begin{pmatrix} Y_{1ij} \\ Y_{2ij} \\ Y_{3ij} \end{pmatrix} \right\} = \begin{pmatrix} 1 & 0.393 & 0.367 \\ 0.393 & 1 & 0.428 \\ 0.367 & 0.428 & 1 \end{pmatrix}. \tag{9.17}$$

From this, it is clear that these three endpoints share a lot of information about the social health of individuals in households. Strong associations are found between the different endpoints, and they all show a similar pattern with age.

9.6 Discussion

Correlated multivariate data are very common in practice. When the outcomes are of different types, there no longer exists a single approach to jointly model the outcomes. Different viewpoints can be adopted. The first viewpoint states that one can rely on large-sample asymptotics, and use conventional normal theory for any endpoint. The second viewpoint states that one should use methods supported by the nature of the data. It is in this latter viewpoint that GLMMs have been proposed by, e.g., Diggle *et al.* (1994). In this chapter, GLMMs were used to jointly model correlated endpoints of mixed types, such as clustered or repeated continuous and discrete endpoints.

A general form of the GLMM was used allowing different link functions and error terms for endpoints of different types. The correlation between continuous and discrete endpoints was modeled via shared or correlated random effects, or via the residual error correlation structure. The clustered (or repeated) correlation structure was also modeled via a random effect, or via specification of the residual error structure. In this way, a flexible modeling framework was obtained to model multiple endpoints of possibly different types.

However, when more than two endpoints are modeled jointly, convergence issues might occur. In this chapter, pseudo-likelihood methods were proposed to enhance convergence. The full marginal likelihood, which involves integrating over the multivariate random effects, was replaced by a product over pairwise bivariate likelihoods. This pseudo-likelihood method yields asymptotically unbiased estimates and valid SEs for the multivariate joint model without many convergence issues.

Chapter 10

Joint analysis of mixed discrete and continuous outcomes via copula models

Beilei Wu, Alexander R. de Leon, and Niroshan Withanage

10.1 Introduction

Practitioners in health and medicine are often confronted with data comprising a mixture of discrete (i.e., nominal, ordinal, or binary) and continuous outcomes. For example, in addition to quantitative health related variables, quality of life outcomes in health studies are often measured or scored on categorical or ordinal scales. The same is true in developmental toxicity studies, where binary and continuous endpoints are commonplace (see Section 10.5). Multivariate modeling of such data often leads to complications in practice due to a relative lack of available models and the difficulties of constructing them.

An obvious, but often inefficient, approach to handling mixed data is to convert one type of variable to another, and then to employ appropriate methods. Although this approach is simple enough and may work in practice, the crude coding or categorization of variables makes it conceptually unattractive and unsatisfactory in many applications. Instead, a model-based alternative is possible by directly or indirectly specifying a joint distribution.

Factorization models directly specify the joint distribution as the product of a conditional distribution of one set of outcomes and a marginal distribution of the other. General location models (GLOMs) (Olkin and Tate, 1961) are based on conditional normal (i.e., Gaussian) distributions, and have received much attention in the literature. They assume conditional normality of continuous components and an arbitrary distribution for discrete components. A reverse factorization entails specifying a latent continuous multivariate distribution from which discrete outcomes are derived by forming discrete classes. Probit-style models for categorical data generated from a multivariate normal distribution of underlying latent variables yield so-called conditional grouped continuous models (CGCMs) (see, e.g., de Leon, 2005). A number of refinements and extensions of these models have since been studied by several authors; e.g., see Chapter 1 for de Leon and Carrière's (2007) general mixed-data model (GMDM), a hybrid of GLOM and CGCM which provides a unified treatment of these models.

While factorization provides a general route for directly specifying mixed-variable distributions, it is not without its shortcomings. Factorization models use a structural approach in classifying outcomes as continuous or discrete to decide the direction of conditioning. A hierarchy in the data is thus induced, with conditioning components treated as intermediate outcomes, and conditioned components as primary responses. As such, factorization models are not invariant to the direction of conditioning taken and the factorization adopted. Variations of GLOM and CGCM have been studied in various applications resulting in models that are not comparable, as parameters have different interpretations depending on the factorization used. It is thus possible for different factorization models to yield very different inferences, especially of associations. Various issues and applications of factorization models are discussed in Chapters 5, 6, and 7.

Indirect approaches to specifying mixed-variable joint distributions have also been studied. One approach that has found widespread adoption in practice introduces shared or correlated random effects to incorporate correlations between outcomes in the resulting joint model. The basic idea in this approach is to use random effects to build in correlation between mixed outcomes. The approach does not resort to factorization, and thus, treats outcomes symmetrically. Its hierarchical structure allows for considerable flexibility in accounting for different measurement levels, delineation of various associations, incorporation of covariate effects, and extension to longitudinal and clustered data settings. However, random effects models also have their drawbacks. For one, correlations may be restricted to lie within artificially narrow ranges; for another, computational difficulties may arise in high-dimensional problems (see de Leon and Carrière Chough, 2010, and McCulloch, 2008, for examples). See Chapters 8 and 9, for discussions of random effects models for mixed data.

This chapter considers mixed-outcome joint models generated by copulas in general, and Gaussian copulas, in particular. The joint models rely on latent variable formulations of discrete outcomes, and generalize conventional normal theory-based generalized linear mixed models (GLMMs) to non-normal outcomes. The models treat mixed outcomes symmetrically and do not resort to factorization. The class is general enough to include previous models studied in the literature as special cases. Section 10.2 introduces the copula approach to joint modeling of disparate outcomes, where latent variable-based joint models for independent and clustered data are introduced. A brief discussion of how associations are incorporated in the models is presented in Section 10.3. Likelihood estimation is outlined in Section 10.4; alternative methods are also briefly mentioned. Section 10.5 is devoted to a case study involving data on teratological effects of ethylene glycol on mice. Finally, Section 10.6 concludes the chapter.

10.2 Joint models via copulas

A recent alternative strategy involves the use of copulas, as discussed by Song (2007), among others. The approach embeds absolutely continuous univariate marginal distribution functions (or margins) $F_{X_1}(\cdot), \cdots, F_{X_P}(\cdot)$, into their corresponding P-dimensional distribution function $F_{X_1,\cdots,X_P}(\cdot)$ via a copula $C(\cdot)$ as

$$F_{X_1,\cdots,X_P}(x_1,\cdots,x_P) = C(F_{X_1}(x_1),\cdots,F_{X_P}(x_P)). \tag{10.1}$$

The distribution $F_{X_1,\cdots,X_P}(\cdot)$ is thus specified via its margins and a copula that "glues" them together. In parametric contexts, the margins need not come from the same parametric family, allowing researchers great flexibility in modeling data of different types. The copula accounts for "dependence" between outcomes in a way that is separate from their marginal specifications.

The choice of an appropriate copula with which to "couple" marginal distributions depends on the suitability of the copula's dependence parameter for describing the data's dependence structure. Gaussian copulas are an important family which has been used in a variety of applications (e.g., Song, 2007). The P-dimensional Gaussian copula is defined as

$$C(u_1,\cdots,u_P) = \Phi_P\left(\Phi^{-1}(u_1),\cdots,\Phi^{-1}(u_P);\widetilde{\mathbf{R}}\right), \tag{10.2}$$

for $u_1,\cdots,u_P \in [0,1]$, where $\Phi^{-1}(\cdot)$ is the inverse function of the standard normal distribution function $\Phi(\cdot)$, and $\Phi_P(\cdot;\widetilde{\mathbf{R}})$ is the P-dimensional standard multivariate normal distribution function (i.e., zero means and unit variances), with correlation matrix $\widetilde{\mathbf{R}}$. For continuous variables $X_1 \sim F_{X_1}(\cdot),\cdots,X_P \sim F_{X_P}(\cdot)$ whose joint distribution function $F_{X_1,\cdots,X_P}(\cdot)$ is specified by Gaussian copula (10.2), we get

$$F_{X_1,\cdots,X_P}(x_1,\cdots,x_P) = \Phi_P\left(\Phi^{-1}(u_1),\cdots,\Phi^{-1}(u_P);\widetilde{\mathbf{R}}\right), \tag{10.3}$$

where $u_1 = F_{X_1}(x_1),\cdots,u_p = F_{X_P}(x_P)$ with $U_1 = F_{X_1}(X_1) \sim \mathrm{U}[0,1],\cdots, U_P = F_{X_P}(X_P) \sim \mathrm{U}[0,1]$ the probability integral transforms (PIT), $\mathrm{U}[0,1]$ is the uniform distribution over $[0,1]$, and $\widetilde{\mathbf{R}}$ is the correlation matrix of so-called normal scores $\Phi^{-1}(U_1), \cdots, \Phi^{-1}(U_P)$. The flexibility and analytical tractability of Gaussian copulas make them a handy tool in many applications. Their popularity is due to the fact that they describe dependence between variables in much the same way that Gaussian distributions do.

While not new, applications of copulas to discrete data (e.g., Nikoloulopoulos and Karlis, 2010, 2009, 2008; Song et al., 2009; Zimmer and Trivedi, 2006; Trégouët et al., 2004; Meester and MacKay, 1994) have only recently been elucidated and clarified (Genest and Nešlehová, 2007). As Genest and Nešlehová (2007) show, a number of complications arise from the direct application of copula models to discrete data. One such complication concerns the failure of the copula to uniquely determine the distribution. Another more practical one involves the interpretability of the dependence parameters. Note that from a modeling perspective, the non-uniqueness of the copula is not an issue, as the parameters of the model are still identifiable and the copula still corresponds to a proper multivariate distribution. Common rank-based association measures like Kendall's tau and Spearman's rho, however, may now depend on the margins (see, e.g., Nešlehová, 2007; Mesfioui and Tajar, 2005), and the range of their possible values may be restricted — severely in some cases — rendering interpretations of such measures problematic.

In what follows, we construct joint models for mixed discrete and continuous outcomes using the Gaussian copula.

10.2.1   Case of independent data

Suppose we have discrete outcomes $X_1, \cdots, X_Q$, and continuous outcomes $Y_1, \cdots, Y_C$. Let X_j have $L_j + 1$ discrete values $x_j^{(0)} < \cdots < x_j^{(L_j)}$, $j = 1, \cdots, Q$. Underlying X_j is Y_j^*, a continuous latent variable whose relationship with X_j is defined by the following threshold model:

$$
X_j = \begin{cases}
x_j^{(L_j)} & \text{, if and only if } Y_j^* > \alpha_j^{(L_j)} \\
x_j^{(\ell_j)} & \text{, if and only if } \alpha_j^{(\ell_j)} < Y_j^* \le \alpha_j^{(\ell_j+1)}, \ \ell_j = 1, \cdots, L_j - 1, \\
x_j^{(0)} & \text{, if and only if } Y_j^* \le \alpha_j^{(1)}
\end{cases} \tag{10.4}
$$

where $\alpha_j^{(1)}, \cdots, \alpha_j^{(L_j)}$, are unknown cutpoints or thresholds. Without loss of generality, we assume that $x_j^{(\ell_j)} = \ell_j$, $\ell_j = 0, \cdots, L_j$.

Given marginal distributions $F_{Y_1^*}(\cdot), \cdots, F_{Y_Q^*}(\cdot)$, and $F_{Y_1}(\cdot), \cdots, F_{Y_C}(\cdot)$, for latent variables and continuous outcomes, respectively, we assume that the joint distribution of $\mathbf{Y}^* = (Y_1^*, \cdots, Y_Q^*)^\top$, and $\mathbf{Y} = (Y_1, \cdots, Y_C)^\top$, is determined by a Gaussian copula with correlation matrix $\widetilde{\mathbf{R}}$. With $P = C + Q$, the joint distribution of $\mathbf{X} = (X_1, \cdots, X_Q)^\top$ and $\mathbf{Y}$ is then

$$
P\left(\begin{array}{l} X_1 = \ell_1, \cdots, X_Q = \ell_Q, \\ Y_1 \le y_1, \cdots, Y_C \le y_C \end{array} \right) = \sum_{\varepsilon_1=0}^{1} \cdots \sum_{\varepsilon_Q=0}^{1} (-1)^{Q + \Sigma_{j=1}^{Q} \varepsilon_j}
$$
$$
\times \Phi_P \left(\begin{array}{l} \Phi^{-1}(u_1^{(\ell_1+\varepsilon_1)}), \cdots, \Phi^{-1}(u_Q^{(\ell_Q+\varepsilon_Q)}), \\ \Phi^{-1}(v_1), \cdots, \Phi^{-1}(v_C) \end{array} ; \widetilde{\mathbf{R}} \right), \tag{10.5}
$$

where $u_j^{(\ell_j+\varepsilon_j)} = F_{Y_j^*}(\alpha_j^{(\ell_j+\varepsilon_j)})$ and $v_k = F_{Y_k}(y_k)$, for $j = 1, \cdots, Q$, and $k = 1, \cdots, C$, with

$$
E(Y_j^*) = \mu_j^*(\mathbf{z}_{1j}, \boldsymbol{\gamma}_j) \quad \text{and} \quad E(Y_k) = \mu_k(\mathbf{z}_{2k}, \boldsymbol{\beta}_k), \tag{10.6}
$$

for vectors $\boldsymbol{\gamma}_j$ and $\boldsymbol{\beta}_k$ of regression coefficients, outcome-specific covariate vectors $\mathbf{z}_{1j}$ and $\mathbf{z}_{2k}$, and link functions $\mu_j^*(\cdot)$ and $\mu_k(\cdot)$ specifying how the covariates are incorporated in the marginal means. For example, if $\mu_j^* = \mathbf{z}_{1j}^\top \boldsymbol{\gamma}_j$ and $\mu_k = \mathbf{z}_{2k}^\top \boldsymbol{\beta}_k$, then we have linear models for the means. For identifiability reasons, we assume that Y_j^* has unit scale (or variance parameter, if it is the scale parameter; in either case, the latent variance is assumed known) and $\boldsymbol{\gamma}_j$ has no intercept term; if X_j is binary (i.e., $L_j = 1$), we may arbitrarily assume the single cutpoint to be zero, and $\boldsymbol{\gamma}_j$ will need to include an intercept term. Note that the margins $F_{Y_1^*}(\cdot), \cdots, F_{Y_Q^*}(\cdot)$, and $F_{Y_1}(\cdot), \cdots, F_{Y_C}(\cdot)$, can be any continuous distributions, thus allowing researchers great flexibility in joint modeling of mixed outcomes. For example, a normal latent distribution for Y_j^* results in a marginal probit model for X_j in the binary case.

The corresponding density of $\mathbf{X}$ and $\mathbf{Y}$ is given by

$$
f_{\mathbf{X},\mathbf{Y}}(\boldsymbol{\ell},\mathbf{y}) = \frac{\phi(t_{C-1|C})\phi(t_{C-2|C:C-1})\cdots\phi(t_{1|C:2})}{\prod_{k=1}^{C-1}\sqrt{1-\widetilde{\rho}_{W_k W_C}^2}\prod_{k=1}^{C-2}\sqrt{1-\widetilde{\rho}_{W_k W_{C-1}|W_C}^2}\cdots\sqrt{1-\widetilde{\rho}_{W_1 W_2|W_C:W_3}^2}}
$$
$$
\times \sum_{\varepsilon_1=0}^{1}\cdots\sum_{\varepsilon_Q=0}^{1}(-1)^{Q+\sum_{j=1}^{Q}\varepsilon_j}\Phi_Q\left(s_{1|C:1}^{(\ell_1+\varepsilon_1)},\cdots,s_{Q|C:1}^{(\ell_Q+\varepsilon_Q)};\widetilde{\mathbf{R}}_{|W_C:W_1}\right)
$$
$$
\times f_{Y_C}(y_C)\prod_{k=1}^{C-1}\frac{f_{Y_k}(y_k)}{\phi(t_k)}, \tag{10.7}
$$

where $\boldsymbol{\ell} = (\ell_1, \cdots, \ell_Q)^\top$ and $\mathbf{y} = (y_1, \cdots, y_C)^\top$, with $f_{Y_k}(\cdot)$ the marginal density of Y_k, $\phi(\cdot)$ the standard normal density, and

$$
t_{C-k|C:C-k+1} = \frac{t_{C-k|C:C-k+2} - \widetilde{\rho}_{W_{C-k}W_{C-k+1}|W_C:W_{C-k+2}}t_{C-k+1|C:C-k+2}}{\sqrt{1-\widetilde{\rho}_{W_{C-k}W_{C-k+1}|W_C:W_{C-k+2}}^2}}
$$
$$
s_{j|C:k}^{(\ell_j+\varepsilon_j)} = \frac{s_{j|C:k+1}^{(\ell_j+\varepsilon_j)} - \widetilde{\rho}_{W_j^* W_k|W_C:W_{k+1}}t_{k|C:k+1}}{\sqrt{1-\widetilde{\rho}_{W_j^* W_k|W_C:W_{k+1}}^2}},
$$

where $W_j^* = \Phi^{-1}\{F_{Y_j^*}(Y_j^*)\}$ and $W_k = \Phi^{-1}\{F_{Y_k}(Y_k)\}$ are the normal scores (latent in the case of W_j^*), $s_j^{(\ell_j+\varepsilon_j)} = \Phi^{-1}(u_j^{(\ell_j+\varepsilon_j)})$, $t_k = \Phi^{-1}(v_k)$, $\widetilde{\rho}_{W_{C-k}W_{C-k+1}|W_C:W_{C-k+2}}$ is the partial correlation between W_{C-k} and W_{C-k+1}, after eliminating $W_{C-k+2}, \cdots, W_C$, $\widetilde{\rho}_{W_j^* W_{C-k}|W_C:W_{C-k+1}}$ is the partial correlation between W_j^* and W_{C-k}, after eliminating $W_{C-k+1}, \cdots, W_C$, and $\widetilde{\mathbf{R}}_{|W_C:W_{C-k+1}} = \widetilde{\mathbf{R}}_{W_1^*:W_{C-k}|W_C:W_{C-k+1}}$ is the partial correlation matrix for $W_1^*, \cdots, W_Q^*, W_1, \cdots, W_{C-k}$, after eliminating $W_{C-k+1}, \cdots, W_C$, for $k = 1, \cdots, C-1$. See de Leon et al. (2012) for more details; see also de Leon and Wu (2011) for the special case $Q = C = 1$.

The approach outlined above contrasts with Song et al.'s (2009) and Song's (2007) direct application of copulas to model discrete outcomes. Our approach adopts a latent variable description of discrete outcomes, and the resulting joint model arises from the joint distribution, constructed via a copula, of continuous outcomes and the latent variables. The joint model is thus indirectly specified through the copula-based joint distribution of continuous outcomes and the latent variables. In addition to its interpretational appeal in medical and health studies, using latent variables to describe discrete outcomes (see Section 10.5) also makes statistical sense, since common regression models for discrete outcomes (e.g., logistic, probit models) have parallel formulations in terms of latent variables (see Teixeira–Pinto and Normand, 2009). In addition, it also leads to familiar association measures for capturing dependence between two discrete outcomes (i.e., polychoric correlation) or between mixed discrete and continuous outcomes (i.e., polyserial correlation). In contrast to Pearson's correlations as measures of association between two discrete outcomes and between mixed discrete and continuous outcomes, polyserial and polychoric correlations are not constrained by marginal probabilities of the discrete outcomes, as they are just the usual pairwise correlations between continuous variables (latent or otherwise). Moreover, the number of polychoric and polyserial correlations remains the same for polychotomous discrete data, certainly not the case for odds ratios.

Finally, note that our approach does not resort to factorization, resulting in a symmetrical treatment of outcomes while preserving the margins, an attractive feature in practice.

10.2.2 Case of clustered data

In this section, we adopt Wu and de Leon's (2012) approach and generalize it to multivariate mixed discrete and continuous clustered data with $Q \geq 2$ and $C \geq 2$. Similar to the previous section, let X_{ij} and Y_{ik} be the respective discrete and continuous outcomes for cluster $i = 1, \cdots, N$, where we assume the same threshold model (10.4) for X_{ij} and its corresponding latent variable Y_{ij}^*. In what follows, we develop copula random effects models which allow for flexible non-normal residual errors and random effects, extending and generalizing previous models studied by de Leon and Wu (2011), Lin et al. (2010), Najita et al. (2009), and Gueorguieva and Agresti (2001), among others. We first assume a shared cluster-specific random effect B_i and construct a joint density for $\mathbf{X}_i = (X_{i1}, \cdots, X_{iQ})^\top$, and $\mathbf{Y}_i = (Y_{i1}, \cdots, Y_{iC})^\top$ as

$$f_{\mathbf{X}_i, \mathbf{Y}_i}(\boldsymbol{\ell}_i, \mathbf{y}_i) = \int_{-\infty}^{+\infty} f_{\mathbf{X}_i, \mathbf{Y}_i | B_i}(\boldsymbol{\ell}_i, \mathbf{y}_i | b) f_{B_i}(b_i) db_i, \qquad (10.8)$$

for $\boldsymbol{\ell}_i = (\ell_{i1}, \cdots, \ell_{iQ})^\top$, where

$$f_{\mathbf{X}_i, \mathbf{Y}_i | B_i}(\boldsymbol{\ell}_i, \mathbf{y}_i | b_i) = \int_{\alpha_1^{\ell_{i1}}}^{\alpha_1^{\ell_{i1}+1}} \cdots \int_{\alpha_Q^{\ell_{iQ}}}^{\alpha_Q^{\ell_{iQ}+1}} f_{\mathbf{Y}_i^*, \mathbf{Y}_i | B_i}(\mathbf{y}_i^*, \mathbf{y}_i | b_i) dy_{iQ}^* \cdots dy_{i1}^*, \qquad (10.9)$$

where $\mathbf{Y}_i^* = (Y_{i1}^*, \cdots, Y_{iQ}^*)^\top$ and $\mathbf{y}_i^* = (y_{i1}^*, \cdots, y_{iQ}^*)^\top$. In (10.8), $f_{B_i}(\cdot)$ is the density of B_i, and $f_{\mathbf{Y}_i^*, \mathbf{Y}_i | B_i}(\cdot | \cdot)$ is the conditional density of $\mathbf{Y}_i^*$ and $\mathbf{Y}_i$, given B_i, and is of the form (10.7) when a Gaussian copula is adopted to construct the conditional distribution of $\mathbf{Y}_i^*$ and $\mathbf{Y}_i$, as in

$$F_{\mathbf{Y}_i^*, \mathbf{Y}_i | B_i}(\mathbf{y}_i^*, \mathbf{y}_i | b_i) = \Phi_P \left(\begin{array}{c} \Phi^{-1}\{w_{i1}^*(b_i)\}, \cdots, \Phi^{-1}\{w_{iQ}^*(b_i)\}, \\ \Phi^{-1}\{w_{i1}(b_i)\}, \cdots, \Phi^{-1}\{w_{iC}(b_i)\} \end{array} ; \widetilde{\mathbf{R}} \right), \qquad (10.10)$$

where $w_{ij}^*(b_i) = F_{Y_{ij}^* | B_i}(y_{ij}^* | b_i)$ and $w_{ik}(b_i) = F_{Y_{ik} | B_i}(y_{ik} | b_i)$, with $F_{Y_{ij}^* | B_i}(\cdot | \cdot)$ and $F_{Y_{ik} | B_i}(\cdot | \cdot)$ the respective (conditional) marginal distributions of Y_{ij}^* and Y_{ik}, for $j = 1, \cdots, Q$, and $k = 1, \cdots, C$. Note that the margins $F_{Y_{ij}^* | B_i}(\cdot | \cdot)$ and $F_{Y_{ik} | B_i}(\cdot | \cdot)$ need not come from the same parametric family, allowing researchers great flexibility in modeling disparate outcomes. Suppose

$$Y_{ij}^* | B_i \sim \text{independent } F_{Y_{ij}^* | B_i}(\cdot | \cdot), \quad Y_{ik} | B_i \sim \text{independent } F_{Y_{ik} | B_i}(\cdot | \cdot),$$

for $i = 1, \cdots, N$, and assuming $E(B_i) = 0$ and $var(B_i) = 1$, with $var(Y_{ij}^* | B_i) = 1$ (assuming the scale parameter is the variance), for identifiability reasons, the conditional mean models, given the random effect B_i, are defined by

$$\mu_{ij}^*(B_i) = E(Y_{ij}^* | B_i) = \mu_{ij}^*(\mathbf{z}_{1ij}, \boldsymbol{\gamma}_j) + \lambda_{1j} B_i, \qquad (10.11)$$

$$\mu_{ik}(B_i) = E(Y_{ik} | B_i) = \mu_{ik}(\mathbf{z}_{2ik}, \boldsymbol{\beta}_k) + \lambda_{2k} B_i, \qquad (10.12)$$

where $\mathbf{z}_{1ij}$ and $\mathbf{z}_{2ik}$ are known covariate vectors (possibly outcome-specific), and $\boldsymbol{\gamma}_j$ and $\boldsymbol{\beta}_k$ are the corresponding unknown regression coefficients, with λ_{1j} and λ_{2k} accounting for difference in scales of Y_{ij}^* and Y_{ik}. The marginal means then follow as $E(Y_{ij}^*) = \mu_{ij}^*(\mathbf{z}_{1ij}, \boldsymbol{\gamma}_j) = \mathbf{z}_{1ij}^\top \boldsymbol{\gamma}_j$, say, and $E(Y_{ik}) = \mu_{ik}(\mathbf{z}_{2ik}, \boldsymbol{\beta}_k) = \mathbf{z}_{2ik}^\top \boldsymbol{\beta}_k$, say, and the marginal variances as $var(Y_{ij}^*) = 1 + \lambda_{1j}^2$ and $var(Y_{ik}) = \sigma_k^2 + \lambda_{2k}^2$, where $\sigma_k^2 = var(Y_{ik} | B_i) > 0$.

Note that in the case of binary X_{ij}, the single cutpoint α_j can be assumed to be 0, provided an intercept term is included in $\boldsymbol{\gamma}_j$. In addition, while the model (and inference) for X_{ij} is done at the latent level, note that this corresponds to a generalized linear mixed model (GLMM) for X_{ij} with a specific link function that depends on the choice of latent distribution. For example, a logistic latent distribution for Y_{ij}^* results in a logistic mixed model for X_{ij}, while a normal latent distribution leads to a probit mixed model. With normal margins for Y_{ij}^* and Y_{ik}, our model specializes to a multivariate version of Najita *et al.*'s (2010) correlated bivariate probit model.

The above approach deviates from the conventional one adopted for specifying GLMMs, where the residual error distribution drives that of the outcomes. Instead of specifying the residual error distribution (usually normal) to construct the conditional joint distribution of outcomes (given random effects), as is usually done in GLMMs, our approach specifies the latter directly by specifying the (conditional) marginal response distributions and coupling them together using the Gaussian copula. The corresponding residual error distributions can then be obtained from the response distribution (given random effects) by transformation methods. The resulting GLMM is thus more flexible and more general than those discussed in Chapters 8 and 9.

Using shared random effects usually yields particularly restrictive association structures for correlated data. The following develops joint models using correlated random effects for the mixed outcomes which allow for greater flexibility in modeling within-cluster and between-outcome associations. For $i = 1, \cdots, N$, define

$$\mu_{ij}^*(B_{1ij}) = \mu_{ij}^*(\mathbf{z}_{1ij}, \boldsymbol{\gamma}_j) + \lambda_{1j}B_{1ij}, \tag{10.13}$$

$$\mu_{ik}(B_{2ik}) = \mu_{ik}(\mathbf{z}_{2ik}, \boldsymbol{\beta}_k) + \lambda_{2k}B_{2ik}, \tag{10.14}$$

where $E(B_{1ij}) = E(B_{2ik}) = 0$, $var(B_{1ij}) = var(B_{2ik}) = 1$, $cov(B_{1ij}, B_{2ik}) = corr(B_{1ij}, B_{2ik}) = \eta_{B_{1j}B_{2k}}$, $cov(B_{1ij}, B_{1ij'}) = corr(B_{1ij}, B_{1ij'}) = \eta_{B_{1j}B_{1j'}}$, and $cov(B_{2ik}, B_{2ik'}) = corr(B_{2ik}, B_{2ik'}) = \eta_{B_{2k}B_{2k'}}$; note that the correlations are constant across clusters. With $\mathbf{B}_{1i} = (B_{1i1}, \cdots, B_{1iQ})^\top$ and $\mathbf{B}_{2i} = (B_{2i1}, \cdots, B_{2iC})^\top$, we have

$$f_{\mathbf{X}_i, \mathbf{Y}_i}(\boldsymbol{\ell}_i, \mathbf{y}_i) = \int_{\mathbb{R}^P} f_{\mathbf{X}_i, \mathbf{Y}_i | \mathbf{B}_{1i}, \mathbf{B}_{2i}}(\boldsymbol{\ell}_i, \mathbf{y}_i | \mathbf{b}_{1i}, \mathbf{b}_{2i}) f_{\mathbf{B}_{1i}, \mathbf{B}_{2i}}(\mathbf{b}_{1i}, \mathbf{b}_{2i}) d\mathbf{b}_{1i} d\mathbf{b}_{2i}, \tag{10.15}$$

where

$$f_{\mathbf{X}_i, \mathbf{Y}_i | \mathbf{B}_{1i}, \mathbf{B}_{2i}}(\boldsymbol{\ell}_i, \mathbf{y}_i | \mathbf{b}_{1i}, \mathbf{b}_{2i}) = \int_{\alpha_1^{\ell_{i1}}}^{\alpha_1^{\ell_{i1}+1}} \cdots \int_{\alpha_Q^{\ell_{iQ}}}^{\alpha_Q^{\ell_{iQ}+1}} f_{\mathbf{Y}_i^*, \mathbf{Y}_i | \mathbf{B}_{1i}, \mathbf{B}_{2i}}(\mathbf{y}_i^*, \mathbf{y}_i | \mathbf{b}_{1i}, \mathbf{b}_{2i}) dy_{iQ}^* \cdots dy_{i1}^*, \tag{10.16}$$

where $f_{\mathbf{B}_{1i}, \mathbf{B}_{2i}}(\cdot)$ is the joint density of $\mathbf{B}_{1i}$ and $\mathbf{B}_{2i}$, and $\mathbb{R}^P$ is the P-dimensional real plane. Note that (10.16) is still of the form (10.7), provided a Gaussian copula is adopted for $F_{\mathbf{Y}_i^*, \mathbf{Y}_i | \mathbf{B}_{1i}, \mathbf{B}_{2i}}(\cdot | \cdot)$ as in (10.10), with (conditional) marginal means (10.13) and (10.14), given correlated random effects $\mathbf{B}_{1i}$ and $\mathbf{B}_{2i}$. Instead of the usual joint normality assumption for $\mathbf{B}_{1i}$ and $\mathbf{B}_{2i}$ as in Gueorguieva and Agresti (2001), it is possible to similarly use the Gaussian copula to build their joint distribution as in Lin *et al.* (2010). This affords flexibility in specifying the marginal distributions of $\mathbf{B}_{1i}$ and $\mathbf{B}_{2i}$; for example, a bridge margin may be assumed for B_{1ij} to facilitate interpretability of marginal effects with a logistic regression model for binary outcome X_{ij} (i.e., a logistic latent distribution for Y_{ij}^*).

10.3 Associations

Copula models usually rely on rank-based association measures, such as Kendall's tau or Spearman's rho (Balakrishnan and Lai, 2009, Chapter 4) to evaluate the strength of dependence between variables, as they are invariant to monotonic transformations (e.g., normal scores). However, unlike with continuous variables for which they provide margin-free measures of the level of dependence, this no longer holds in the discrete case (e.g., Genest and Nešlehová, 2007). Denuit and Lambert (2005) and Mesfioui and Tajar (2005) adopt a "continuous-ation" approach as a possible remedy. Interpretability of the association measures is another issue since their range varies, and there is thus a need to re-scale them.

Directly applying copulas to model multivariate discrete data, as in Song *et al.* (2009), suffers from the same problem many multivariate discrete distributions do: correlations and dependence parameters are unnaturally constrained to ensure the propriety of the joint probabilities. However, merely using latent variables for the discrete variables and indirectly applying copula at the latent level do not always work. This is seen in the recent work of Li and Wong (2011), who adopt the multivariate Gumbel copula to specify a joint model for correlated binary data, via a threshold model as in (10.4); however, as shown by Nikoloulopoulos (2012), the dependence parameters of the Gumbel copula are constrained to ensure the positivity of the copula density. In contrast, our use of the Gaussian copula is advantageous in that the matrix of normal correlations is not constrained; the only requirement is that the matrix be positive definite. Thus, using latent variable-based marginal models together with the Gaussian copula results in margin-free and unconstrained dependence measures (albeit measured at the latent level).

10.3.1   Case of independent data

Consider first the case of independent data. The correlation matrix $\widetilde{\mathbf{R}}$ contains three types of correlations: correlation $\widetilde{\rho}_{W_j^* W_{j'}^*} = corr(W_j^*, W_{j'}^*)$ between latent normal scores W_j^* and $W_{j'}^*$ based on latent variables Y_j^* and $Y_{j'}^*$, correlation $\widetilde{\rho}_{W_j^* W_k} = corr(W_j^*, W_k)$ between a latent normal score W_j^* and normal score W_k based on continuous outcome Y_k, and correlation $\widetilde{\rho}_{W_k W_{k'}} = corr(W_k, W_{k'})$ between normal scores based on Y_k and $Y_{k'}$. Bodnar *et al.* (2010) show that any pair of variables (latent and otherwise) from $\mathbf{Y}^*$ and $\mathbf{Y}$, hence any pair from $\mathbf{X}$ and $\mathbf{Y}$, are independent if and only if the corresponding correlation between their normal scores (latent and otherwise) is 0. The correlations in $\widetilde{\mathbf{R}}$ are called normal or dependence correlation coefficients (Bodnar *et al.*, 2010; Klaassen and Wellner, 1997), when variables are observable, as is the case with $\widetilde{\rho}_{W_k W_{k'}}$. Since $\mathbf{Y}^*$ is latent and unobservable and because $\widetilde{\rho}_{W_j^* W_{j'}^*}$ and $\widetilde{\rho}_{W_j^* W_k}$ are analogous to polychoric and polyserial correlations $\rho_{Y_j^* Y_{j'}^*} = corr(Y_j^*, Y_{j'}^*)$ and $\rho_{Y_j^* Y_k} = corr(Y_j^*, Y_k)$, respectively, we refer to $\widetilde{\rho}_{Y_j^* Y_{j'}^*}$ as a polychoric normal correlation coefficient, and to $\widetilde{\rho}_{Y_j^* Y_k}$ as a polyserial normal correlation coefficient. From the nonlinearity of normal quantile transforms, it follows from Theorem 6.1 of Klaassen and Wellner (1997) that

$$\rho_{Y_j^* Y_{j'}^*} < |\widetilde{\rho}_{W_j^* W_{j'}^*}|, \quad \rho_{Y_j^* Y_k} < |\widetilde{\rho}_{W_j^* W_k}|, \quad \rho_{Y_k Y_{k'}} < |\widetilde{\rho}_{W_k W_{k'}}|. \tag{10.17}$$

Given $\widetilde{\rho}_{W_j^* W_{j'}^*}$, $\widetilde{\rho}_{W_j^* W_k}$, and $\widetilde{\rho}_{W_k W_{k'}}$, it is possible to obtain $\rho_{Y_j^* Y_{j'}^*}$, $\rho_{Y_j^* Y_k}$, and $\rho_{Y_k Y_{k'}}$, as

$$\rho_{Y_j^* Y_{j'}^*} = \psi_{jj'}^{**}(\widetilde{\rho}_{W_j^* W_{j'}^*}), \quad \rho_{Y_j^* Y_k} = \psi_{jk}^*(\widetilde{\rho}_{W_j^* W_k}), \quad \rho_{Y_k Y_{k'}} = \psi_{kk'}(\widetilde{\rho}_{W_k W_{k'}}), \tag{10.18}$$

for some functions $\psi_{jj'}^{**}(\cdot)$, $\psi_{jk}^*(\cdot)$, and $\psi_{kk'}(\cdot)$. Kugiumtzis and Bora–Senta (2010) use piece-wise linear approximations based on truncated standard bivariate normal variables to obtain $\psi_{jj'}^{**}(\cdot)$, $\psi_{jk}^*(\cdot)$, and $\psi_{kk'}(\cdot)$. While easy to implement, it may occasionally yield a non-positive definite correlation matrix of $\rho_{Y_j^* Y_{j'}^*}$, $\rho_{Y_j^* Y_k}$, and $\rho_{Y_k Y_{k'}}$; this arises mainly due to such correlations needing to satisfy certain admissible ranges.

Because $\mathbf{W}^*$ and $\mathbf{W}$ are monotonic transformations of $\mathbf{Y}^*$ and $\mathbf{Y}$, respectively, it follows that the Kendall's tau measures are such that $\widetilde{\tau}_{W_j^* W_{j'}^*} = \tau(W_j^*, W_{j'}^*) = \tau_{Y_j^* Y_{j'}^*} = \tau(Y_j^*, Y_{j'}^*)$, $\widetilde{\tau}_{W_j^* W_k} = \tau(W_j^*, W_k) = \tau_{Y_j^* Y_k} = \tau(Y_j^*, Y_k)$, and $\widetilde{\tau}_{W_k W_{k'}} = \tau(W_k, W_{k'}) = \tau_{Y_k Y_{k'}} = \tau(Y_k, Y_{k'})$. Assuming a Gaussian copula-generated joint distribution as in (10.5) and (10.7) for $\mathbf{Y}^*$ and $\mathbf{Y}$, these measures are easy to calculate given $\widetilde{\mathbf{R}}$:

$$\widetilde{\tau}_{W_j^* W_{j'}^*} = \frac{2}{\pi} \sin^{-1}(\widetilde{\rho}_{W_j^* W_{j'}^*}), \quad \widetilde{\tau}_{W_j^* W_k} = \frac{2}{\pi} \sin^{-1}(\widetilde{\rho}_{W_j^* W_k}), \quad \widetilde{\tau}_{W_k W_{k'}} = \frac{2}{\pi} \sin^{-1}(\widetilde{\rho}_{W_k W_{k'}}); \tag{10.19}$$

they can also capture the full range of possible associations in $\mathbf{Y}^*$ and $\mathbf{Y}$, making them quite attractive in practice. A similar approach may be adapted to Spearman's correlation rho $\rho_S = 6\sin^{-1}(\sin(\pi\tau/2)/2)/\pi$.

10.3.2   Case of clustered data

For marginal correlations in the random effects models described in Section 10.2.2, let Y_{ijh}^* and Y_{ikh} be discrete outcome $j = 1, \cdots, Q$, and continuous outcome $k = 1, \cdots, C$, respectively, for subject $h = 1, \cdots, n_i$, in cluster $i = 1, \cdots, N$. Note that we previously suppressed the index "h" for notational convenience. For joint model (10.8) with shared random effects and assuming $var(Y_{ij}^*|B_i) = 1$ for all i, j, the correlations in $\widetilde{\mathbf{R}}$ (including polychoric normal and polyserial normal correlations for binary-outcome pairs and mixed binary-continuous outcome pairs, respectively) are now conditional on random effect B_i. The marginal between-subject and within-subject correlations are then found to be

$$corr(Y_{ijh}^*, Y_{ijh'}^*) = \frac{\lambda_{1j}^2}{1+\lambda_{1j}^2}, \quad corr(Y_{ijh}^*, Y_{ij'h}^*) = \frac{\rho_{Y_j^*Y_{j'}^*|B_i} + \lambda_{1j}\lambda_{1j'}}{\sqrt{(1+\lambda_{1j}^2)(1+\lambda_{1j'}^2)}}, \quad (10.20)$$

$$corr(Y_{ikh}, Y_{ik'h}) = \frac{\rho_{Y_kY_{k'}|B_i}\sigma_k\sigma_{k'} + \lambda_{2k}\lambda_{2k'}}{\sqrt{(\sigma_k^2+\lambda_{2k}^2)(\sigma_{k'}^2+\lambda_{2k'}^2)}}, \quad corr(Y_{ikh}, Y_{ikh'}) = \frac{\lambda_{2k}^2}{\sigma_k^2+\lambda_{2k}^2}, \quad (10.21)$$

$$corr(Y_{ijh}^*, Y_{ij'h'}^*) = \frac{\lambda_{1j}\lambda_{1j'}}{\sqrt{(1+\lambda_{1j}^2)(1+\lambda_{1j'}^2)}}, \quad corr(Y_{ikh}, Y_{ik'h'}) = \frac{\lambda_{2k}\lambda_{2k'}}{\sqrt{(\sigma_k^2+\lambda_{2k}^2)(\sigma_{k'}^2+\lambda_{2k'}^2)}}, (10.22)$$

$$corr(Y_{ijh}^*, Y_{ikh}) = \frac{\rho_{Y_j^*Y_k|B_i}\sigma_k + \lambda_{1j}\lambda_{2k}}{\sqrt{(1+\lambda_{1j}^2)(\sigma_k^2+\lambda_{2k}^2)}}, \quad corr(Y_{ijh}^*, Y_{ikh'}) = \frac{\lambda_{1j}\lambda_{2k}}{\sqrt{(1+\lambda_{1j}^2)(\sigma_k^2+\lambda_{2k}^2)}}, \quad (10.23)$$

where $\rho_{Y_j^*Y_{j'}^*|B_i} = corr(Y_{ijh}^*, Y_{ij'h}^*|B_i)$, $\rho_{Y_kY_{k'}|B_i} = corr(Y_{ikh}, Y_{ik'h}|B_i)$, and $\rho_{Y_j^*Y_k|B_i} = corr(Y_{ijh}^*, Y_{ikh}|B_i)$; note that these correlations depend only on the outcome type, and are constant over subjects and clusters. Observe that

$$corr(Y_{ijh}^*, Y_{ikh'}) = \sqrt{corr(Y_{ijh}^*, Y_{ijh'}^*)}\sqrt{corr(Y_{ikh}, Y_{ikh'})},$$

so that having a shared random effect yields a particularly restrictive association structure for the mixed outcomes from different subjects.

Expressions in (10.20), (10.21), and (10.22) remain unchanged in joint model (10.8) with correlated random effects, and with $\widetilde{\mathbf{R}}$ now containing conditional correlations $\rho_{Y_j^*Y_{j'}^*|\mathbf{B}_{1i}\mathbf{B}_{2i}} = corr(Y_{ijh}^*, Y_{ij'h}^*|\mathbf{B}_{1i}, \mathbf{B}_{2i})$ and $\rho_{Y_kY_{k'}|\mathbf{B}_{1i}\mathbf{B}_{2i}} = corr(Y_{ikh}, Y_{ik'h}|\mathbf{B}_{1i}, \mathbf{B}_{2i})$; correlations in (10.23) are modified accordingly as

$$corr(Y_{ijh}^*, Y_{ikh}) = \frac{\rho_{Y_j^*Y_k|\mathbf{B}_{1i}\mathbf{B}_{2i}}\sigma_k + \eta_{B_{1j}B_{2k}}\lambda_{1j}\lambda_{2k}}{\sqrt{(1+\lambda_{1j}^2)(\sigma_k^2+\lambda_{2k}^2)}}, \quad (10.24)$$

$$corr(Y_{ijh}^*, Y_{ikh'}) = \frac{\eta_{B_{1j}B_{2k}}\lambda_{1j}\lambda_{2k}}{\sqrt{(1+\lambda_{1j}^2)(\sigma_k^2+\lambda_{2k}^2)}}, \quad (10.25)$$

where $\rho_{Y_j^*Y_k|\mathbf{B}_{1i}\mathbf{B}_{2i}} = corr(Y_{ijh}^*, Y_{ikh}|\mathbf{B}_{1i}, \mathbf{B}_{2i})$. Noting that

$$corr(Y_{ijh}^*, Y_{ikh'}) = \eta_{B_{1j}B_{2k}}\sqrt{corr(Y_{ijh}^*, Y_{ijh'}^*)}\sqrt{corr(Y_{ijh}, Y_{ijh'})},$$

suggesting that a more flexible association structure is possible with a model with correlated random effects. Wu and de Leon (2012) obtain the same correlations, albeit in the case of bivariate mixed discrete and continuous clustered outcomes (i.e., $C = Q = 1$).

Note that correlations between outcomes are not modeled directly in our approach; instead, correlations are incorporated for normal scores, which are transformations of the original outcomes (or latent variables in the case of discrete data assumed to have latent structure). In practice, the matrix of correlations between normal scores (both latent and observed) is used as proxy for the outcome correlations. Alternatively, Kugiumtzis and Bora–Senta's (2010) piecewise-linear approximation may be similarly used to recover the latter from the former. We can also use nonparametric rank-based measures like Kendall's tau to gauge associations.

10.4   Likelihood estimation

Consider first the case of independent data and model (10.7). Let $\mathbf{x}_i = (x_{i1}, \cdots, x_{iQ})^\top$ and $\mathbf{y}_i = (y_{i1}, \cdots, y_{iC})^\top$ denote the observed vectors of discrete and continuous outcomes for subject $i = 1, \cdots, N$; in addition, suppose $\mathbf{z}_{i1j}$ and $\mathbf{z}_{i2k}$ represent subject i's vectors of covariates for discrete outcome $j = 1, \cdots, Q$, and continuous outcome $k = 1, \cdots, C$, respectively. Putting $\boldsymbol{\Theta}$ as the vector of parameters, the log-likelihood contribution of subject i takes the following form:

$$
\begin{aligned}
\ell_i(\boldsymbol{\Theta}) &= \sum_{\boldsymbol{\ell}} \mathrm{I}\{\mathbf{x}_i = \boldsymbol{\ell}\} \log \left\{ \sum_{\varepsilon_1 = 0}^{1} \cdots \sum_{\varepsilon_Q = 0}^{1} (-1)^{Q + \Sigma_{j=1}^{Q} \varepsilon_j} \Phi_Q \left(s_{1|C:1}^{(\ell_1 + \varepsilon_1)}, \cdots, s_{Q|C:1}^{(\ell_Q + \varepsilon_Q)}; \widetilde{\mathbf{R}}_{|W_C:W_1} \right) \right\} \\
&+ \sum_i \left(\sum_{k=1}^{C} \log f_{Y_k}(y_{ik}) + \sum_{k=1}^{C-1} \log \left\{ \frac{\phi(t_{i,C-k|C:C-k+1})}{\phi(t_{ik})} \right\} \right) - \frac{N}{2} \sum_{k=1}^{C-1} \log(1 - \widetilde{\rho}_{W_k W_C}^2) \\
&- \frac{N}{2} \sum_{k=1}^{C-2} \log(1 - \widetilde{\rho}_{W_k W_{C-1}|W_C}^2) - \cdots - \frac{N}{2} \log(1 - \widetilde{\rho}_{W_1 W_2|W_C:W_3}^2).
\end{aligned}
\tag{10.26}
$$

where $\mathrm{I}\{\cdot\}$ is the indicator function. Here, $\boldsymbol{\Theta}$ contains marginal parameters (including cutpoints for $X_1, \cdots, X_Q$, and regression parameters $\boldsymbol{\gamma}_j$ and $\boldsymbol{\beta}_k$, $j = 1, \cdots, Q$, $k = 1, \cdots, C$) as well as the (conditional and marginal) correlations in (10.7). Note that the correlations in $\boldsymbol{\Theta}$ are a one-to-one re-parameterization of the marginal correlations in $\widetilde{\mathbf{R}}$. To see this, consider the case $C = Q = 2$. The likelihood function in this case involves marginal correlations $\widetilde{\rho}_{W_1 W_2}$, $\widetilde{\rho}_{W_1^* W_2}$, and $\widetilde{\rho}_{W_2^* W_2}$, and conditional correlations $\widetilde{\rho}_{W_1^* W_1|W_2}, \widetilde{\rho}_{W_2^* W_1|W_2}$, and $\widetilde{\rho}_{W_1^* W_2^*|W_1}$. Noting that

$$
\widetilde{\rho}_{W_1^* W_1} = \widetilde{\rho}_{W_1^* W_2} \widetilde{\rho}_{W_1 W_2} + \widetilde{\rho}_{W_1^* W_1|W_2} \sqrt{(1 - \widetilde{\rho}_{W_1^* W_2}^2)(1 - \widetilde{\rho}_{W_1 W_2}^2)},
\tag{10.27}
$$

$$
\widetilde{\rho}_{W_2^* W_1} = \widetilde{\rho}_{W_2^* W_2} \widetilde{\rho}_{W_1 W_2} + \widetilde{\rho}_{W_2^* W_1|W_2} \sqrt{(1 - \widetilde{\rho}_{W_2^* W_2}^2)(1 - \widetilde{\rho}_{W_1 W_2}^2)},
\tag{10.28}
$$

$$
\begin{aligned}
\widetilde{\rho}_{W_1^* W_2^*} &= \widetilde{\rho}_{W_1^* W_2} \widetilde{\rho}_{W_2^* W_2} + \widetilde{\rho}_{W_1^* W_1|W_2} \widetilde{\rho}_{W_2^* W_1|W_2} \sqrt{(1 - \widetilde{\rho}_{W_1^* W_2}^2)(1 - \widetilde{\rho}_{W_2^* W_2}^2)} \\
&+ \widetilde{\rho}_{W_1^* W_2^*|W_1} \sqrt{(1 - \widetilde{\rho}_{W_1^* W_1|W_2}^2)(1 - \widetilde{\rho}_{W_2^* W_1|W_2}^2)(1 - \widetilde{\rho}_{W_1^* W_2}^2)(1 - \widetilde{\rho}_{W_2^* W_2}^2)},
\end{aligned}
\tag{10.29}
$$

it is clear that $\widetilde{\mathbf{R}}$ can be recovered from $\boldsymbol{\Theta}$.

For the Gaussian copula model with shared random effects described in Section 10.2.2 in the case of clustered data, the (marginal) log-likelihood contribution of subject i is

$$
\ell_i(\boldsymbol{\Theta}) = \log \int_{-\infty}^{+\infty} L_i(\boldsymbol{\Theta}|b) f_{B_i}(b) db,
\tag{10.30}
$$

with $L_i(\boldsymbol{\Theta}|b) = \exp\{\ell_i(\boldsymbol{\Theta}|b)\}$ the (conditional) likelihood contribution of subject i, given the shared random effect B_i, where $\ell_i(\boldsymbol{\Theta}|b)$ takes the form (10.26) and $\boldsymbol{\Theta}$ contains the additional parameters $\lambda_{1j}, \lambda_{2k}$, $i = 1, \cdots, Q$, $k = 1, \cdots, C$, with $\widetilde{\mathbf{R}}$ now a conditional correlation matrix, given B_i. In the correlated random effects model, $\ell_i(\boldsymbol{\Theta})$ becomes

$$\ell_i(\boldsymbol{\Theta}) = \log \int_{\mathbb{R}^P} L_i(\boldsymbol{\Theta}|\mathbf{b}_{1i},\mathbf{b}_{2i})f_{\mathbf{B}_{1i},\mathbf{B}_{2i}}(\mathbf{b}_{1i},\mathbf{b}_{2i})d\mathbf{b}_{1i}d\mathbf{b}_{2i}, \tag{10.31}$$

where $L_i(\boldsymbol{\Theta}|\mathbf{b}_{1i},\mathbf{b}_{2i}) = \exp\{\ell_i(\boldsymbol{\Theta}|\mathbf{b}_{1i},\mathbf{b}_{2i})\}$, and $\ell_i(\boldsymbol{\Theta}|\mathbf{b}_{1i},\mathbf{b}_{2i})$ again takes the form of (10.26). Parameter vector $\boldsymbol{\Theta}$ contains, in addition to those in (10.30) above, the normal correlations $\tilde{\eta}_{B_{1j}B_{1j'}}$, $\tilde{\eta}_{B_{1j}B_{2k}}$, and $\tilde{\eta}_{B_{2k}B_{2k'}}$ $(j \neq j', k \neq k')$, between the random effects, assuming a Gaussian copula is used to construct the random effects distribution. The correlation matrix $\tilde{\mathbf{R}}$ is again conditional on the random effects $\mathbf{B}_{1i}$ and $\mathbf{B}_{2i}$.

The maximum likelihood estimate (MLE) $\hat{\boldsymbol{\Theta}}$ is obtained by maximizing $\ell(\boldsymbol{\Theta}) = \sum_{i=1}^N \ell_i(\boldsymbol{\Theta})$ using an iterative technique such as Newton–Raphson or quasi-Newton methods. It can be easily verified that $\hat{\boldsymbol{\Theta}}$ is consistent and asymptotically multivariate normal with mean $\boldsymbol{\Theta}$ and covariance matrix given by the inverse of the Fisher information matrix $E\{-h(\boldsymbol{\Theta})\} = E\{s(\boldsymbol{\Theta})s^\top(\boldsymbol{\Theta})\}$, where $s(\boldsymbol{\Theta}) = \partial\ell(\boldsymbol{\Theta})/\partial\boldsymbol{\Theta}$ is the score function and $h(\boldsymbol{\Theta}) = \partial^2\ell(\boldsymbol{\Theta})/\partial\boldsymbol{\Theta}\partial\boldsymbol{\Theta}^\top$ is the Hessian matrix. Standard errors (SEs) for $\hat{\boldsymbol{\Theta}}$ are calculated from diagonals of $\{s(\hat{\boldsymbol{\Theta}})s^\top(\hat{\boldsymbol{\Theta}})\}^{-1}$ or of $-h^{-1}(\hat{\boldsymbol{\Theta}})$, provided either matrix is invertible. Note that estimation can be done quite easily using standard software and statistical packages, such as PROC NLMIXED in SAS, which automatically includes SEs in its output. This is convenient in practice, as it obviates the need for deriving and coding of score and Hessian functions.

One drawback of MLE in this case however, has to do with the computation of the rectangle probabilities in (10.26), which involves repeated multidimensional integration; the same is true of (10.30), and especially of (10.31), where likelihood evaluations can take on prohibitive computational costs for large P, when integrating out the random effects from the conditional likelihood. Consequently, full likelihood inference might be difficult for high-dimensional settings; see Nikoloulopoulos and Karlis (2009). A possible remedy which achieves important computational economies is the use of composite likelihood methods; see, e.g., Varin et al. (2011) and Varin (2008). Chapter 9 includes an illustration of the method applied to mixed-data GLMMs.

A related method is Joe and Xu's (1996) method of inference function for margins (IFM), which estimates marginal parameters solely from margins and only uses the copula as basis for estimating association parameters. The method is particularly suited to our setting, where the marginal models and the dependence between outcomes are specified separately. Computational and statistical performance (i.e. bias and efficiency) of these methods has been shown to range from acceptably good to excellent by, among others, de Leon and Wu (2011) for copula-based joint models similar to ones considered here, and by de Leon and Carrière (2007) and de Leon (2005) for the GMDM and CGCM, respectively.

10.5 Analysis of ethylene glycol toxicity data

The purpose of developmental toxicity studies is to determine toxicity levels of chemical and pharmaceutical substances with the goal of properly regulating their use. They are typically done on laboratory animals, mostly pregnant mice, which are exposed to increasing dose levels of toxicants whose adverse effects (e.g., low birth weight, congenital malformations) on offspring are then observed. The U. S. National Toxicology Program conducted a series of such studies involving five different chemicals, one of which is ethylene glycol (EG), a common chemical used as antifreeze, brake fluid, paint solvent, etc., in a variety of industrial applications. The main concern about EG is its potentially toxic effects on humans, for whom its ingestion can be fatal.

The data we consider were previously analyzed by a number of authors, among them Gueorguieva and Agresti (2001), Regan and Catalano (2002, 1999a, 1999b), Catalano and Ryan (1992), and Fitzmaurice and Laird (1995); they are also analyzed in Chapters 8 and 9. In the study, a total of 94 pregnant mice, called dams, were exposed to EG at daily dose levels 0, 0.75, 1.5, and 3g/kg over a period of development of major fetal organs. The dams were then sacrificed, and their live fetuses were examined for various defects, including fetal weight and fetal malformations. Details of the study design are given by Price *et al.* (1985).

Two outcomes — believed to be indicative of toxicity — are considered: X_{ih}, a binary outcome indicating the absence ($X_{ih} = 0$) or presence ($X_{ih} = 1$) of malformations for fetus $h = 1, \cdots, n_i$, in litter $i = 1, \cdots, N$; and Y_{ih}, a continuous outcome representing fetal weight of fetus $h = 1, \cdots, n_i$, in litter $i = 1, \cdots, N$. The primary interest is the simultaneous effects of dose z_i on both outcomes. Summary statistics for fetal malformations and fetal weight for the $N = 1,028$ live fetuses from the 94 litters are displayed in Table 8.1 of Chapter 8. It is apparent that fetal weight decreases, on average, with increasing dose, with the mean weight ranging from 0.704g at the highest dose, to 0.972g at dose 0. Similarly, the proportion of malformed fetuses monotonically increases with dose, from less than 1% at dose 0, up to about 57% at dose 3g/kg.

In our analyses, we first ignore the clustering in the data by adopting model (10.7) with $C = Q = 1$, and assume that all $N = 1,028$ live fetuses are independent, notwithstanding that some of them came from the same dams. This will have consequences in the results as discussed in Section 10.5.1. We then incorporate the clustered nature of the data and employ the random effects models in Section 10.2.2. More details of the analyses, especially concerning quantitative risk assessment, are discussed in Wu and de Leon (2012).

Table 10.1 *Full and marginal (profile) likelihood estimates, their SEs, and t-values for EG data, ignoring clustering and assuming independence, and using the robit-normal model.*

Effect		Marginal			Full		
		Est	SE	t	Est	SE	t
Malformation							
Intercept	γ_1	−1.8846	0.0862	−21.86	−2.0596	0.1129	−18.24
Dose	γ_2	0.7549	0.0450	16.78	0.8353	0.0540	15.47
Degrees-of-freedom	ν	8	—	—	8	—	—
Fetal weight							
Intercept	β_1	0.9497	0.0048	197.85	0.9504	0.0052	182.77
Dose	β_2	−0.0909	0.0033	−27.55	−0.0911	0.0032	−28.50
Error SD	σ	0.1112	0.0029	38.34	0.1111	0.0024	46.29
Correlation							
Normal biserial	$\widetilde{\rho}$	−0.4397	0.0424	−10.37	−0.4386	0.0386	−11.36
Biserial	ρ	−0.3925	—	—	−0.3909	—	—
Kendall's tau	$\widetilde{\tau}$	−0.29	0.0301	−9.63	−0.2892	0.0274	−10.55

10.5.1 Analysis ignoring clustering

In what follows, we drop the index h and assume (X_i, Y_i) are independent, $i = 1, \cdots, N$. We assume a latent variable $Y_i^* \sim t_1(\mu_i^*, 1, \nu)$ (i.e., the univariate t-distribution with mean μ_i^*, unit scale, and degrees of freedom ν; see Liu, 2004) underlying fetal malformation indicator X_i such that $X_i = I\{Y_i^* > 0\}$, and consider the following marginal linear models relating the means μ_{1i} of Y_i^* and μ_{2i} of fetal weight $Y_i \sim N(\mu_{1i}, \sigma^2)$ to dose z_i:

$$\mu_i^* = \gamma_1 + \gamma_2 z_i = \mathbf{z}_i^\top \boldsymbol{\gamma}, \qquad \mu_i = \beta_1 + \beta_2 z_i = \mathbf{z}_i^\top \boldsymbol{\beta}, \qquad (10.32)$$

where $\mathbf{z}_i = (1, z_i)^\top$. A joint model to glue the marginal models in (10.32) is specified by a bivariate Gaussian copula with correlation $\widetilde{\rho} = corr[\Phi^{-1}\{F_{Y_i^*}(Y_i^*)\}, \Phi^{-1}\{F_{Y_i}(Y_i)\}]$. The choice of a marginal robit regression of X_i on z_i — using a t-latent distribution for Y_i^* — results in inferences that are robust to the presence of outliers. Liu (2004) also shows that robit models approximate both logit (with $\nu \approx 7$) and probit (with large ν) regressions, and thus provide a general approach to binary regression modeling. Note that this is exactly joint model (10.7) for $C = Q = 1$, with a normal continuous outcome and a t-latent variable for the binary outcome. From (10.7), we get the joint density of X_i and Y_i as

$$
f_{X_i,Y_i}(\ell,y) =
\begin{cases}
\frac{1}{\sigma}\phi\left(\frac{y-\mu_i}{\sigma}\right)\Phi\left(\frac{\frac{\tilde{\rho}}{\sigma}(y-\mu_i)-\Phi^{-1}(u_i)}{\sqrt{1-\tilde{\rho}^2}}\right) & , \text{if } \ell = 1 \\
\frac{1}{\sigma}\phi\left(\frac{y-\mu_i}{\sigma}\right)\Phi\left(\frac{\Phi^{-1}(u_i)-\frac{\tilde{\rho}}{\sigma}(y-\mu_i)}{\sqrt{1-\tilde{\rho}^2}}\right) & , \text{if } \ell = 0
\end{cases}
, \tag{10.33}
$$

where $u_i = P(Y_i^* \leq 0) = \int_{-\infty}^{0} f_{Y_i^*}(y^*; \mathbf{z}_{1i}, \boldsymbol{\gamma}, v)dy^*$, the cumulative distribution function of $t_1(\mu_i^*, 1, v)$ at 0. Given data $\{\ell_i, y_i, z_i\}$, $i = 1, \cdots, N$, the log-likelihood function is obtained from (10.26) as

$$
\begin{aligned}
\ell(\boldsymbol{\Theta}) = & -N\log\sigma + \sum_{i=1}^{N}\log\phi\left(\frac{y_i-\mu_i}{\sigma}\right) + \sum_{\forall\ell_i=1}\log\Phi\left(\frac{\frac{\tilde{\rho}}{\sigma}(y_i-\mu_i)-\Phi^{-1}(u_i)}{\sqrt{1-\tilde{\rho}^2}}\right) \\
& + \sum_{\forall\ell_i=0}\log\Phi\left(\frac{\Phi^{-1}(u_i)-\frac{\tilde{\rho}}{\sigma}(y_i-\mu_i)}{\sqrt{1-\tilde{\rho}^2}}\right),
\end{aligned} \tag{10.34}
$$

where $\boldsymbol{\Theta}^{\top} = (\boldsymbol{\gamma}^{\top}, \boldsymbol{\beta}^{\top}, \sigma, \tilde{\rho}, v)$.

Table 10.2 *PLEs, their SEs, and t-values for EG mice data, accounting for clustering, and using the robit-normal mixed model with shared cluster-level random effects, with only linear dose effect and with both linear and quadratic dose effects.*

Effect		Linear only			Linear & quadratic		
		Est	SE	t	Est	SE	t
Malformation							
Intercept	γ_1	−2.3974	0.1769	−13.55	−3.9621	0.4195	−9.44
Dose	γ_2	0.9301	0.0861	10.8	3.2463	0.4933	6.58
Dose^2	γ_3	—	—	—	−0.6258	0.1263	−4.95
Scale factor	λ_1	−0.6212	0.0822	−7.55	−0.5731	0.0926	−6.19
Degrees-of-freedom	v	8	—	—	4	—	—
Fetal weight							
Intercept	β_1	0.9586	0.0125	76.87	0.9789	0.0139	70.57
Dose	β_2	−0.0886	0.0077	−11.46	−0.1606	0.0261	−6.15
Dose^2	β_3	—	—	—	0.0239	0.0084	2.83
Scale factor	λ_2	0.0848	0.0064	13.22	0.0829	0.0069	12.01
Error SD	σ	0.0749	0.0018	41.61	0.0748	0.0018	41.56
Correlation							
Conditional normal biserial	$\tilde{\rho}$	−0.133	0.0553	−2.4	−0.1458	0.0564	−2.58
Conditional biserial	ρ	−0.1094	—	—	−0.12	—	—

To avoid problems associated with estimating the degrees of freedom v (Kotz and Nadarajah, 2004, pp. 235–236), we employ the method of profile likelihood (Song et al., 2007). The method entails maximizing the log-likelihood at fixed grid points $v \in (2, M]$, for some suitably large constant M. Our estimates of γ_1, γ_2, β_1, β_2, σ, and $\tilde{\rho}$, correspond to those obtained for $\hat{v}$, at which the profile likelihood is maximum on $(2, M]$. This method is easy to implement using dense grid points on $(2, M]$, and is more computationally efficient than full MLE. Asymptotic properties analogous to those for MLEs can be similarly established.

Alternatively, we adopt the IFM method to estimate marginal parameters γ_1, γ_2, β_1, β_2, σ, and v, from the respective marginal log-likelihoods, and use the joint log-likelihood only to estimate the correlation $\widetilde{\rho}$. For the binary part, we use the method of (marginal) profile likelihood by maximizing the (marginal) t-profile log-likelihood at fixed grid points $v \in (2, M]$; estimates of the continuous data parameters are easily obtained by least squares. To estimate $\widetilde{\rho}$, we maximize the (joint) log-likelihood evaluated at the marginal estimates. Asymptotic results similar to those for joint estimates apply as well to these estimates.

To remove the constraints on $\widetilde{\rho} \in (-1, 1)$, it was re-parameterized as Fisher's z-transformation $\log\{(1+\widetilde{\rho})/(1-\widetilde{\rho})\}/2$.

From Table 10.1, we can see that the full and marginal likelihood estimates are very close, especially for outcome fetal weight. Estimates of the regression coefficient γ_2 for dose in the robit regression of fetal malformation on dose indicates that the probability of malformation increases with increasing dose. Similarly, estimates of the regression coefficient β_2 for dose in the linear model of fetal weight on dose indicate that the mean fetal weight decreases with increasing dose. Note that SEs for both sets of estimates are very close. Estimates of normal biserial correlation coefficient $\widetilde{\rho}$ suggest a moderate negative association between fetal weight and the latent variable underlying fetal malformation. This suggests that fetal weight and fetal malformation are moderately negatively associated. The estimate of Kendall's tau is also displayed along with its SE obtained by the delta method. The t-values suggest that both dose coefficients for fetal malformation and fetal weight are significantly positive and negative, respectively.

Comparing these estimates against previous values reported by Fitzmaurice and Laird (1995) and Catalano and Ryan (1992), we see that our estimates, both jointly and marginally obtained, are relatively close in magnitude. The closeness of estimates is not surprising, since the degrees of freedom estimate of 8 in both cases is large enough for the t-latent distribution to approximate a standard normal latent distribution. However, SEs are noticeably smaller than earlier estimates; this is explained by the fact that we ignored the clustered nature of the data, and assumed them to be independent, effectively assuming that there are more subjects providing independent information than is in fact the case, resulting in underestimated SEs. Note that without clustering, our effective sample size is 1,028, but with clustering, we have a sample size of only 94, about a tenth of the former.

10.5.2   Analysis accounting for clustering

In this section, we reanalyze the EG data by incorporating the clustering in the data via the models in Section 10.2.2. We first assume a shared cluster-level random effect B_i, and suppose a latent variable $Y_{ih}^*|B_i \sim t_v(\mu_i^*(B_i), 1)$, such that $X_{ih} = I\{Y_{ih}^* > 0\}$, and $Y_{ih}|B_i \sim N(\mu_i(B_i), \sigma^2)$, where the conditional means are given by the following (conditional) linear mixed models

$$\mu_i^*(B_i) = z_i^\top \gamma + \lambda_1 B_i = \mu_i^* + \lambda_1 B_i, \qquad \mu_i(B_i) = z_i^\top \beta + \lambda_2 B_i = \mu_i + \lambda_2 B_i, \quad (10.35)$$

for $h = 1, \cdots, n_i$, $i = 1, \cdots, N$. We assume litters (i.e., dams) are independent, and $B_1, \cdots, B_N \overset{iid}{\sim} N(0, 1)$, where "$\overset{iid}{\sim}$" means "independent identically distributed." To assemble the conditional joint model $f_{X_{ih}, Y_{ih}|B_i}(\cdot|\cdot)$ (given B_i), we adopt the Gaussian copula with (conditional) biserial normal correlation $\widetilde{\rho} = corr[\Phi^{-1}\{F_{Y_i^*|B_i}(Y_i^*|B_i)\}, \Phi^{-1}\{F_{Y_i|B_i}(Y_i|B_i)\}]$. Note that $f_{X_{ih}, Y_{ih}|B_i}(\cdot|\cdot)$ is exactly model (10.9) with $C = Q = 1$, and is of the form (10.33); hence, we refer to the model as the robit-normal mixed model. As in the robit-normal model in Section 10.5.1, the latent mixed model in (10.35) is equivalent to a mixed effects logistic regression model for X_{ih} for $v \approx 7$; with large v, it yields a mixed effects probit regression model. Note that this generalizes Najita et al.'s (2010) correlated bivariate probit model by allowing for non-normal residual errors. It reduces to Najita et al.'s (2010) model when a normal latent distribution is assumed for Y_i^*.

Several numerical methods (e.g., Gauss–Hermite quadrature, importance sampling) are available in standard statistical packages, such as PROC NLMIXED in SAS, for evaluating the marginal log-likelihood function in (10.30). The same profile likelihood method outlined in Section 10.5.1 is used in this case to avoid problems with estimating the degrees of freedom v.

The resulting estimates for the robit-normal mixed model for the EG mice data are reported in Table 10.2. We can see that the estimate of γ_2, in the robit regression of fetal malformation on dose, indicates that the conditional malformation probability increases with increasing dose. Similarly, the estimate of β_2, in the linear mixed model of fetal weight on dose, suggests that the mean fetal weight decreases with increasing dose. The dose coefficient for fetal malformation and weight are both statistically significant. The estimate of correlation $\tilde{\rho}$ suggests that a slightly negative association exists between fetal weight and the latent variable underlying fetal malformation. Kugiumtzis and Bora–Senta's (2010) piecewise linear approximation method yields an estimated conditional biserial correlation $\hat{\rho} = \widehat{corr(Y_{ih}^*, Y_{ih}|B_i)}$ between fetal malformation and fetal weight of -0.1094 for the model with only linear dose effect, and -0.12 for that with both linear and quadratic dose effects. Estimates of marginal correlations in (10.20)–(10.23) are reported in Table 10.4. Note that $var(Y_{ih}^*|B_i) = v/(v-2) \neq 1$, since we assumed unit scale in this case.

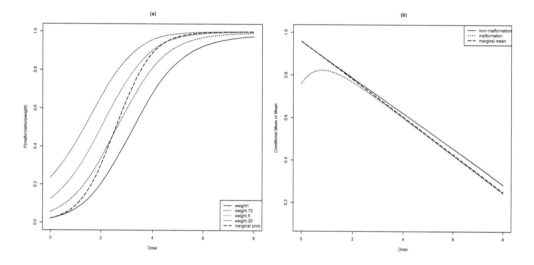

Figure 10.1 *Plots of estimated conditional (a) fetal malformation probability (given fetal weight) and (b) mean fetal weight (given malformation indicator), as functions of dose, for the robit-normal mixed model with shared cluster-level random effects and with linear dose effect. The estimated marginal fetal malformation probability and marginal mean fetal weight, both as functions of dose, are also shown for comparison.*

From the joint model, we can obtain the conditional distribution of fetal malformation as a function of dose, given fetal weight, or vice versa. To see this, we obtained the conditional probability of malformation at fixed doses, given fetal weight, as

$$
P(X_{ih} = \ell | Y_{ih} = y; z_i) = \begin{cases} \int_{-\infty}^{+\infty} \Phi\left(\dfrac{\frac{\tilde{\rho}}{\sigma}\{y - \mu_i(b_i)\} - \Phi^{-1}\{u_{ih}^*(b_i)\}}{\sqrt{1 - \tilde{\rho}^2}} \right) \phi(b_i) db_i & \text{, if } \ell = 1 \\[3mm] \int_{-\infty}^{+\infty} \Phi\left(\dfrac{\Phi^{-1}\{u_{ih}^*(b_i)\} - \frac{\tilde{\rho}}{\sigma}\{y - \mu_i(b_i)\}}{\sqrt{1 - \tilde{\rho}^2}} \right) \phi(b_i) db_i & \text{, if } \ell = 0 \end{cases},
$$

where $\mu_i(b_i) = \mathbf{z}_i^\top \boldsymbol{\beta} + \lambda_1 b_i$, $u_{ih}(b_i) = P(Y_{ih}^* \leq 0 | B_i = b_i)$ and $\mathbf{z}_i = (1, z_i)^\top$; similarly, we get conditional mean fetal weight, as a function of dose, as

$$
E(Y_{ih} | X_{ih} = \ell; z_i) = \begin{cases} \int_{-\infty}^{+\infty} \left(\mu_i^*(b_i) + \dfrac{\tilde{\rho}\sigma}{1 - u_{ih}^*(b_i)} \phi[\Phi^{-1}\{u_{ih}^*(b_i)\}] \right) \phi(b_i) db_i & \text{, if } \ell = 1 \\[3mm] \int_{-\infty}^{+\infty} \left(\mu_i(b_i) - \dfrac{\tilde{\rho}\sigma}{u_{ih}^*(b_i)} \phi[\Phi^{-1}\{u_{ih}^*(b_i)\}] \right) \phi(b_i) db_i & \text{, if } \ell = 0 \end{cases},
$$

where $\mu_i^*(b_i) = \mathbf{z}_i^\top \boldsymbol{\alpha} + \lambda_2 b_i$. We plotted the estimated conditional malformation probability as a function of dose level $z = 0, 0.75, 1.5, 3\text{g/kg}$, at each fixed fetal weight $y = 0.25, 0.5, 0.75, 1\text{g}$, in

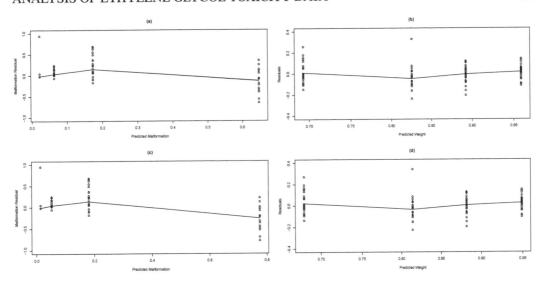

Figure 10.2 *LOESS-smoothed plots of marginal residuals against predicted malformation (a and c) and against predicted weight (b and d) using mixed models with shared and correlated random effects, and linear dose effect. Plots (a) and (b) are based on the shared random effects model, while plots (c) and (d) are based on the correlated random effects model.*

Figure 10.1a, and the conditional mean fetal weight, as a function of dose, given malformation indicator $X = 0$ or $X = 1$, in Figure 10.1b. Observe that for given fetal weight, the estimated conditional probability of malformation increases with the dose; we also find that larger fetuses are less likely to have malformations. Given the malformation indicator, the conditional mean fetal weight decreases with dose. For comparisons, we include in Figure 10.1 the estimated marginal probability of malformation and marginal mean fetal weight (shown as dashed lines), as functions of dose.

The fit of the joint model can be assessed by plotting residuals against predicted values. Figures 10.2a and 10.2b plot average malformation residuals (Fitzmaurice and Laird, 1995) and fetal weight residuals against predicted marginal values. LOESS-smoothed curves are added to detect any apparent trends. Some evidence of a curvilinear trend in the plots suggests that a quadratic effect of dose should be considered. The results for the model with a quadratic dose effect is also displayed in Table 10.3, including estimates of marginal correlations. The t-values of the quadratic dose coefficients are found to be statistically significant, indicating that it is necessary to incorporate the effect in the model. Figures 10.3a and 10.3b support this, as no apparent trend can be seen in the residual plots.

We now consider a conditional joint model with correlated cluster-level random effects and a logistic-latent distribution for Y_{ih}^*, which implies a marginal logistic regression model for the binary outcome. Our model generalizes Lin *et al.*'s (2010) and Gueorguieva and Agresti's (2001) models by allowing for non-normal residual errors. Given correlated random effects B_{1i} and B_{2i}, we have the following mixed models for the conditional means $\mu_i^*(B_{1i})$ and $\mu_i(B_{2i})$:

$$\mu_i^*(B_{1i}) = \mathbf{z}_i^\top \boldsymbol{\gamma} + \lambda_1 B_{1i} = \mu_i^* + \lambda_1 B_{1i}, \quad \mu_i(B_{2i}) = \mathbf{z}_i^\top \boldsymbol{\beta} + \lambda_2 B_{2i} = \mu_i + \lambda_2 B_{2i}, \quad (10.36)$$

for $h = 1, \cdots, n_i$, $i = 1, \cdots, N$. For ease of interpretation and to allow the marginal model for the binary outcome to have a logistic structure, we assume a bridge distribution for B_{1i} as follows:

$$f_{B_{1i}}(b_{1i}) = \frac{\sin(\phi\pi)}{2\pi\{\cosh(\phi b_{1i}) + \cos(\phi\pi)\}}, \quad (10.37)$$

for any real number b_{1i}, where ϕ is the attenuation parameter. A discussion of the bridge distribution and its nice marginalization properties is given in Lin *et al.* (2010). With $B_{2i} \sim N(0, 1)$, a

Table 10.3 *MLEs, their SEs, and t-values for EG mice data, accounting for clustering, and using the logit-normal mixed model with correlated cluster-level random effects, with only linear dose effect and with both linear and quadratic dose effects.*

Effect		Linear only			Linear & quadratic		
		Est	SE	t	Est	SE	t
Malformation							
Intercept	γ_1	−4.2648	0.3989	−10.94	−5.85	0.6951	−8.42
Dose	γ_2	1.8344	0.2046	8.97	5.0137	0.9079	5.52
Dose^2	γ_3	—	—	—	−1.0042	0.2451	−4.1
Scale factor	λ_1	2.0101	0.2065	7.72	1.6443	0.2309	7.12
Fetal weight							
Intercept	β_1	0.9489	0.0133	71.35	0.9792	0.0148	66.07
Dose	β_2	−0.0899	0.0075	−12.02	−0.1668	0.026	−6.42
Dose^2	β_3	—	—	—	0.0278	0.0081	3.45
Scale factor	λ_2	0.0942	0.0081	11.65	0.0807	0.0059	13.62
Error SD	σ	0.0746	0.0017	43.13	0.0746	0.0017	43.09
Correlations							
Normal correlation for random effects	$\tilde{\eta}$	−0.7221	0.057	−12.68	−0.5877	0.077	−7.64
Correlation for random effects	η	−0.6931	—	—	−0.559	—	—
Conditional normal biserial	$\tilde{\rho}$	−0.2085	0.0548	−3.8	−0.2136	0.0551	−3.88
Conditional biserial	ρ	−0.1736	—	—	−0.1772	—	—

NOTE: *The logit-normal mixed model above has correlated random effects having marginal bridge (for binary outcome) and normal (for continuous outcome) distributions.*

Table 10.4 *Estimates of marginal (including tetrachoric and biserial) correlations for the robit-normal mixed model with shared random effects and both linear and quadratic dose effects, and for the logit-normal mixed model with correlated cluster-level random effects with both linear and quadratic dose effects.*

Correlation	Robit-normal shared	Logit-normal correlated
$corr(Y_{ih}^*, Y_{ih})$	−0.3533	−0.1624
$corr(Y_{ih}^*, Y_{ih'}^*)$	0.141	0.0118
$corr(Y_{ih}, Y_{ih'})$	0.5508	0.5392
$corr(Y_{ih}^*, Y_{ih'})$	−0.2787	−0.0429

joint density for B_{1i} and B_{2i} is constructed using the Gaussian copula with correlation coefficient $\tilde{\eta} = corr[\Phi^{-1}\{F_{B_{1i}}(B_{1i})\}, \Phi^{-1}\{F_{B_{2i}}(B_{2i})\}]$. Note that given B_{1i} and B_{2i}, the conditional joint model $f_{X_{ih}, Y_{ih}|B_{1i}, B_{2i}}(\cdot|\cdot)$ is still of the form (10.33), albeit with a logistic latent distribution for Y_{ih}^*. This latter specification implies that X_{ih} is modeled via a logistic mixed model.

This model is fit using PROC NLMIXED in SAS, and the results are displayed in Table 10.3. The positive estimated dose effect for fetal malformation and the negative estimated dose effect on fetal weight are consistent with previous analyses, including ours above based on a joint model with shared cluster-level random effects.

LOESS-smoothed residual plots for the model are shown in Figures 10.2c and 10.2d. Note that these suggest the need to incorporate a quadratic term in dose for the joint model. Results of fitting this model are reported in Table 10.3, with the additional coefficients γ_3 and β_3 for the quadratic dose

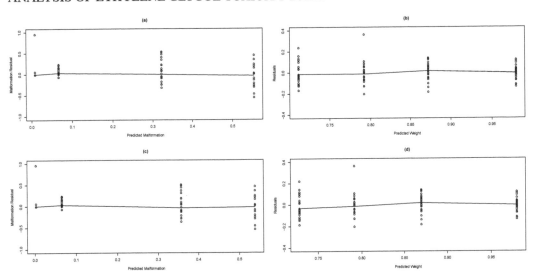

Figure 10.3 *LOESS-smoothed plots of marginal residuals against predicted malformation (a and c) and against predicted weight (b and d) using mixed models with shared and correlated random effects, and quadratic, in addition to linear, dose effect. Plots (a) and (b) are based on the shared random effects model, while plots (c) and (d) are based on the correlated random effects model.*

Table 10.5 *AIC, BIC and LR values for the robit-normal mixed model with shared cluster-level random effects (with and without quadratic dose effect) and for the logit-normal mixed model with correlated cluster-level random effects (with and without quadratic dose effect).*

| Model fit | Robit-normal shared | | Logit-normal correlated | |
criterion	Linear only	With quadratic	Linear only	With quadratic
LR	$-1,429$	$-1,453$	$-1,306$	$-1,322$
AIC	$-1,413$	$-1,433$	$-1,288$	$-1,300$
BIC	$-1,392$	$-1,408$	$-1,265$	$-1,272$

effects, which are found to be statistically significant. Corresponding LOESS-smoothed residual plots are displayed in Figures 10.3c and 10.3d. It is clear that including the quadratic dose effect has improved the fit of the joint model.

Using the estimate of $\widetilde{\rho} = corr[\Phi^{-1}\{F_{Y_i^*|B_{1i}}(Y_i^*|B_{1i})\}, \Phi^{-1}\{F_{Y_i|B_{2i}}(Y_i|B_{2i})\}]$ for the logit-normal mixed model with correlated random effects and with both linear and quadratic effects of dose, the piecewise linear approximation method can again be used to obtain estimate of the conditional biserial correlations $\rho = corr(Y_{ih}^*, Y_{ih}|B_{1i}, B_{2i})$ between fetal malformation and fetal weight; see Table 10.3, where the estimate of $\eta = corr(B_{1i}, B_{2i})$, obtained by piecewise linear approximation based on the estimate of $\widetilde{\eta}$, is shown as well. The estimate of ρ is used to obtain estimates of the marginal correlations. These are displayed in Table 10.4. Note that $var(Y_{ih}^*|B_i) = \pi^2/3 \neq 1$, since we assumed unit scale in this case.

Finally, we report in Table 10.5 the Akaike's information criterion (AIC) and Bayesian information criterion (BIC) values, along with the likelihood ratio statistic (LR=$-2 \times$ log-likelihood), for the models considered. Note that the shared random effects robit-normal mixed model with a quadratic dose effect has smaller LR, AIC and BIC values than that with only the linear dose effect. Similarly, the correlated random effects logit-normal mixed model with the quadratic dose effect has much smaller LR, AIC and BIC values than that without the quadratic dose effect. This indicates that the models incorporating a quadratic term for dose provide better fits than those with only the linear dose effect.

10.6 Discussion

In this chapter, we developed a Gaussian copula-based joint regression model for mixed multivariate discrete and continuous outcomes. We also introduced random effects to incorporate within-cluster associations in clustered data settings. The marginal regression models are specified using GLMMs linking the outcomes' means to covariates, conditional on random effects. An attractive feature of the joint models is their use of copulas to separately model dependencies between outcomes, thereby preserving the outcomes' distinct marginal properties. They thus offer a flexible alternative to conventional approaches that generally rely on the assumption that outcomes, or some suitable transformations of them, follow a Gaussian distribution. The models adopt a latent variable description of discrete variables, a quite appealing formulation of discrete outcomes in developmental toxicology, for example, where a latent formulation can provide a natural description of the underlying biological process. It also makes statistical sense, as common regression models for discrete (i.e., binary) responses (e.g., logistic, probit) can be equivalently formulated in terms of latent variables. In addition, it also leads to familiar association measures for capturing dependence between two discrete outcomes (i.e., polychoric correlation) or between mixed discrete and continuous outcomes (i.e., polyserial correlation).

The analysis of correlated mixed outcomes raises many difficult statistical issues. The approach we outlined in the chapter makes exclusive use of the Gaussian copula to describe the mixed-outcome joint model, as in Song *et al.* (2009). Even though Gaussian copulas are mathematically and computationally tractable and have been used in a variety of applications, they may not work in all cases. Other copula families could be considered as well, and the choice of the copula could then be investigated via various model selection criteria. As well, our joint model may not be appropriate for mixed continuous and count outcomes. The same is true if the outcomes comprise nominal in addition to continuous and count measurements. While suitable for ordinal outcomes, the notion of continuous latent variables underlying nominal outcomes may not be appropriate.

Chapter 11

Analysis of mixed outcomes in econometrics: Applications in health economics

David M. Zimmer

11.1 Introduction

Passed by the U. S. Congress in 1965, the original Medicare program did not include coverage for prescription drugs. Although this omission did not initially generate much political attention, rapid growth in spending on pharmaceuticals during subsequent decades fostered political pressure to include prescription drug coverage under the umbrella of Medicare. The discussions became especially heated in the early 2000s as drug spending, as a percentage of total medical spending, increased from 8.6% in 1993 to almost 13% by 2003 (*OECD Health Data, 2005*). In response, the Bush Administration signed the Medicare Modernization Act (MMA) in 2003, representing the largest overhaul in Medicare's 38-year history.

The MMA introduced optional prescription drug coverage, beginning in 2006, widely known as Medicare Part D. Initially, the program experienced lower-than-expected enrollment, largely due to confusion about enrollment options. As these confusions dissipated, however, the program experienced rapid growth. By mid 2006, more than 90% of seniors had some form of prescription drug coverage, compared with 67% before 2006 (Khan and Kaestner, 2009; Levy and Weir, 2010; Engelhardt and Gruber, 2010).

Justification for passage of the MMA rested, in part, on two widely-held beliefs. The first belief is that unhealthy individuals need additional financial protection because unhealthy individuals incur more drug expenses. Despite this seemingly intuitive reasoning, empirical evidence confirming a positive link between health problems and drug expenses is tenuous. Presumably, the link could be investigated by examining the conditional density

$$f(\text{drug spending}|\text{health problem}).$$

Yet, individuals presumably purchase health care services in order to improve their health status, which implies that conditioning might run in the opposite direction. Indeed, the standard model in health economics views health care services as inputs into the production of health (Grossman, 1972).

Ambiguity regarding conditioning suggests that a more appropriate model would assume the form

$$f(\text{drug spending}, \text{health problem}).$$

Then relationships between the two outcomes could be analyzed from the joint model. In practice, however, specifying such a joint density encounters difficulties due to how medical spending and health status are typically measured in household surveys. Specifically, this chapter uses data from the Medical Expenditures Panel Survey (MEPS), the most widely-used household survey for U. S. healthcare analysis. The MEPS records medical spending as annual dollars spent (both by patients and their insurers), and thus, should be treated as a continuous measure. The presence of a health problem, on the other hand, is measured as a discrete dummy indicator for whether the subject reports "fair or poor" health.

The second widely-held belief used to justify passage of the MMA is that increased usage of prescription drugs reduces spending on other, potentially costlier, health care services, and therefore, increased drug consumption might reduce aggregate medical spending. Achieving such cost-offsets has long been a goal of policymakers, especially with Medicare spending projected to consume all tax revenues by as early as the year 2050 (Deb *et al.*, 2009; Zhang *et al.*, 2009; Goldman *et al.*, 2007; Chandra *et al.*, 2007). Presumably, the link between drug usage and nondrug spending could be analyzed by examining the conditional density

$$f(\text{nondrug spending}|\text{drug usage}).$$

However, the conditioning could be reversed, because in order for a patient to consume a prescription drug, he usually must first acquire a prescription from a physician. Thus, at least some nondrug spending might precede drug usage.

This lack of obvious conditioning direction suggests moving to a joint density

$$f(\text{nondrug spending}, \text{drug usage}).$$

But similar to above, while MEPS records nondrug spending as annual dollars spent, and thus, should be treated as continuous, drug usage is recorded as the number of annual prescriptions (including refills), which follows a discrete count distribution.

In sum, empirical confirmation of the two widely-held beliefs upon which passage of the MMA was based both appear to call for joint modeling of mixed continuous-discrete outcomes. In addition to the fact that economic theory rarely provides justification for clearly-defined causal relationships suitable for univariate marginal analysis, economists might wish to analyze joint models for two other reasons. First, joint analysis offers efficiency improvements in contrast to separate analysis of each outcome. Such efficiency gains are especially important when working with small sample sizes. Second, research conclusions often hinge upon measures of dependence between two variables, including subtle issues like asymmetric dependence and tail dependence. Univariate (or conditional univariate) models are not suited for assessing such relationships.

One approach for forming a joint distribution of mixed outcomes is to factorize it into a conditional distribution of one outcome given the other, and a marginal distribution for the conditioning variable. However, as already noted, in analyzing the above-mentioned questions regarding MMA, it is desirable to avoid such conditional relationships. More generally, such a factorization introduces a hierarchy in outcomes, with the conditioning outcome treated as an intermediate variable (de Leon and Carrière Chough, 2010). Such a hierarchy might not be appropriate in certain applied settings. See Chapter 9 for a discussion of certain types of hierarchical models.

Other methods for modeling mixed distributions have appeared in the statistics literature, but to date those methods have received little attention in economics research, either due to computational complexity or lack of generality. One such approach, the quadratic exponent model, accommodates any mixture of outcomes from an exponential family (Sammel *et al.*, 1997). Another approach, introduced by Regan and Catalano (2002), extends the correlated probit model for binary outcomes to incorporate continuous outcomes. Yet another approach, often called the Plackett–Dale approach, derives a mixed density by defining a global cross-ratio between the outcomes (Molenberghs *et al.*, 2001). Song (2007) elaborates on these and other methods.

This chapter models mixed outcomes using two approaches, random effects models and copula functions, that share three features common to model classes most commonly estimated by economists. First, both approaches are fully parametric. Second, both methods rely on full-information maximum likelihood estimation (MLE) methods. Third, both methods use regression-type structures, in which marginal distributions depend upon a set of explanatory covariates. In developing these models, this chapter seeks to investigate two questions about medical spending and drug usage: (1) Does drug spending correlate with health problems? (2) Does drug consumption correlate with reduced nondrug spending? Investigating both questions requires models for mixed outcomes. See Chapter 8 for an in depth treatment of random effects models, and Chapter 10 for an exposition on copula models for mixed data.

The remainder of the chapter is as follows. Section 11.2 presents a version of a random effects model that has appeared elsewhere in the economics literature. Section 11.3 presents an alternative approach based on copula functions. Section 11.4 applies these two modeling approaches to analyze the link between drug spending and health problems. Section 11.5 applies the models to investigate whether drug usage correlates with reduced spending on nondrug services. The final section concludes the chapter.

11.2 Random effects models

Let Y_{1i} measure individual i's annual drug spending, and consider a sample of N individuals, with $\mathbf{Y}_1 = (Y_{11}, \cdots, Y_{1N})^\top$. Then, Y_{1i} follows a continuous density parametrically specified as $f_1(y_{1i}; \mathbf{x}_i^\top \boldsymbol{\beta}_1)$, where $\mathbf{x}_i$ represents a $k \times 1$ vector of explanatory variables with a $k \times 1$ vector $\boldsymbol{\beta}_1$ of estimable coefficients, where k denotes the number of explanatory variables, including a constant. The density might also include ancillary terms, such as shape parameters, which are omitted for notational brevity. Variables in $\mathbf{x}_i$ include person-specific characteristics expected to influence drug spending, including age, gender, race, and marital status.

Let Y_{2i} be a discrete indicator for the presence of a health problem, with $\mathbf{Y}_2 = (Y_{21}, \cdots, Y_{2N})^\top$. Then, Y_{2i} follows a discrete density parametrically specified as $f_2(y_{2i}; \mathbf{x}_i^\top \boldsymbol{\beta}_2)$. The same explanatory variables impact both drug spending and the presence of a health problem, although this need not be the case in general.

The previous two paragraphs are couched in terms of the first of the two above-mentioned questions: Does drug spending correlate with health problems? For the second question, which concerns links between nondrug spending and drug usage, Y_{1i} is a continuous measure of annual nondrug spending, and Y_{2i} is a discrete measure of the annual number of prescribed medicines (including refills). The presentation in the following subsections applies to both questions.

11.2.1 Specification of joint distribution

One approach to specifying a mixed outcome joint distribution introduces shared random effects in the marginal distributions. Thus, the marginal distributions for Y_{1i} and Y_{2i} are specified as $f_1(y_{1i}; \mathbf{x}_i^\top \boldsymbol{\beta}_1, \eta_i)$ and $f_2(y_{2i}; \mathbf{x}_i^\top \boldsymbol{\beta}_2, \eta_i)$, where Y_{1i} and Y_{2i} are assumed to be conditionally independent, given the shared random effect η_i. Since η_i is not observed, but assumed to be independent and identically distributed, with known distribution given by $h(\eta_i)$, it can be integrated out of the marginal densities:

$$f(y_{1i}, y_{2i}; \boldsymbol{\beta}_1, \boldsymbol{\beta}_2) = \int f_1(y_{1i}; \mathbf{x}_i^\top \boldsymbol{\beta}_1, \eta_i) f_2(y_{2i}; \mathbf{x}_i^\top \boldsymbol{\beta}_2, \eta_i) h(\eta_i) d\eta_i, \tag{11.1}$$

with the integral calculated over the support of η_i. The joint density then follows from (11.1).

The integral in (11.1) does not, in general, integrate to an analytical form. It can, however, be approximated by Monte Carlo integration by randomly simulating values of η_i from its distribution $h(\cdot)$, and averaging over all simulations, i.e.,

$$\widetilde{f}(y_{1i}, y_{2i}; \boldsymbol{\beta}_1, \boldsymbol{\beta}_2) = \frac{1}{S} \sum_{s=1}^{S} f_1(y_{1i}; \mathbf{x}_i^\top \boldsymbol{\beta}_1, \widetilde{\eta}_{si}) f_2(y_{2i}; \mathbf{x}_i^\top \boldsymbol{\beta}_2, \widetilde{\eta}_{si}), \tag{11.2}$$

where $\widetilde{\eta}_{si}$ denotes a simulated value of η_i.

Numerous options exist for how the random effect term η_i enters into each marginal. In estimates reported below, these terms enter as

$$f_1(y_{1i}; \mathbf{x}_i^\top \boldsymbol{\beta}_1 + \lambda_1 \tilde{\eta}_{si}) \quad \text{and} \quad f_2(y_{2i}; \mathbf{x}_i^\top \boldsymbol{\beta}_2 + \lambda_2 \tilde{\eta}_{si}). \tag{11.3}$$

In the presentation of this model in the statistics literature, the coefficients λ_1 and λ_2 are included to adjust for differences in scales between the two outcomes (McCulloch, 2008). Of course, the coefficients serve a similar purpose in economic applications, but in the economic literature, they are called loading factors, and are interpreted as capturing dependence between outcomes. Each cannot be separately identified (Deb and Trivedi, 2006), so estimates below restrict $\lambda_2 = 1$, so that λ_1 provides a measure of dependence. Economists refer to this specification as a one-factor model, while statisticians refer to it as a shared random effects model, as dependence between the two outcomes derives from one latent term η_i. Alternatively, one could estimate what economists call a two-factor model, and what statisticians call a correlated random effects model, by simulating two latent terms, say, $\tilde{\eta}_{1si}$ and $\tilde{\eta}_{2si}$, from a bivariate normal distribution with an estimable correlation term that captures dependence (Munkin and Trivedi, 1999). This chapter does not pursue this approach for reasons discussed below.

11.2.2 Maximum likelihood estimation

With the simulated joint density, the likelihood function is formed as

$$L(\boldsymbol{\beta}_1, \boldsymbol{\beta}_2, \lambda_1; \mathbf{y}_1, \mathbf{y}_2) = \sum_{i=1}^{N} \log \tilde{f}(y_{1i}, y_{2i}; \boldsymbol{\beta}_1, \boldsymbol{\beta}_2, \lambda_1), \tag{11.4}$$

and the likelihood function is maximized with respect to $\boldsymbol{\Omega} = (\boldsymbol{\beta}_1^\top, \boldsymbol{\beta}_2^\top, \lambda_1)^\top$. This maximization proceeds numerically using a modified iterative Newton–Raphson algorithm, with numerically approximated first and second derivatives. Economists refer to this estimation method as maximum simulated likelihood estimation (MSLE) (Bhat et al., 2010).

Letting $\mathbf{g}(\widehat{\boldsymbol{\Omega}}) = \partial \log L / \partial \boldsymbol{\Omega}|_{\boldsymbol{\Omega} = \widehat{\boldsymbol{\Omega}}}$ denote the gradient vector, and

$$\mathbf{H}(\widehat{\boldsymbol{\Omega}}) = \left. \frac{\partial^2}{\partial \boldsymbol{\Omega} \partial \boldsymbol{\Omega}^\top} \log L(\boldsymbol{\Omega}) \right|_{\boldsymbol{\Omega} = \widehat{\boldsymbol{\Omega}}}$$

the Hessian matrix, both evaluated at the converged parameter values; the covariance matrix is calculated using the robust sandwich formula

$$var(\widehat{\boldsymbol{\Omega}}) = \mathbf{H}^{-1}(\widehat{\boldsymbol{\Omega}}) \left\{ \frac{N}{N-1} \sum_{i=1}^{N} \mathbf{g}_i(\widehat{\boldsymbol{\Omega}}) \mathbf{g}_i(\widehat{\boldsymbol{\Omega}})^\top \right\} \mathbf{H}^{-1}(\widehat{\boldsymbol{\Omega}}). \tag{11.5}$$

The square roots of the diagonal elements of this matrix provide standard errors (SEs) for the MSLEs.

11.2.3 Quasi-random number generation

The final detail involves the number S of simulation repeats to use in the numerical integration. Gourieroux and Monfort (1996) show that if $S, N \to \infty$ and $\sqrt{N}/S \to 0$, then the MSLE is asymptotically equivalent to the conventional MLE. Aside from an endorsement for large S, this asymptotic result does not provide guidance for choosing the actual number of simulations. In practice, researchers typically continue to estimate the model with increasingly larger numbers of simulations until converged parameter estimates and the maximized log-likelihood value appear to stabilize.

Recent applications of MSLE have achieved stable convergence using fewer simulations by employing quasi-random draws, as opposed to traditional pseudo-random draws, based on Halton sequences (Bhat, 2001). To form a Halton sequence, begin with a prime number, say 2; note that Halton sequences for non-primes are not unique. Divide the unit interval into that many equal parts, with the cutpoint, in this example $1/2$, serving as the first number in the Halton sequence. Next divide each part into two more parts, with the cutpoints, in this example $1/4$ and $3/4$, appended to the Halton sequence. At this point, the Halton sequence is $1/2, 1/4, 3/4$. The process continues, so that for prime number 2, the first ten numbers in the sequence are $\{1/2, 1/4, 3/4, 1/8, 5/8, 3/8, 7/8, 1/16, 9/16, 5/16\}$. For prime number 3, the first 10 numbers are $\{1/3, 2/3, 1/9, 4/9, 7/9, 2/9, 5/9, 8/9, 1/27, 10/27\}$. See Train (2003, Chapter 9), for a formal exposition on Halton sequences.

The standard practice is to discard the first 20 elements of a Halton sequence, because those early elements tend to be correlated across Halton sequences. The simulator in (11.2) calls for S draws for each observation $i = 1, \cdots, N$. This can be achieved by generating a sequence of length $S \times (N + 20)$, and after discarding the first 20 elements, each observation has S simulated values. After generating a uniformly distributed Halton sequence, applying a quantile function, such as the inverse normal, produces simulated values from a desired distribution.

To visualize Halton sequences, consider Figure 11.1, which shows $1,000$ uniform draws for two variates. The left panel shows pseudo-random draws using the standard random number generator embedded in any statistical software package. The right panel shows Halton draws based on the prime number 2.

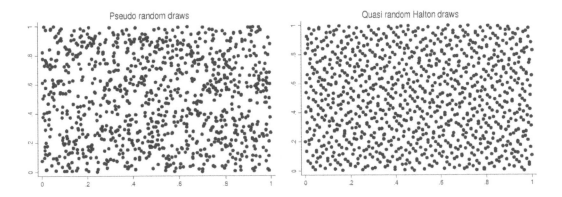

Figure 11.1 *Simulated values for two independent uniform variates.*

The cost of Halton sequences is largely philosophical, in that Halton draws, despite their quasi-random moniker, are clearly not random. The benefit is that by offering more even coverage of the unit interval, MSLE procedures require fewer simulation draws to achieve convergence (Bhat, 2001).

Results reported below use $S = 1,000$ quasi-random Halton draws, which appeared sufficient to ensure stable convergence. Increasing the number of simulations did not change the results.

11.2.4 Drawbacks of random effects models

The random effects approach introduces correlation through the inclusion of random effects into each marginal distribution. In general, the bounds of permissible correlation are narrower than those for Pearson correlation. The bounds depend on the parametric forms of the marginal distributions, as well as on how the random effects enter each marginal (McCulloch, 2008). For similar reasons, it is not straightforward to interpret the magnitude of estimated correlations. However, Deb and Trivedi (2006) argue that those narrower bounds do not often present problems in person-level microeconomic applications, where socioeconomic control variables tend to absorb a large proportion of statistical variation, leaving remaining conditional correlations relatively small in magnitude. Similarly, difficulty in interpreting the magnitude of estimated correlations also does not pose many problems, because microeconomists rarely care what these values are; rather, they care more about whether a correlation is positive or negative.

MLE of the random effects approach requires numerical integration, which can be time consuming for large data sets, and for models with many parameters to be estimated. The applications in this chapter use data sets of approximate size $N = 5,000$, and with 5 covariates in each marginal, plus ancillary parameters and a loading factor. With $S = 1,000$ simulations, these models required about an hour to converge on a standard desktop computer. Convergence times could be shortened by providing analytical first and/or second derivatives.

Another drawback of the random effects approach is that, in contrast to the copula approach discussed in the following section, it can be difficult in a random effects model to condition the loading factor λ_1 on observed covariates. A possible solution is to use a shared random effect, or two-factor, model with two random effects drawn from, say, a bivariate normal distribution with a correlation term that depends on observable covariates. In practice, however, since coefficients to be estimated are buried within a correlation parameter, which is, itself, attached to latent terms being numerically integrated out, estimation requires a large number simulations, especially for large sample sizes. Attempts to estimate such models with an appropriate number of simulations for this chapter quickly exhausted the available memory on a standard desktop computer.

Finally, the exposition in this section focuses on bivariate applications. In principle, extensions to higher dimensions should be straightforward, although higher-dimension numerical integration techniques do pose some complications, especially in terms of simulation errors (Caflisch, 1998).

11.3   Copula models

Consider a sample of N observations on two outcome variables, Y_{1i} and Y_{2i}, with $\mathbf{Y}_1 = (Y_{1i}, \cdots, Y_{1N})^\top$ and $\mathbf{Y}_2 = (Y_{2i}, \cdots, Y_{2N})^\top$. The same notation as in the previous section applies here, where $f_1(y_{1i}; \mathbf{x}_i^\top \boldsymbol{\beta}_1)$ denotes the density of the continuous outcome, and $f_2(y_{2i}; \mathbf{x}_i^\top \boldsymbol{\beta}_2)$ denotes that of the discrete outcome.

11.3.1   Copula basics

An alternative parametric approach to specifying the joint density uses copula functions (Song, 2000; Joe, 1997; Joe, 1993; Hutchinson and Lai, 1990; Marshall and Olkin, 1988; Sklar, 1959), generalized to mixed discrete-continuous settings by de Leon and Wu (2011) and Song et al. (2009). A bivariate copula is a bivariate cumulative distribution function (CDF) with both univariate margins distributed as $U(0, 1)$. Consider the bivariate CDF $F(y_{1i}, y_{2i})$, with univariate marginal CDFs $F_1(y_{1i})$ and $F_2(y_{2i})$, and inverse probability transforms (i.e., quantile functions) $F_1^{-1}(\cdot)$ and $F_2^{-1}(\cdot)$, where, for succinctness, we suppress conditioning on covariates. Then, $y_{1i} = F_1^{-1}(u_{1i}) \sim F_1(\cdot)$ and $y_{2i} = F_2^{-1}(u_{2i}) \sim F_2(\cdot)$, where u_{1i} and u_{2i} are uniformly distributed variates. The transforms of uniform variates are distributed as $F_1(\cdot)$ and $F_2(\cdot)$. Hence,

$$F(y_{1i}, y_{2i}) \;\; = \;\; F\{F_1^{-1}(u_{1i}), F_2^{-1}(u_{2i})\} \;\; = \;\; C(u_{1i}, u_{2i}; \theta) \qquad (11.6)$$

is the copula associated with the CDF, where θ captures the dependence. By this line of reasoning, the copula satisfies

$$F(y_{1i}, y_{2i}) \;\; = \;\; C(F_1(y_{1i}), F_2(y_{2i}); \theta). \qquad (11.7)$$

Note that, as is evident by the use of uppercase letters, both of the marginal CDFs $F_1(\cdot)$ and $F_2(\cdot)$, as well as the copula itself, are expressed in terms of CDFs, rather than densities. The copula in (11.7) is unique, only if the marginal distributions are absolutely continuous.

It is worth noting that for discrete outcomes, the associated copula representation is not unique. Such non-uniqueness arises from the fact that the CDF of a discrete random variable does not map such a variable to the entire unit interval $[0, 1]$, and thus, the copula $C(\cdot; \theta)$ satisfying (11.7) need not be uniform over rectangles. See Joe (1997, p. 14), for a detailed discussion of this issue. This result, however, does not create serious problems from a modeling viewpoint. While a copula is not unique for a joint distribution of discrete variables, it is unique on $\text{Ran}(F_1) \times \text{Ran}(F_2)$, where $\text{Ran}(G)$ denotes the range of all possible values of the CDF $G(\cdot)$ (see Nelsen, 1999, p. 15).

11.3.2 Specification of joint distribution

The copula approach provides a simple recipe for forming joint distributions. By plugging the known marginal CDFs into a copula function $C(\cdot)$, the right-hand side of (11.7) gives a parametric representation of the unknown joint CDF on the left-hand side, from which the corresponding density is obtained, a necessary requirement of MLE. For continuous outcomes, the density is determined as $\partial^2 C(F_1(y_{1i}), F_2(y_{2i}); \theta)/\partial y_{1i}\partial y_{2i}$. For discrete outcomes, the density comes from differencing as

$$P(Y_{1i} = y_{1i}, Y_{2i} = y_{2i}) = C(F_1(y_{1i}), F_2(y_{2i}); \theta) - C(F_1(y_{1i} - 1), F_2(y_{2i}); \theta)$$
$$-C(F_1(y_{1i}), F_2(y_{2i} - 1); \theta) + C(F_1(y_{1i} - 1), F_2(y_{2i} - 1); \theta). \quad (11.8)$$

For mixed continuous-discrete outcomes, the joint density combines differentiating and differencing for the continuous and discrete outcomes, respectively, as in

$$f(y_{1i}, y_{2i}; \boldsymbol{\beta}_1, \boldsymbol{\beta}_2, \theta) = f_1(y_{1i}; \mathbf{x}_i^\top \boldsymbol{\beta}_1) \left\{ \frac{\partial}{\partial y_{1i}} C(F_1(y_{1i}), F_2(y_{2i}); \theta) \right.$$
$$\left. - \frac{\partial}{\partial y_{1i}} C(F_1(y_{1i}), F_2(y_{2i} - 1); \theta) \right\}. \quad (11.9)$$

See Song et al. (2009) for a general version of this formula extended to higher dimensions.

Because different copulas imply different dependence structures between outcome variables, a crucial modeling decision is what parametric form the copula $C(\cdot; \theta)$ should assume. In the two empirical applications presented below, it is important to remain agnostic about the direction of correlation. In the first application, which considers the relationship between drug spending and health problems, positive correlation would be expected if unhealthier subjects seek drugs, but negative correlation would be expected if drugs improve health. In the second application, which investigates the relationship between nondrug spending and drug usage, positive correlation would be expected if subjects consume drug and nondrug services together, but the cost-offset hypothesis posits a negative relationship between these two measures. This consideration rules out many commonly-used copulas, including those by Clayton (1978) and Gumbel (1960), both of which allow only positive dependence.

Estimates presented below use the Gaussian copula, which allows both negative and positive dependence. The Gaussian copula takes the form

$$C(F_1(y_{1i}), F_2(y_{2i}); \theta) = \Phi_2 \left(\Phi^{-1}\{F_1(y_{1i})\}, \Phi^{-1}\{F_2(y_{2i})\}; \theta \right), \quad (11.10)$$

where $\Phi_2(\cdot; \theta)$ denotes the standard bivariate normal CDF with zero means, unit variances, and correlation θ; and $\Phi^{-1}(\cdot)$ denotes the standard normal quantile function. The following graphs show simulated values from the Gaussian copula with standard normal marginal distributions. As Figure 11.2 shows, the Gaussian copula has symmetric dependence consistent with the normal distribution. The models are also estimated using the Frank copula, which also allows symmetric positive and negative dependence, albeit of a slightly different shape than in the Gaussian copula. Results were nearly identical to those presented below.

The copula approach presents an attractive feature in that it allows researchers to specify dependence conditionally on covariates. For example, specifying

$$\theta_i = \tanh(\mathbf{x}_i^\top \boldsymbol{\gamma}), \quad (11.11)$$

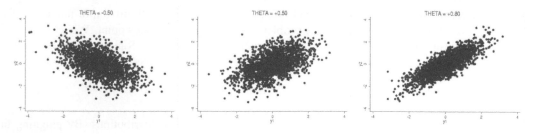

Figure 11.2 *Simulated values from bivariate Gaussian copulas with correlation* $\theta = -0.5, 0.5,$ *and* 0.8.

with estimable coefficients $\boldsymbol{\gamma}$ allows an investigation of how specific covariates affect the magnitude of dependence. The hyperbolic tangent operator constrains $\theta \in (-1, 1)$, as required by the Gaussian copula.

11.3.3   Maximum likelihood estimation

With the joint density specified using a copula, the likelihood function is formed as

$$L(\boldsymbol{\beta}_1, \boldsymbol{\beta}_2, \theta; \mathbf{y}_1, \mathbf{y}_2) \quad = \quad \sum_{i=1}^{N} \log\, f(y_{1i}, y_{2i}; \boldsymbol{\beta}_1, \boldsymbol{\beta}_2, \theta) \tag{11.12}$$

where $f(y_{1i}, y_{2i}; \boldsymbol{\beta}_1, \boldsymbol{\beta}_2, \theta)$ is given by (11.9). The likelihood function is maximized with respect to $\boldsymbol{\Omega} = (\boldsymbol{\beta}_1^\top, \boldsymbol{\beta}_2^\top, \theta)^\top$; in the case where $\theta_1 = \tanh(\mathbf{x}_1^\top \boldsymbol{\gamma})$, θ is replaced with $\boldsymbol{\gamma}$. Similar to the random effects model, maximization proceeds numerically using a modified iterative Newton–Raphson algorithm with numerically approximated first and second derivatives, and the Hessian matrix and gradient vector are analogously defined.

11.3.4   Drawbacks of copula models

The copula approach allows researchers to investigate intricate dependence patterns by selecting specific copula functions. For a detailed discussion of dependence structures associated with specific copula functions, see for example, Trivedi and Zimmer (2007). However, having access to such a wide variety of copulas presents a potential pitfall, in that models based on copulas that do not show dependence patterns present in the data can lead to poor out-of-sample predictions. Indeed, recent evidence suggests that the 2008 housing crisis that began in Florida stemmed, at least in part, from misspecified copula functions (Zimmer, 2012).

A second disadvantage of the copula approach involves extensions beyond bivariate settings. Constructing m dimensional copulas with $m(m-1)/2$ dependence parameters is difficult (Nelsen, 2006, pp. 105–108). Earlier attempts to construct flexible higher dimensional copulas either failed to perform well in applications, or required complex simulation-based estimates, which negates the main advantage of copula estimation (Joe, 1994, 1990; Husler and Reiss, 1989). A more recent development, called vine copula (Aas *et al.*, 2009), has received attention in statistics, but almost none in economics. The method specifies a higher-dimensional distribution as a chain of lower-dimensional conditional copulas and marginal distributions. While the approach can accommodate wide varieties of dependence patterns, it also requires certain outcomes to be conditional on other outcomes, which, as noted in the introduction section, is not always ideal in economic applications.

Finally, the copula approach has a somewhat mechanical drawback in that while copulas are expressed in terms of CDFs, MLE requires converting to densities. For bivariate applications, the conversion follows from (11.9). Although not prohibitive in most applications, the random effects approach, by starting with density representations, avoids this conversion altogether.

11.4 Application to drug spending and health status

Justification for passage of the Medicare Modernization Act (MMA) rested, in part, on the assumption that unhealthy individuals incur more drug expenses, and therefore need financial protection. To assess the validity of this assumption, it is important to verify the existence of a link between drug spending and health status. In addition, because the MMA offered elderly individuals subsidized access to private drug coverage, a more complete assessment of the impact of the legislation requires analyzing whether the link between drug spending and health status depends on private insurance status. Similar applications of mixed outcome models to analyze insurance-related questions are discussed in Erhardt and Czado (2012) and Czado *et al.* (2011).

Linear regression-based methods provide a reasonable starting point in this setting. We have

$$\text{drug spending}_i \ = \ \beta_0 + \beta_1 \times \text{health status}_i + \varepsilon_i \tag{11.13}$$

where ε_i is a homogeneous error term, and i indexes the individual. Such models pose two complications. First, medical spending is highly skewed, with high density at low values and a fat upper tail. Second, the causal relationship might be reversed, with drug spending affecting health status. Manning *et al.* (1982) provide a detailed discussion of this second issue. Together, these concerns cloud interpretation of the main coefficient of interest β_1.

Were it only for the first concern regarding skewness, the coefficients could be consistently estimated by moving to some nonlinear specification. But the second concern regarding reverse causality calls for a more elaborate model. Drawing inspiration from Murteira and Lourenco (2011), this section considers a joint model of the form $f(y_1, y_2)$, for outcomes Y_1 and Y_2, where Y_1 measures annual person-level drug spending, and Y_2 provides a binary measure of health, equaling 1, if the person reports "fair or poor" health, and 0, otherwise.

11.4.1 Data

Data come from the Medical Expenditures Panel Survey (MEPS), collected and published by the Agency for Healthcare Research and Quality, a unit of the U. S. Department of Health and Human Services. The MEPS provides a nationally representative sample of the U. S. noninstitutionalized population, and contains detailed individual-level information on socioeconomic characteristics, medical insurance status, and health care spending. Data come from the 2004 and 2005 years of the survey, the two years before Medicare Part D became active.

The sample focuses on Medicare enrollees, 65 years of age and older, who recorded at least $100 in annual drug spending. Two reasons motivate focusing on those who spend at least $100. First, as presented by Pohlmeier and Ulrich (1995), one economic process governs whether an individual incurs positive spending, and a second economic process determines the magnitude of spending. The reason is that, in the U. S. healthcare system, patients typically initiate first contact with a medical care provider, but after contact is made, the provider usually determines the intensity of treatment. These two processes cannot be explained by one statistical distribution, not even one that accommodates zero inflation (Deb and Trivedi, 1997). The standard approach in models of health care demand, which is followed here by focusing on positive spenders, is to treat the two economic processes as stochastically independent.

The second reason for focusing on positive spenders is that they, naturally, represent the biggest concern from a government budget perspective. Indeed, as widely reported by the U. S. Healthcare Financing Administration, approximately 10% of Medicare enrollees account for 70% of program spending.

Although MEPS does include a limited longitudinal dimension, data from the two years are stacked and treated as a pooled cross section. The final estimation sample consists of 5,096 Medicare enrollees. Table 11.1 reports sample means for several variables partitioned according to health. As the table indicates, subjects reporting "fair or poor" health spend more on prescription drugs compared to their healthy counterparts. Mean spending for both groups is large, which reflects the

Table 11.1 *Means and standard deviations (SDs) for sample of subjects 65 or older drawn from the 2004 and 2005 waves of MEPS, who spent at least $100 annually on prescription drugs.*

	Better than fair health N = 3,819		Fair or poor health N = 1,277	
	Mean	SD	Mean	SD
Drug spending (in $1000s)	17.10	15.21	25.57	20.61
Age (in 10s)	7.46	0.64	7.50	0.62
Female	0.58	0.49	0.56	0.50
Black	0.09	0.28	0.15	0.36
Married	0.58	0.49	0.55	0.50

highly skewed distribution of drug spending, but it also reflects the simple fact that American seniors spend heavily on prescription drugs. Unhealthy subjects are also slightly older and more likely to be black. Gender and marital status appear comparable across health status.

Figure 11.3 shows empirical density plots for the two outcome variables, drug spending and health status.

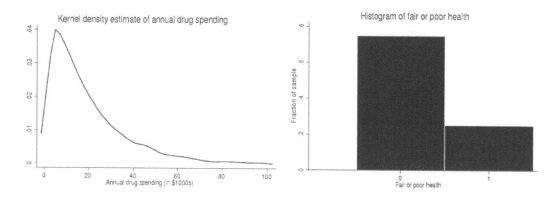

Figure 11.3 *Empirical density of drug spending and histogram of health status.*

Drug spending follows a continuous, highly-skewed distribution, as shown in the left panel of Figure 11.3. The skewed empirical density does not, by itself, eliminate the possibility of modeling based on some symmetric distribution, because the inclusion of explanatory variables in a regression-based model might account for the remaining variation so that the distribution becomes relatively symmetric. In practice, however, drug spending is determined by many unmeasured factors, such that it is unrealistic to expect regression effects to eliminate the skewness. The other outcome variable, which indicates whether the subject reports fair or poor health, is a binary discrete measure, as explained by the histogram in the right panel of Figure 11.3. It is clear from this that more than 20% of subjects report being in fair or poor health.

11.4.2   Specification of marginal and joint distributions

Estimation for the random effects and copula-based models requires specification of the distributions of the two outcome measures, drug spending and health status. For drug spending, although skewed distributional shapes often lend themselves well to log-normal transformations, recent evidence in health economics extols the improved fit offered by gamma-based models applied to non-logged medical spending (Wooldridge, 2010, pp. 740–742; Buntin and Zaslavsky, 2004). Thus, the

Table 11.2 *Estimates and their SEs based on random effects and copula models for drug spending and health status.*

| | Drug spending | | | | Fair or good health | | | |
| | Random effects | | Copula | | Random effects | | Copula | |
	Est	SE	Est	SE	Est	SE	Est	SE
Constant	2.079	0.165	2.412	0.156	−1.704	0.345	−1.153	0.243
Age (in 10s)	0.045	0.021	0.031	0.020	0.104	0.044	0.067	0.031
Female	0.085	0.027	0.085	0.026	−0.102	0.058	−0.072	0.041
Black	−0.042	0.042	−0.058	0.040	0.508	0.085	0.354	0.060
Married	−0.083	0.028	−0.072	0.027	−0.054	0.059	−0.036	0.042
σ (shape)	1.608	0.041	1.364	0.024	—	—	—	—
λ_1 (factor loading)	0.373	0.023	—	—	—	—	—	—

density is specified by

$$f_1(y_{1i}) \quad = \quad \frac{y^{\sigma-1}}{\mu_i^\sigma \Gamma(\sigma)} \exp\left(-\frac{y_{1i}}{\mu_i}\right), \tag{11.14}$$

where $\mu_i = \exp(\mathbf{x}_i^\top \boldsymbol{\beta}_1)$ (or $\mu_i = \exp(\mathbf{x}_i^\top \boldsymbol{\beta}_1 + \lambda_1 \eta_i)$ in the random effects specification) is the mean parameter, σ denotes a shape parameter, and $\Gamma(\cdot)$ is the gamma function.

The other outcome measure, health status, follows a probit specification, given by

$$P(Y_{2i} = 1) \quad = \quad \Phi(\mathbf{x}_i^\top \boldsymbol{\beta}_2), \tag{11.15}$$

or $\Phi(\mathbf{x}_i^\top \boldsymbol{\beta}_2 + \eta_i)$ in the random effects model, where $\Phi(\cdot)$ denotes the CDF of the standard normal distribution.

In the random effects model, the density of the binary outcome is $f_2(y_{2i}) = \Phi\{(2y_{2i}-1)(\mathbf{x}_i^\top \boldsymbol{\beta}_2 + \eta_i)\}$, and the joint density is formed from (11.2). In the copula model, Song *et al.* (2009) show that, for the Gaussian copula, the joint density in (11.9) simplifies to

$$f(y_{1i}, y_{2i}) \quad = \quad \begin{cases} (1 - C_{1i}) f_1(y_{1i}|\mathbf{x}_i^\top \boldsymbol{\beta}_1) & , \text{if } y_{2i} = 0 \\ C_{1i} f_1(y_{1i}; \mathbf{x}_i^\top \boldsymbol{\beta}_1) & , \text{if } y_{2i} = 1 \end{cases}, \tag{11.16}$$

where

$$C_{1i} \quad = \quad \Phi\left(\frac{\mathbf{x}_i^\top \boldsymbol{\beta}_2 - \theta \Phi^{-1}\{F_1(y_{1i})\}}{\sqrt{1-\theta^2}}\right).$$

11.4.3   Results

Table 11.2 presents estimates from the random effects and copula models. The two approaches produce similar qualitative estimates for the coefficients of the explanatory variables. Drug spending and health problems both appear to increase with age. Females spend more on drugs, but report worse health. Blacks report worse health, but spend similar amounts on drugs compared to their non-black counterparts. Married subjects spend less on drugs, but report similar health compared to non-married subjects.

Conditional dependence between the two marginal distributions is captured by the loading factor in the random effects model, and by the dependence parameter in the copula specification. The latter is estimated as $\widehat{\theta} = 0.28$ (SE $= 0.018$). Although the magnitudes of these two estimates are not directly comparable, since each enters its joint density in a different fashion, both estimates are

Table 11.3 *Tests on comparison of model fits of random effects, copula, and separate models, for drug spending and health status.*

Null hypothesis	t-statistic	p-value
H_0 : Copula not superior to separate models	7.48	< 0.01
H_0 : Random effects not superior to separate models	7.86	< 0.01
H_0 : Random effects not superior to copula	3.04	< 0.01

Table 11.4 *Estimates and their SEs based on copula model for drug spending and health status, with dependence conditioned on insurance status.*

	Drug spending		Fair or poor health		Dependence	
	Est	SE	Est	SE	Est	SE
Constant	2.420	0.156	−1.141	0.243	0.266	0.023
Age (in 10s)	0.030	0.020	0.065	0.031	—	—
Female	0.084	0.026	−0.070	0.041	—	—
Black	−0.058	0.040	0.352	0.060	—	—
Married	−0.072	0.027	−0.035	0.042	—	—
Private drug insurance	—	—	—	—	0.068	0.040
σ (shape)	1.364	0.024	—	—	—	—

positive and precisely estimated. The implication is that unmeasured traits that lead a person to be in worse health also correlate with higher drug spending.

Log-likelihood values for the random effects and copula models are $-22,723.96$ and $-22,743.56$, respectively. However, standard model comparison tests of likelihood-based models, such as the standard likelihood ratio, Wald, and Lagrange multiplier tests, are not appropriate for non-nested models. Thus, we used Vuong's (1989) test in Table 11.3, to compare the fit of the random effects and copula models, as well as a model that estimates each marginal separately (effectively setting the loading and dependence parameters equal to zero). Both the random effects and copula models provide superior fit compared to separate marginal estimation, which is not surprising given the precisely estimated loading/dependence terms. Furthermore, the random effects approach appears to provide a better fit than the copula model.

Estimates in Table 11.2 omit a potentially important link between health status and drug spending, and one that has important policy implications concerning MMA: private drug coverage. Yet, it is not appropriate to include drug coverage as an additional explanatory variable, because subjects might acquire insurance in anticipation of drug needs or health status. In economic jargon, drug coverage might be endogenous with respect to both drug spending and health status. A possible solution is to specify the dependence parameter conditional on insurance status, as in (11.11). Table 11.4 presents copula estimates for which the dependence parameter depends on private drug coverage. Estimates suggest that while overall dependence is positive, as evidenced by the estimate of the constant (i.e., 0.266), this dependence increases when the subject has private drug coverage. The resulting log-likelihood value is $-22,742.14$.

Table 11.5 *Magnitude of dependence, as measured by Pearson's correlation coefficient, between drug spending and health status, for each insurance status.*

Insurance status	Correlation
Private coverage = 0	0.26
Private coverage = 1	0.32

Table 11.5 translates the dependence numbers in Table 11.4 into Pearson's correlation coefficients for each insurance status. For subjects without private drug coverage, the correlation equals 0.26, but this increases to 0.32, for subjects with private coverage. Supporters of Medicare Part D used a similar argument during debates over the law. Specifically, individuals with private drug coverage appear to show stronger links between poor health and drug spending. And if Part D, which subsidizes seniors' purchases of private drug plans, allows more seniors access to necessary drugs, then the law should offer unhealthy seniors financial protection from large medical expenses.

11.5   Application to nondrug spending and drug usage

By as early as the year 2050, Medicare expenditures are expected to consume all federal tax revenues. In the absence of additional revenue sources, policymakers must find a way to slow projected growth in medical expenditures. To that end, another justification for passage of the MMA was based on the theory that increased consumption of prescription drugs pay for themselves through reduced spending on other, possibly more costly, health care services. The net result should be reduced aggregate medical expenditures. Health economists refer to this as the "cost offset hypothesis."

Similar to the application in the previous section, it does not make sense to model nondrug spending conditional on drug usage, as in

$$\text{nondrug spending}_i \;=\; \beta_0 + \beta_1 \times \text{drug usage}_i + \varepsilon_i, \tag{11.17}$$

because the direction of causality might run in the opposite direction. Indeed, to obtain a prescription, a patient usually first visits a physician, which implies that at least some nondrug spending must precede drug usage. Therefore, this section seeks to model the two jointly.

11.5.1   Data

The data come from a subset of individuals used in the previous application who report at least $100 in nondrug medical spending, and who report consuming at least 1 prescribed medicine. Reasons for considering positive spenders and users were discussed in the previous section. The sample contains 3,542 unique individuals. The two outcome variables of interest are total annual medical spending not including expenses on prescription drugs, and total number of annual prescribed medicines including refills.

The density plot and histogram in Figure 11.4 indicate that the two outcomes follow different distributions, each of which shows significant skewness. As discussed above, it is not realistic to expect covariates in a regression-based model to eliminate skewness.

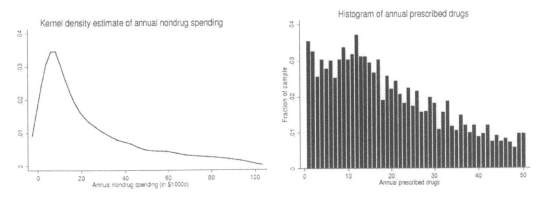

Figure 11.4 *Empirical density of nondrug spending and histogram of drug consumption.*

Table 11.6 *Estimates and their SEs based on random effects and copula models for nondrug spending and drug usage.*

	Nondrug spending				Drug usage			
	Random effects		Copula		Random effects		Copula	
	Est	SE	Est	SE	Est	SE	Est	SE
Constant	0.606	0.268	2.309	0.205	2.162	0.157	2.163	0.156
Age (in 10s)	0.073	0.030	0.106	0.020	0.106	0.020	0.107	0.020
Female	−0.037	0.039	−0.068	0.034	0.058	0.026	0.056	0.026
Black	−0.263	0.062	−0.078	0.054	0.068	0.041	0.082	0.041
Married	0.080	0.041	0.044	0.035	−0.060	0.027	−0.062	0.027
σ (shape)	1.608	0.118	1.101	0.023	—	—	2.043	0.053
α (overdispersion)	—	—	—	—	0.478	0.013	2.043	0.053
λ_1 (factor loading)	0.111	0.014	—	—	—	—	—	—

11.5.2   Specification of marginal and joint distributions

To accommodate skewness, nondrug spending is modeled to follow a gamma distribution. Thus, the density is specified as in (11.14), i.e.,

$$f_1(y_{1i}) = \frac{y^{\sigma-1}}{\mu_i^\sigma \Gamma(\sigma)} \exp\left(-\frac{y_{1i}}{\mu_i}\right). \tag{11.18}$$

The second marginal describing prescribed medicines is assumed to have a discrete negative binomial setup, which has been employed in previous studies of health care consumption based on counted outcomes (Cameron and Trivedi, 1986). The negative binomial density is

$$f_2(y_{2i}; \mathbf{x}_i^\top \boldsymbol{\beta}_2) = \frac{\Gamma(y_{2i}+\alpha^{-1})}{\Gamma(y_{2i}+1)\Gamma(\alpha^{-1})} \left(\frac{\alpha^{-1}}{\alpha^{-1}+\mu_i}\right)^{1/\alpha} \left(\frac{\mu_i}{\alpha^{-1}+\mu_i}\right)^{y_{2i}}, \tag{11.19}$$

where the mean is $\mu_i = \exp(\mathbf{x}_i^\top \boldsymbol{\beta}_2)$ (or $\mu_i = \exp(\mathbf{x}_i^\top \boldsymbol{\beta}_2 + \eta_i)$ in the random effects specification) and α, an overdispersion term, measures the extent to which the conditional variance exceeds the conditional mean.

Although the negative binomial distribution is the standard distribution in the analysis of the number of contacts with health care providers, other count models exist. As evident in Figure 11.4, the count outcome does not display a disproportionately large mass on any particular value, in part because the sample includes only positive users. However, alternative specifications that allow "inflation" on particular values might provide better fits to the data when such mass points exist (Czado *et al.*, 2007).

For the random effects model, the joint density is formed from (11.2). For the copula approach, again based on the Gaussian copula, the joint density in (11.9) simplifies to

$$f(y_{1i}, y_{2i}) = f_1(y_{1i}; \mathbf{x}_i^\top \boldsymbol{\beta}_1) \left\{ \Phi\left(\frac{\Phi^{-1}\{F_2(y_{2i})\} - \theta\Phi^{-1}\{F_1(y_{1i})\}}{\sqrt{1-\theta^2}}\right) \right.$$
$$\left. -\Phi\left(\frac{\Phi^{-1}\{F_2(y_{2i}-1)\} - \theta\Phi^{-1}\{F_1(y_{1i})\}}{\sqrt{1-\theta^2}}\right) \right\}. \tag{11.20}$$

11.5.3   Results

Table 11.6 shows estimates from random effects and copula models. Nondrug spending and drug usage both appear to increase with age. Females use more drugs than males, but the copula model finds that females spend less on nondrug services. Blacks consume more drugs, but the random

Table 11.7 *Tests on comparison of model fits of random effects, copula, and separate models, for nondrug spending and drug usage.*

Null hypothesis	t-statistic	p-value
H_0 : Copula not superior to separate models	5.64	< 0.01
H_0 : Random effects not superior to separate models	2.93	< 0.01
H_0 : Random effects not superior to copula	0.47	0.32

Table 11.8 *Estimates and their SEs based on copula model for nondrug spending and drug usage, with dependence conditioned on insurance status.*

	Nondrug spending		Drug usage		Dependence	
	Est	SE	Est	SE	Est	SE
Constant	2.313	0.205	2.161	0.156	0.135	0.021
Age (in 10s)	0.105	0.026	0.107	0.020	—	—
Female	−0.066	0.034	0.055	0.026	—	—
Black	−0.081	0.054	0.082	0.041	—	—
Married	0.046	0.035	−0.061	0.027	—	—
Private drug insurance	—	—	—	—	0.068	0.035
σ (shape)	1.101	0.023	—	—	—	—
α (overdispersion)	—	—	2.039	0.053	—	—

effects model finds that they spend less on nondrug services. Married subjects appear to consume fewer drugs, but spend more on nondrug services.

The main estimates of interest — the loading factor in the random effects model and the dependence parameter in the copula specification — both suggest positive dependence (i.e., $\hat{\theta} = 0.159, \text{SE} = 0.016$), indicating that unmeasured factors that induce higher drug usage also lead to higher spending on nondrug medical services. This finding appears to draw into question the existence of cost offsets.

Log-likelihood values for the random effects and copula models are $-28,493.47$ and $-28,487.87$, respectively. Vuong's (1989) tests of model fit appear in Table 11.7. Test results confirm that both joint models outperform non-joint separate modeling of each outcome. In contrast to results from the first application, which show that the random effects model outperformed its copula counterpart, in this second application, the copula slightly outperforms the random effects specification, although the test suggests that this improvement is not statistically significant.

Similar to the previous example, these results omit the potentially important role of private drug coverage. Table 11.8 presents results for the copula model, where the dependence parameter is specified as a function of private drug coverage. Subjects with private drug coverage appear to

Table 11.9 *Magnitude of dependence, as measured by Pearson's correlation coefficient, between nondrug spending and drug usage, for each insurance status.*

Insurance status	Correlation
Private coverage $= 0$	0.13
Private coverage $= 1$	0.20

show stronger dependence than their counterparts without such coverage. Table 11.9 translates these numbers into Pearson's correlation coefficients. The correlation between nondrug spending and drug usage for subjects with private drug coverage is 0.13, but this number increases to 0.20, an increase of more than 50%, for those with private coverage.

Opponents of Medicare Part D have used such arguments to attack the reform as contributing to rising health care costs. First, in contrast to the cost offset hypothesis, higher drug usage appears to correlate with higher nondrug spending. Furthermore, private drug coverage appears to strengthen, not weaken, this positive association.

11.6   Discussion

This chapter specifies random effects and copula models to analyze two mixed outcome applications relevant to passage of the MMA. The first application investigates the link between drug spending and health problems, and how private insurance status impacts that link. Results indicate a positive correlation between drug spending and health problems, with private drug insurance increasing the magnitude of this association. Proponents of the MMA have used arguments similar to results produced by the first application to argue that the law offers financial protection to seniors with health problems.

The second application studies the link between nondrug spending and drug usage. Results find a positive correlation, with private drug coverage increasing the magnitude of this link. Opponents of the MMA have relied on arguments similar to these findings to argue that the law contributes to increasing medical expenses, which places further strain on federal budgets.

Chapter 12

Sparse Bayesian modeling of mixed econometric data using data augmentation

Helga Wagner and Regina Tüchler

12.1 Introduction

Multidimensional data of different types are routinely collected in economic and social surveys. In econometric analyses of these data, regression models are typically used for the outcome variables. However, outcome components are usually studied using separate models. The data-inherent dependencies are thereby neglected. Recently, researchers have tried to resolve this problem, and joint modeling of mixed data has become increasingly important. An overview of the variety of models for mixed data is given in Chapter 1.

Joint models for data of mixed types have been proposed in the Bayesian econometrics literature as treatment models, to deal with endogeneity of a binary covariate on a continuous outcome. Correlation between the binary treatment variable and the potential outcomes under both treatments, is modeled using a latent variable specification for treatment, via a latent utility. For the two potential outcomes and the latent utility, regression models are specified with a multivariate error distribution. Koop and Poirier (1997) specify a joint trivariate normal distribution for the latent utility and the potential outcomes, whereas Chib (2007) specifies the model through bivariate Student t-distributions for the latent utility and each potential continuous outcome. Another more and more popular option to model dependence in multivariate models is via copulas. Pitt *et al.* (2006) develop a method for Bayesian estimation of regression models joined by Gaussian copulas which can also be used for outcomes of mixed type; an extension to non-elliptical copulas is proposed recently by Smith and Khaled (2012). Very general models for clustered mixed outcomes are proposed in Dunson (2000), where each outcome is modeled as a function of an unobserved underlying variable, which is assumed to be related to covariates and latent variables through a generalized linear model. The multivariate latent variables are assumed to have an effect on each underlying variable, which then induces dependence in the outcomes. Dunson (2006) considers a hierarchical data structure, where outcomes of different types are measured for subjects within groups. Outcomes of one subject then share a latent factor with a semiparametrically modeled density, which is allowed to differ across groups. Chapter 13 reviews additional Bayesian literature and addresses as well the problem of missing data.

In this chapter, we focus on marginal regression models with fixed or random effects, for which data augmentation with latent auxiliary variables leads to a representation as a linear Gaussian model in the augmented data. The resulting joint model for all response components is a multivariate regression model. Dependence between different components can be specified either directly through the correlations of the corresponding error terms, which is the route we follow for fixed effects models, or via the correlations of the random effects for random effects models. A related model is considered in Komárek and Komárková (2012), where model-based clustering of longitudinal outcomes of mixed types is performed using mixtures of generalized linear mixed models (GLMMs) with correlated random effects.

An important issue in any econometric model is that of sparse modeling to avoid overfitting. The most convenient way to specify a joint model for several outcomes is to start with a very general model, where all potentially relevant regressors are included in the model, and then to use variable selection to identify those covariates, which have non-zero effects. In our specification of the joint model as a multivariate linear model for the augmented data, variable selection and covariance selection of random effects can be easily incorporated.

We discuss model specifications for cross-sectional, as well as for longitudinal data, in Section 12.2, where data augmentation and modeling of dependence structure is described. In Section 12.3, we consider the case of a bivariate response with a binary and a normal component. Model specification and Markov Chain Monte Carlo (MCMC) estimation with simultaneous variable selection, is described in detail and illustrated in an example with simulated data. An application for cross-sectional data is presented in Section 12.4, where the goal is to jointly investigate material deprivation (binary) and income of households (continuous). An application to longitudinal data from marketing with count and continuous responses is given in Section 12.5.

12.2 Model specification

Our strategy to join models for responses of different types consists in combining marginal regression models for the different responses. For cross-sectional data, we use fixed effects models, and for longitudinal data, we employ random effects models. The advantage of our approach is that marginal models have their usual interpretation, and additionally, we can incorporate variable selection and covariance selection, which allow sparse modeling.

For cross-sectional data, let $\mathbf{Y}_i = (Y_i^1, \cdots, Y_i^K)^\top$ denote a multivariate response variable, which is observed for N subjects. The components Y_i^k, $k = 1, \cdots, K$, may be either continuous (i.e., normal), binary, ordered categorical variables, or Poisson counts. Let Y_i^k denote the observation of the kth component measured for subject i. To relate the mean $\mu_i^k = E(Y_i^k)$ to the linear predictor η_i^k, we use a distinct link function $g_k(\mu_i^k) = \eta_i^k$, $k = 1, \cdots, K$, for each component depending on its type. For Poisson components, we use the log-link-function

$$\mu_i^k = \exp(\eta_i^k). \tag{12.1}$$

For binary and ordinal components, we consider the logit link function

$$\mu_i^k = \frac{\exp(\eta_i^k)}{1 + \exp(\eta_i^k)}, \tag{12.2}$$

whereas for normal components, we use the identity link

$$\mu_i^k = \eta_i^k, \tag{12.3}$$

and assume a constant variance, i.e., $Y_i^k \sim N(\mu_i^k, \sigma_k^2)$.

For the linear predictor η_i^k of Y_i^k, we consider the regression specification

$$\eta_i^k = \mathbf{x}_i^\top \boldsymbol{\beta}^k,$$

where $\mathbf{x}_i$ is a design vector of dimension $d \times 1$, and $\boldsymbol{\beta}^k$ is the vector of regression coefficients. We assume that the same covariates are used for each of the K response components, but regression coefficients are of course allowed to differ. Covariates could be different for each response component, but as we incorporate variable selection, a convenient modeling strategy is to include all covariates at hand in the linear predictor of each outcome, and learn from the data which covariates actually have non-zero effects.

For longitudinal data, we assume that the multivariate response

$$\mathbf{Y}_i = \begin{pmatrix} \mathbf{Y}_i^1 \\ \vdots \\ \mathbf{Y}_i^K \end{pmatrix} \tag{12.4}$$

is measured on $t = 1, \cdots, T_i$ occasions. The observation of the kth component measured for subject i at time point t, is denoted by Y_{it}^k. Depending on the data type, we use the link functions given in (12.1)–(12.3), and consider the following random effects specification for the linear predictors $\boldsymbol{\eta}_i^k$ of the sequence $\mathbf{Y}_i^k$:

$$\boldsymbol{\eta}_i^k = \mathbf{X}_i \boldsymbol{\beta}_i^k.$$

Here $\mathbf{X}_i$ is a design matrix of dimension $T_i \times d$, and $\boldsymbol{\beta}_i^k$ are normally distributed random effects. As for cross-sectional data, we assume that the same covariates are used for each of the K response components, and that random effects may differ between components. Dependence between repeated measurements is captured by the random effects $\boldsymbol{\beta}_i^k$, shared by all measurements of one response component. Note that we start with a very general model specification, where all predictor variables are included and their effects are specified as random. Variable and covariance selection allows to learn from the data which of the coefficients are non-zero, and whether effects are fixed rather than random.

For simplicity, we replace the superscripts k by n for a normal response, by b for a binary response, by o for an ordinal response, and by c for a Poisson count in the following sections.

12.2.1 Data augmentation

To specify the joint model, we use a latent variable formulation for the discrete outcomes, where the distribution of the latent variable is continuous and is approximated by a mixture of normal components. This data augmentation provides much flexibility, and allows us to perform Bayesian estimation and model selection. For simplicity, we describe the data augmentation steps for the different types of discrete data for cross-sectional data. For longitudinal data, these steps have to be carried out for each observation.

Data augmentation for a binary response Y_i^b is based on the interpretation of logit models in terms of latent utilities; see McFadden (1974). The difference U_i of the utility for choosing category 1 and the utility for the baseline category 0 is modeled as

$$U_i = \mathbf{x}_i^\top \boldsymbol{\beta}^b + \varepsilon_i. \tag{12.5}$$

If ε_i follows a standard logistic distribution, the logit model results for Y_i^b if $Y_i^b = \mathrm{I}\{U_i > 0\}$. Frühwirth–Schnatter and Frühwirth (2010) show that the standard logistic distribution can be represented as a scale mixture of 6 normal distributions with fixed scale parameters and component weights

$$f_\varepsilon(\varepsilon) \approx \sum_{r=1}^{6} \phi\left(\frac{\varepsilon}{s_r}\right) \pi_r,$$

where $\phi(\cdot)$ denotes the standard normal density.

As the weights π_r and variances s_r^2 are fixed in this approximation, only the component indicators r have to be introduced to achieve a representation of the logit model as a latent Gaussian model. Conditional on the mixture component r_i, we define the auxiliary variables

$$\widetilde{Y}_i^b = \mathbf{x}_i^\top \boldsymbol{\beta}^b + \widetilde{\varepsilon}_i^b, \quad \widetilde{\varepsilon}_i^b \sim N(0, s_{r_i}^2).$$

For a response Y_i^o taking values in ordered categories, the same data augmentation scheme can be used. If the discrete response has $P + 1$ ordered categories, labeled $p = 0, \cdots, P$, additional $P - 1$ latent thresholds $\gamma_2, \cdots, \gamma_P$, have to be introduced. With $\gamma_0 = -\infty$, $\gamma_1 = 0$, and $\gamma_{P+1} = +\infty$, the ordinal response Y_i^o is related to the latent utility by

$$Y_i^o = p \quad \text{if and only if} \quad \widetilde{Y}_i^o \in (\gamma_p, \gamma_{p+1}).$$

For a Poisson count Y_i^c, a representation through continuous latent variables can be obtained using the properties of a Poisson process; see Frühwirth–Schnatter et al. (2009) and Frühwirth–Schnatter and Wagner (2006). If Y_i^c is the number of jumps of a Poisson process with intensity μ_i^c in the time interval $[0,1]$, the inter-arrival times between these jumps are independent, and follow an exponential distribution with mean $1/\mu_i^c$. Depending on the value of Y_i^c, either one or two latent variables are required. For each count observation, the inter-arrival time $\tau_{i,1}^c$ between the last jump before and the first jump after the end of the unit interval is introduced. If $Y_i^c > 0$, an additional arrival time $\tau_{i,2}^c$, corresponding to the last jump before 1, is introduced. The latent variable representation of the Poisson model is then given as

$$-\log \tau_{i,1}^c = \eta_i^c + \varepsilon_{i,1}^c, \tag{12.6}$$
$$-\log \tau_{i,2}^c = \eta_i^c + \varepsilon_{i,2}^c. \tag{12.7}$$

The error term $\varepsilon_{i,1}^c$ follows a Type I extreme value distribution, and $\varepsilon_{i,2}^c$ has the same distribution as the negative logarithm of a gamma random variable, with integer shape parameter $v = Y_i^c$. These distributions can also be approximated by mixtures of normal components $N(m_r, s_r^2)$. The number of components necessary to obtain an accurate approximation depends on v; see Frühwirth–Schnatter et al. (2009) for details.

Conditioning on the mixture component $r_{i,j}$ of the variables $\tau_{i,j}$, we define the auxiliary variables $\widetilde{Y}_{i,j}^c = -\log \tau_{i,j}^c - m_{r_{i,j}}$, for $j = 1,2$, and we denote by $\widetilde{\mathbf{Y}}_i^c$ the vector which is obtained by stacking the auxiliary variables for subject i. Note that $\widetilde{\mathbf{Y}}_i^c$ has either one or two components, depending on the value of Y_i^c. Defining $\widetilde{\mathbf{X}}_i^c$ as the corresponding design matrix, i.e., $\widetilde{\mathbf{X}}_i^c = \mathbf{x}_i^\top$ or $\widetilde{\mathbf{X}}_i^c = (\mathbf{x}_i, \mathbf{x}_i)^\top$, the conditional Gaussian model in the auxiliary variables can be written as

$$\widetilde{\mathbf{Y}}_i^c = \widetilde{\mathbf{X}}_i^c \boldsymbol{\beta}_i^c + \widetilde{\boldsymbol{\varepsilon}}_i^c, \quad \widetilde{\boldsymbol{\varepsilon}}_i^c \sim \begin{cases} N(0, s_{r_{i,1}}^2) & , \text{if } Y_i^c = 0 \\ N_2(\mathbf{0}, diag(s_{r_{i,1}}^2, s_{r_{i,2}}^2)) & , \text{otherwise} \end{cases}.$$

12.2.2　Modeling dependence for cross-sectional data

Data augmentation as described above yields a representation of the marginal regression models for discrete responses as linear Gaussian regression models in the auxiliary variables $\widetilde{Y}$. For cross-sectional data, we join the marginal models by specifying a multivariate normal distribution for the error terms. After stacking all auxiliary variables and the continuous responses for subject i into the vector $\widetilde{\mathbf{Y}}_i$, the covariates into an appropriate design matrix $\widetilde{\mathbf{X}}_i$, and the regression coefficients into the vector $\boldsymbol{\beta}$, the joint model can be written as

$$\widetilde{\mathbf{Y}}_i = \widetilde{\mathbf{X}}_i \boldsymbol{\beta} + \widetilde{\boldsymbol{\varepsilon}}_i, \quad \widetilde{\boldsymbol{\varepsilon}}_i \sim N_{K'}(\mathbf{0}, \boldsymbol{\Sigma}_i). \tag{12.8}$$

Note that the dimension of $\widetilde{\mathbf{Y}}_i$, and hence, of $\boldsymbol{\Sigma}_i$, depends on the types of the response components. For multivariate responses, where all K components are binary, ordinal or normal, $\widetilde{\mathbf{Y}}_i$ has $K' = K$ elements; if one component Y_i^c is a count variable, the vector $\widetilde{\mathbf{Y}}_i$ has dimension $K' = K$, if $Y_i^c = 0$, and $K' = K + 1$, otherwise.

To make the model identifiable, we impose a structure on the elements of $\boldsymbol{\Sigma}_i$; in particular, we assume correlations between the jth and kth components to be equal across subjects. Hence, the covariance between the latent utility of a binary or ordered variable and a normal response, is specified as

$$cov(\widetilde{Y}_i^b, Y_i^n) = \rho_{bn} s_{r_i} \sigma_n,$$

where $|\rho_{bn}| < 1$. For count response $Y_i^c = 0$, the covariance between the corresponding auxiliary variable and the normal response, is given as

$$cov(\widetilde{Y}_i^c, Y_i^n) = \rho_{cn} s_{r_{i,1}} \sigma_n,$$

and for $Y_i^c > 0$, the covariance matrix of the two auxiliary variables and the normal response, is given as

$$cov(\widetilde{\mathbf{Y}}_i^c, Y_i^n) = \begin{pmatrix} s_{r_{i,1}}^2 & 0 & \rho_{cn} s_{r_{i,1}} \sigma_n \\ 0 & s_{r_{i,2}}^2 & \rho_{cn} s_{r_{i,2}} \sigma_n \\ \rho_{cn} s_{r_{i,1}} \sigma_n & \rho_{cn} s_{r_{i,2}} \sigma_n & \sigma_n^2 \end{pmatrix}.$$

Thus — as in any relevant analysis, at least one of the observed counts is greater than zero — the correlation coefficient ρ_{cn} has to be restricted to the interval $|\rho_{cn}| < 1/\sqrt{2}$ to guarantee positive definiteness of the covariance matrix, for $Y_i^c > 0$.

Note that for the augmented data, the joint model is a seemingly unrelated regressions (SUR) model. For SUR models, it is known that point estimates using feasible generalized least squares, where dependence is taken into account, are equivalent to separate ordinary least squares (OLS) estimates for each regression model, if the same regressors are used in all models; see e.g., Wooldridge (2010). Hence, we would also expect little differences for Bayes point estimates, under uninformative priors, using a joint model compared to separate modeling of each component. However, joint inference involving regression coefficients for different response components requires their joint posterior distribution, which is only available from a joint model.

12.2.3   Modeling dependence for longitudinal data

For longitudinal data, we specify a regression model for the response vector $\mathbf{Y}_i$. We could, in principle, follow the same approach as for cross-sectional data, and model the dependence by specifying a correlation structure on the covariance matrix of the vector $\widetilde{\boldsymbol{\varepsilon}}_i$ of error terms. However, the panel structure of the data allows us to specify the marginal model for each response component as a random effects model. Heterogeneity between subjects with respect to covariate effects, is thereby captured.

Using data augmentation, the resulting marginal model for the kth response component is a linear Gaussian random effects model in the auxiliary variables $\widetilde{\mathbf{Y}}_i^k$:

$$\widetilde{\mathbf{Y}}_i^k = \widetilde{\mathbf{X}}_i^k \boldsymbol{\beta}_i^k + \widetilde{\boldsymbol{\varepsilon}}_i^k, \quad k = 1, \cdots, K, \tag{12.9}$$

where the error term $\widetilde{\boldsymbol{\varepsilon}}_i^k$ is distributed as $N(\mathbf{0}, \boldsymbol{\Sigma}_i^k)$. For a normal response, no data augmentation is needed, and hence, $\widetilde{\mathbf{Y}}_i^n = \mathbf{Y}_i^n$, $\widetilde{\mathbf{X}}_i^n = \mathbf{X}_i$, and $\boldsymbol{\Sigma}_i^n = \sigma_n^2 \mathbf{I}$. Note that we suppress the dimension of the normal distribution for the error term for simplicity. It is T_i for normal, binary or ordinal components, but depends on the observed values for count observations and can thus range from T_i (if $Y_{it}^c = 0$, for all $t = 1, \cdots, T_i$) to $2T_i$ (if $Y_{it}^c > 0$, for all $t = 1, \cdots, T_i$).

Dependence between repeated measurements is described by the random effects $\boldsymbol{\beta}_i^k$, shared by all measurements of one response component. To take dependence across the components into account, we assume that the random effects $\boldsymbol{\beta}_i = (\{\boldsymbol{\beta}_i^1\}^\top, \cdots, \{\boldsymbol{\beta}_i^K\}^\top)^\top$ of one subject, follow the multivariate normal distribution

$$\boldsymbol{\beta}_i \sim N_{dK}(\boldsymbol{\beta}, \mathbf{Q}), \tag{12.10}$$

with mean $\boldsymbol{\beta}$ and covariance matrix $\mathbf{Q}$.

The joint model for all outcome components is obtained by combining the separate models (12.9) into the following Gaussian random effects model:

$$\widetilde{\mathbf{Y}}_i = \widetilde{\mathbf{X}}_i \boldsymbol{\beta}_i + \widetilde{\boldsymbol{\varepsilon}}_i, \quad \widetilde{\boldsymbol{\varepsilon}}_i \sim N(\mathbf{0}, \boldsymbol{\Sigma}_i), \tag{12.11}$$

where $\widetilde{\mathbf{X}}_i$ and $\boldsymbol{\Sigma}_i$ are block diagonal matrices with entries $\widetilde{\mathbf{X}}_i^1, \cdots, \widetilde{\mathbf{X}}_i^K$, and $\boldsymbol{\Sigma}_i^1, \cdots, \boldsymbol{\Sigma}_i^K$, respectively. Note that (12.10)–(12.11) specify the model with centered parameterization. Again, the dimension of the error term is dropped as it depends on the type of components. It is KT_i if all components are normal, binary or ordinal, but can be higher if some components are count data with positive values. The equivalent non-centered parameterization is given as

$$\widetilde{\mathbf{Y}}_i \;=\; \widetilde{\mathbf{X}}_i \boldsymbol{\beta} + \widetilde{\mathbf{X}}_i \mathbf{C} \mathbf{Z}_i + \widetilde{\boldsymbol{\varepsilon}}_i, \quad \widetilde{\boldsymbol{\varepsilon}}_i \sim N(\mathbf{0}, \boldsymbol{\Sigma}_i), \quad \mathbf{Z}_i \sim N_{dK}(\mathbf{0}, \mathbf{I}), \tag{12.12}$$

where $\mathbf{C}$ denotes the lower-triangular Cholesky factors of $\mathbf{Q}$, i.e., $\mathbf{Q} = \mathbf{C}\mathbf{C}^\top$, see Meng and van Dyk (1998). The representation of the original model as a conditional Gaussian model in non-centered parameterization allows simple MCMC schemes and incorporation of variable and covariance selection.

12.2.4   Prior distributions

Bayesian model specification is completed by assigning prior distributions to the model parameters. For cross-sectional data, a joint prior has to be specified for $\boldsymbol{\beta}$, the correlations $\boldsymbol{\rho}$, and the standard deviations $\boldsymbol{\sigma}$ of the normal components. We use a prior of the structure

$$p(\boldsymbol{\beta}, \boldsymbol{\rho}, \boldsymbol{\sigma}) \;=\; p(\boldsymbol{\beta}) p(\boldsymbol{\rho}) p(\boldsymbol{\sigma}).$$

As we deal with a linear Gaussian model in the auxiliary variables, we propose to use conjugate priors for $\boldsymbol{\beta}$, e.g., uninformative normal priors $\boldsymbol{\beta} \sim N(\mathbf{0}, \mathbf{B}_0)$ to estimate the unrestricted model.

To perform variable selection, we use mixture priors with a spike concentrated at zero and a slab component, which has its mass spread over a range of plausible values; see Ishwaran and Rao (2005), George and McCulloch (1997), and Mitchell and Beauchamp (1988), for different variants of spike and slab priors. To specify a spike and slab prior, we introduce a vector of indicators $\boldsymbol{\delta} = (\delta_1, \cdots, \delta_d)$, where $\delta_j = 1$, if and only if β_j is allocated to the slab component.

Denoting by $\boldsymbol{\beta}_{\boldsymbol{\delta}}$ the vector comprising those elements of $\boldsymbol{\beta}$ with corresponding indicator $\delta_j = 1$, we consider priors of the following structure:

$$p(\boldsymbol{\beta} | \boldsymbol{\delta}) \;=\; p_{slab}(\boldsymbol{\beta}_{\boldsymbol{\delta}}) \prod_{j:\delta_j=0} \Delta_0(\beta_j), \tag{12.13}$$

$$p(\boldsymbol{\delta}) \;=\; \text{Beta}(p_{\boldsymbol{\delta}} + 1, dK - p_{\boldsymbol{\delta}} + 1). \tag{12.14}$$

Here, $\Delta_0(\beta_j)$ denotes the Dirac measure at zero, $\text{Beta}(\cdot)$ is the beta function, and $p_{\boldsymbol{\delta}} = \sum_j \delta_j$. The prior on $\boldsymbol{\delta}$ is proposed in Smith and Kohn (2002), and results from assuming prior independence of the elements of $\boldsymbol{\delta}$ and the hierarchical specification

$$p(\delta_j = 1) \;=\; \omega, \quad \omega \sim \mathscr{B}(1,1),$$

where $\mathscr{B}(1,1)$ is a beta distribution.

For the slab component, we use a normal prior $N(\mathbf{b}_{0,\boldsymbol{\delta}}, \mathbf{B}_{0,\boldsymbol{\delta}})$, e.g., an independence slab with

$$\mathbf{b}_{0,\boldsymbol{\delta}} \;=\; \mathbf{0} \quad \text{and} \quad \mathbf{B}_{0,\boldsymbol{\delta}} \;=\; B_0 \mathbf{I},$$

or a fractional slab with fraction f, which is given as

$$\mathbf{b}_{0,\boldsymbol{\delta}} \;=\; (\widetilde{\mathbf{X}}_{\boldsymbol{\delta}}^\top \widetilde{\mathbf{X}}_{\boldsymbol{\delta}})^{-1} \widetilde{\mathbf{X}}_{\boldsymbol{\delta}}^\top \widetilde{\mathbf{Y}} \quad \text{and} \quad \mathbf{B}_{0,\boldsymbol{\delta}} \;=\; \frac{1}{f} (\widetilde{\mathbf{X}}_{\boldsymbol{\delta}}^\top \widetilde{\mathbf{X}}_{\boldsymbol{\delta}})^{-1}. \tag{12.15}$$

Here $\widetilde{\mathbf{Y}}$ is the vector obtained by stacking the observations $\widetilde{\mathbf{Y}}_i$ for all subjects, and $\widetilde{\mathbf{X}}_{\boldsymbol{\delta}}$ is the corresponding design matrix with only those regressors for which the corresponding indicator variable takes the value 1, see e.g., Frühwirth−Schnatter and Tüchler (2008) and Malsiner−Walli and Wagner (2011), for details on this prior.

For the correlations $\rho_{k\ell}$, we use independent normal priors $N(r_{0,k\ell}, R_{0,k\ell})$, restricted to the appropriate range. The priors for the variances of normal components are also assumed to be independent, and are specified in terms of $\theta = \log \sigma_n \sim N(d_0, D_0)$. Alternatively, independent inverse gamma priors on σ_n^2, i.e., $\sigma_n^2 \sim \mathscr{G}^{-1}(s_0/2, S_0/2)$, could be used.

For longitudinal data, we specify a joint normal prior for $\boldsymbol{\beta}$ and $\mathbf{C}$. To perform joint variable and covariance selection, we define a further indicator vector $\boldsymbol{\gamma}$ of dimension $dK(dK+1)/2$, to select elements in $\mathbf{C}$, and specify a joint spike and slab prior for $\boldsymbol{\beta}$ and $\mathbf{C}$ as

$$p(\boldsymbol{\beta}, \mathbf{C}|\boldsymbol{\delta}, \boldsymbol{\gamma}) = p_{slab}(\boldsymbol{\beta_\delta}, \mathbf{C_\gamma}) \prod_{j:\delta_j=0} \Delta_0(\beta_j) \prod_{j:\gamma_j=0} \Delta_0(C_j), \qquad (12.16)$$

$$p(\boldsymbol{\delta}) = \mathrm{Beta}(p_{\boldsymbol{\delta}}+1, dK - p_{\boldsymbol{\delta}}+1), \qquad (12.17)$$

$$p(\boldsymbol{\gamma}) = \mathrm{Beta}\left(p_{\boldsymbol{\gamma}}+1, \frac{dK(dK+1)}{2} - p_{\boldsymbol{\gamma}}+1\right), \qquad (12.18)$$

where $p_{\boldsymbol{\gamma}} = \sum_j \gamma_j$. The slab distribution can again be specified as a proper normal prior or a fractional prior, see Wagner and Tüchler (2010) for more details. Note that the ordering of the covariates influences the total number of free elements in $\mathbf{C}$ in prior (12.16), see Frühwirth–Schnatter and Tüchler (2008).

12.3 Logit-normal model

To exemplify our method, we describe in detail one particular model of the general model class defined above. We consider cross-sectional data for a bivariate response with binary and continuous components, and specify a marginal logit regression model for the binary, and a Gaussian regression model for the continuous response.

12.3.1 Model specification

Let $\mathbf{Y}^b$ and $\mathbf{Y}^n$ denote the vectors of binary and normal responses, respectively. Further, let $\widetilde{\mathbf{Y}}^b$ denote the auxiliary variables generated for the binary observations. Conditional on the component indicator r_i, we specify a bivariate normal distribution for the auxiliary variable $\widetilde{Y}_i^b$ and the normal response Y_i^n as

$$\widetilde{\mathbf{Y}}_i = \begin{pmatrix} \widetilde{Y}_i^b \\ Y_i^n \end{pmatrix} = \begin{pmatrix} \mathbf{x}_i^\top \boldsymbol{\beta}^b \\ \mathbf{x}_i^\top \boldsymbol{\beta}^n \end{pmatrix} + \widetilde{\boldsymbol{\varepsilon}}_i, \quad \widetilde{\boldsymbol{\varepsilon}}_i \sim N_2(\mathbf{0}, \boldsymbol{\Sigma}),$$

where

$$\boldsymbol{\Sigma} = \begin{pmatrix} s_{r_i}^2 & s_{r_i}\sigma\rho \\ s_{r_i}\sigma\rho & \sigma^2 \end{pmatrix},$$

ρ is the correlation between the binary and the normal responses, and σ^2 is the variance of the normal response.

Thus, conditional on $r_i = r$, the standardized residuals $\widetilde{\varepsilon}_i^b/s_r$ and ε_i^n/σ, of the two marginal models are combined via a standard Gaussian copula, so that their joint cumulative distribution function (CDF) is given as

$$F(\widetilde{\varepsilon}_i^b, \varepsilon_i^n|r) = \Phi_2\left(\frac{\widetilde{\varepsilon}_i^b}{s_r}, \frac{\varepsilon_i^n}{\sigma}; \rho\right).$$

Here, $\Phi_2(\cdot; \rho)$ denotes the CDF of a standard bivariate normal distribution with correlation ρ. However, marginalizing over r, the standard logistic and the standard normal errors are joined by a mixture of Gaussian copulas. That is, the marginal bivariate distribution is given as

$$F(\widetilde{\varepsilon}_i^b, \varepsilon_i^n) = \sum_{r=1}^{6} \pi_r \Phi_2\left(\frac{\widetilde{\varepsilon}_i^b}{s_r}, \frac{\varepsilon_i^n}{\sigma}; \rho\right),$$

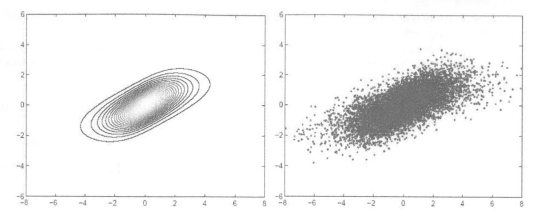

Figure 12.1 *Contour (left) and scatter (right) plots of* 10,000 *simulated data from the bivariate distribution with standard logistic and standard normal margins and* $\rho = 0.7$.

whereas a Gaussian copula of the standard logistic and the standard normal errors, which is used in Song *et al.* (2009), would be

$$F(\widetilde{\varepsilon}_i^b, \varepsilon_i^n) \quad = \quad \Phi_2\left(\Phi^{-1}\left\{\sum_{r=1}^{6}\pi_r\Phi\left(\frac{\widetilde{\varepsilon}_i^b}{s_r}\right)\right\}, \frac{\varepsilon_i^n}{\sigma};\rho\right).$$

Figure 12.1 (left panel) displays a contour plot of the bivariate density with standard logistic and standard normal margins resulting from the mixture of Gaussian copulas with $\rho = 0.7$. The right panel shows a scatter plot of $n = 10,000$ data generated from this bivariate distribution.

The likelihood contribution of the bivariate observation $\mathbf{y}_i$ is given as

$$p(\mathbf{y}_i) \quad = \quad p(y_i^n)p(y_i^b|y_i^n) \quad = \quad \phi\left(\frac{y_i^n - \mathbf{x}_i^\top\boldsymbol{\beta}^n}{\sigma}\right)p(y_i^b|y_i^n).$$

It can easily be determined from the conditional normal distribution of $\widetilde{Y}_i^b|Y_i^n \sim N(m_{i,r}, s_{i,r})$, with conditional mean and standard deviation

$$m_{i,r} \quad = \quad \mathbf{x}_i^\top\boldsymbol{\beta}^b + s_r\rho\frac{\varepsilon_i^n}{\sigma}, \tag{12.19}$$

$$s_{i,r} \quad = \quad s_r\sqrt{1-\rho^2}. \tag{12.20}$$

Hence, the likelihood contribution of y_i^b, conditional on y_i^n and r, is given as

$$p(y_i^b|y_i^n,r) \quad = \quad \int_{\gamma(y_i^b)}^{\gamma(y_i^b+1)}\phi\left(\frac{\widetilde{y}_i^b - m_{i,r}}{s_{i,r}}\right)d\widetilde{y}_i^b,$$

with $\gamma(0) = -\infty$, $\gamma(1) = 0$, and $\gamma(2) = +\infty$. Marginalizing over the mixture components, we get

$$p(\mathbf{y}_i) \quad = \quad \begin{cases} \phi\left(\frac{y_i^n - \mathbf{x}_i^\top\boldsymbol{\beta}^n}{\sigma}\right)\left\{1 - \sum\Phi\left(\frac{m_{i,r}}{s_{i,r}}\right)\pi_r\right\} & \text{, if } y_i^b = 0 \\ \phi\left(\frac{y_i^n - \mathbf{x}_i^\top\boldsymbol{\beta}^n}{\sigma}\right)\sum\Phi\left(\frac{m_{i,r}}{s_{i,r}}\right)\pi_r & \text{, if } y_i^b = 1 \end{cases}. \tag{12.21}$$

12.3.2 *Priors*

Priors are specified, as discussed in Section 12.2.4, as $p(\boldsymbol{\beta},\theta,\rho) = p(\boldsymbol{\beta})p(\theta)p(\rho)$, with $\theta = \log\sigma$. For the regression parameters, we recommend to use a proper normal prior, as in logit models, a proper prior with moments up to order ℓ guarantees the existence of posterior moments of the same order (Rossi, 1996). Using a fractional prior might lead to an improper posterior when separation occurs in the binary data (Wagner and Duller, 2012).

To perform variable selection, we assign independent spike and slab priors with a Dirac spike at zero, and a $N(0, B_0)$-slab to all regression coefficients. We specify the spike and slab priors, with indicators δ_j, hierarchically as

$$p(\beta_j) = (1 - \delta_j)\Delta_0(\beta_j) + \delta_j \phi\left(\frac{\beta_j}{\sqrt{B_0}}\right).$$

We assume that for both the continuous and binary responses, the intercept is not subject to selection, and hence, we use independent normal priors (i.e., we set $\delta_j = 1$) for the corresponding elements of $\boldsymbol{\beta}$.

12.3.3 MCMC estimation

The joint distribution of observations, latent variables, and parameters, is given as

$$p(\mathbf{y}, \widetilde{\mathbf{y}}^b, \mathbf{r}, \boldsymbol{\delta}, \boldsymbol{\beta}, \rho, \theta) = p(\mathbf{y}|\boldsymbol{\beta}, \rho, \theta)p(\widetilde{\mathbf{y}}^b|\boldsymbol{\beta}, \rho, \theta, \mathbf{r}, \mathbf{y})p(\boldsymbol{\beta}|\boldsymbol{\delta})p(\boldsymbol{\delta})p(\rho)p(\theta)p(\mathbf{r}),$$

where $\mathbf{r} = (r_1, \cdots, r_n)$ denotes the vector of component indicators.

To estimate the model parameters, we use an MCMC scheme with the following steps:

Step 1. Sample the component indicators $\mathbf{r}$ from $p(\mathbf{r}|\boldsymbol{\delta}, \boldsymbol{\beta}, \rho, \theta, \widetilde{\mathbf{y}})$.

Step 2. Sample (ρ, θ) and the latent utilities $\widetilde{\mathbf{y}}^b$:

Step 2a. Sample ρ and θ together using an MH-step from the posterior $p(\rho, \theta|\boldsymbol{\delta}, \boldsymbol{\beta}, \mathbf{r}, \mathbf{y})$.

Step 2b. For $i = 1, \cdots, n$, sample the latent utilities $\widetilde{y}_i^b$ from the posterior $p(\widetilde{y}_i^b|\boldsymbol{\delta}, \boldsymbol{\beta}, \rho, \theta, r_i, \mathbf{y}_i)$.

Step 3. Sample $(\boldsymbol{\delta}, \boldsymbol{\beta})$ from the normal posterior $p(\boldsymbol{\delta}|\rho, \theta, \widetilde{\mathbf{y}}, \mathbf{r})p(\boldsymbol{\beta}|\boldsymbol{\delta}, \rho, \theta, \widetilde{\mathbf{y}}, \mathbf{r})$:

Step 3a. Sample each element δ_j from $p(\delta_j|\boldsymbol{\delta}_{\backslash j}, \rho, \theta, \widetilde{\mathbf{y}}, \mathbf{r})$, where $\boldsymbol{\delta}_{\backslash j}$ denotes all elements of $\boldsymbol{\delta}$ except δ_j.

Step 3b. Sample $\boldsymbol{\beta}$ from $p(\boldsymbol{\beta}|\boldsymbol{\delta}, \rho, \theta, \widetilde{\mathbf{y}}, \mathbf{r})$.

To perform this MCMC scheme, starting values are required for $\rho, \theta, \widetilde{\mathbf{y}}^b$, and $\boldsymbol{\beta}$. For ρ and θ, starting values can be drawn from the prior distribution or set to 0; for the latent variables $\widetilde{\mathbf{y}}_i$, starting values can be drawn from the standard logistic distribution; and for $\boldsymbol{\beta}$, the OLS estimates $(\mathbf{X}^\top \mathbf{X})^{-1} \mathbf{X}^\top \widetilde{\mathbf{y}}$ of the joint regression model are used. When using spike and slab priors, we recommend to start with the unrestricted model, i.e., setting $\delta_j = 1$ for all j.

Details of the sampling steps are given in the following subsections.

12.3.3.1 Sampling component indicators

The posterior for the component indicators $\mathbf{r}$ is formally given as

$$p(\mathbf{r}|\boldsymbol{\beta}_{\boldsymbol{\delta}}, \rho, \theta, \widetilde{\mathbf{y}}^b, \mathbf{y}) \propto p(\widetilde{\mathbf{y}}^b|\boldsymbol{\beta}_{\boldsymbol{\delta}}, \rho, \theta, \mathbf{r}, \mathbf{y})p(\mathbf{r}) = \prod_{i=1}^n p(\widetilde{y}_i^b|\boldsymbol{\beta}_{\boldsymbol{\delta}}, \rho, \theta, r_i, \mathbf{y}_i)p(r_i). \quad (12.22)$$

As we use a mixture of 6 normal components, each with mean 0 and standard deviations s_r, the posterior probabilities $P(r_i = r)$, $r = 1, \cdots, 6$, are given as

$$P(r_i = r) \propto \phi\left(\frac{\widetilde{y}_i^b - m_{i,r}}{s_{i,r}}\right)\pi_r,$$

where the conditional means $m_{i,r}$ and standard deviations $s_{i,r}$ are given in (12.19)–(12.20).

12.3.3.2   Sampling variance and correlation parameters

We sample (ρ, θ) jointly not from its full conditional but from the conditional posterior, marginal-ized over the latent utilities $\widetilde{\mathbf{Y}}^b$, and hence, our sampler is a partially collapsed Gibbs sampler (Van Dyk and Park, 2009). The marginalized posterior for (ρ, θ) is given as

$$p(\rho, \theta | \mathbf{y}, \boldsymbol{\beta}_{\boldsymbol{\delta}}, \mathbf{r}) \propto p(\rho)p(\theta)\prod_{i=1}^{n} p(\mathbf{y}_i | \boldsymbol{\beta}_{\boldsymbol{\delta}}, \mathbf{r}, \rho, \theta),$$

where the contribution of $\mathbf{y}_i$ in the likelihood is given in (12.21). As this posterior is not of closed form, we use a MH-step with a tailored proposal to sample (ρ, θ). The proposal is a bivariate Student t-distribution, with 10 degrees of freedom. The mean of this t-distribution is the maximum likelihood estimate (MLE) after a few maximizing iterations of the likelihood $\prod_{i=1}^{n} p(\mathbf{y}_i | \boldsymbol{\beta}_{\boldsymbol{\delta}}, \mathbf{r}, \theta, \rho)$, and the covariance matrix is the inverse Hessian at this point. In our application, we use 10 iterations to get a proposal with an acceptance rate of 93%.

12.3.3.3   Sampling latent utilities

Conditional on the continuous response Y_i^n, the model parameters $(\boldsymbol{\beta}_{\boldsymbol{\delta}}, \rho, \theta)$, and the component indicator r_i, the distribution of $\widetilde{Y}_i^b$ is the normal distribution

$$\widetilde{Y}_i^b | \boldsymbol{\beta}_{\boldsymbol{\delta}}, \rho, \theta, y_i^n \sim N(m_{i,r}, s_{i,r}^2),$$

with mean and standard deviation given in (12.19)–(12.20). Further, conditioning on Y_i^b truncates this normal distribution to the interval $(-\infty, 0)$, for $Y_i^b = 0$, and to the interval $(0, +\infty)$, for $Y_i^b = 1$. Hence, $\widetilde{Y}_i^b$ is drawn from the $N(m_{i,r}, s_{i,r}^2)$ truncated to $(-\infty, 0)$, for $Y_i^b = 0$, and truncated to $(0, +\infty)$, for $Y_i^b = 1$.

12.3.3.4   Sampling indicator variables and regression coefficients

The indicator variables are sampled one at a time from the posterior marginalized over $\boldsymbol{\beta}$, given as

$$p(\delta_j = 1 | \boldsymbol{\delta}_{\backslash j}, \rho, \theta, \mathbf{r}, \widetilde{\mathbf{y}}) = \frac{1}{1 + \frac{p(\delta_j = 0, \boldsymbol{\delta}_{\backslash j})}{p(\delta_j = 1, \boldsymbol{\delta}_{\backslash j})} R_j}, \quad R_j = \frac{p(\widetilde{\mathbf{y}} | \delta_j = 0, \boldsymbol{\delta}_{\backslash j}, \rho, \theta, \mathbf{r})}{p(\widetilde{\mathbf{y}} | \delta_j = 1, \boldsymbol{\delta}_{\backslash j}, \rho, \theta, \mathbf{r})}.$$

The posterior for δ_j involves the conditional marginal likelihoods of two heteroscedastic linear regression models, with design matrices differing only by inclusion/exclusion of the jth column of the matrix $\mathbf{X}$. The conditional marginal likelihood of a linear regression model is available in closed form as

$$p(\widetilde{\mathbf{y}} | \boldsymbol{\delta}, \rho, \theta) \propto \frac{|\mathbf{B}_{\boldsymbol{\delta}}|^{1/2}}{|\mathbf{B}_{0,\boldsymbol{\delta}}|^{1/2}} \exp\left\{-\frac{1}{2}(\widetilde{\mathbf{y}}^{\top}\boldsymbol{\Sigma}^{-1}\widetilde{\mathbf{y}} - \mathbf{b}_{\boldsymbol{\delta}}^{\top}\mathbf{B}_{\boldsymbol{\delta}}^{-1}\mathbf{b}_{\boldsymbol{\delta}} + \mathbf{b}_{0,\boldsymbol{\delta}}^{\top}\mathbf{B}_{0,\boldsymbol{\delta}}^{-1}\mathbf{b}_{0,\boldsymbol{\delta}})\right\},$$

where $\mathbf{B}_{\boldsymbol{\delta}}$ and $\mathbf{b}_{\boldsymbol{\delta}}$ are the moments of the normal posterior

$$\mathbf{B}_{\boldsymbol{\delta}} = (\mathbf{X}_{\boldsymbol{\delta}}^{\top}\boldsymbol{\Sigma}^{-1}\mathbf{X}_{\boldsymbol{\delta}} + \mathbf{B}_{0,\boldsymbol{\delta}}^{-1})^{-1},$$

$$\mathbf{b}_{\boldsymbol{\delta}} = \mathbf{B}_{\boldsymbol{\delta}}(\mathbf{X}_{\boldsymbol{\delta}}^{\top}\boldsymbol{\Sigma}^{-1}\widetilde{\mathbf{y}} + \mathbf{B}_{0,\boldsymbol{\delta}}^{-1}\mathbf{b}_{0,\boldsymbol{\delta}}),$$

and $\mathbf{X}_{\boldsymbol{\delta}}$ is the appropriate design matrix, including those regressors, for which the corresponding indicator variable takes the value 1. Regression coefficients β_j for which the corresponding indicator $\delta_j = 0$, are set to zero, and the remaining elements $\boldsymbol{\beta}_{\boldsymbol{\delta}}$ are sampled from the normal posterior $N(\mathbf{b}_{\boldsymbol{\delta}}, \mathbf{B}_{\boldsymbol{\delta}})$.

12.3.4 Simulation example

We illustrate inference for the joint logit-normal model in an example with simulated data. We generated $n = 1,000$ bivariate data with binary and normal components. The latent utilities were generated from the mixture of 6 normal components, which is used as an approximation of the logistic distribution in the MCMC scheme. We set the correlation coefficient $\rho = 0.7$, and the variance for the normal component to $\sigma^2 = 0.1$. Covariates $x_1, \cdots, x_4$, were generated as independent $N(0,1)$ variables, and the vector of regression coefficients which includes also the intercept, is $\boldsymbol{\beta}^b = (-0.8, 0.3, 0.3, 0, 0)^\top$ for the binary component, and $\boldsymbol{\beta}^n = (0.5, 0.05, 0, 0.05, 0)^\top$ for the normal component. We use a spike and slab prior, with independent $N(0,5)$-slabs for the regression coefficients, a $N(0,1)$-prior on θ, and a $N(0,1)$ distribution truncated to $(-1,1)$ as a prior for ρ. MCMC as described in Section 12.3.3 is performed for $10,000$ iterations after a burn-in of $2,000$. The first $1,000$ draws of the burn-in are drawn from the unrestricted model.

Table 12.1 gives the model averaged posterior means of the regression coefficients and the estimated posterior inclusion probabilities for the regression effects. Effects with inclusion probabilities larger than 0.5 are in bold. No posterior inclusion probabilities are reported for the intercepts, as they are not subject to selection.

Table 12.1 *Posterior means of regression coefficients and estimated posterior inclusion probabilities from the joint model for the simulation example. Effects with inclusion probabilities larger than 0.5 are in bold.*

Effect	Binary component $\widehat{\beta}$	Binary component $P(\delta_j = 1)$	Normal component $\widehat{\beta}$	Normal component $P(\delta_j = 1)$
Intercept	−0.69	—	0.51	—
x_1	**0.33**	**1.00**	**0.05**	**1.00**
x_2	**0.21**	**0.83**	−0.01	0.38
x_3	−0.01	0.12	**0.05**	**1.00**
x_4	0.00	0.05	−0.00	0.11

To illustrate the impact of joint modeling, Table 12.2 gives the results obtained from variable selection on the marginal logit and normal regression models, respectively. Results are similar for most effects; however, for x_2, the non-zero effect on the binary response is detected in the joint model, but not in the marginal logistic model. We see that in this example, joint modeling is necessary to find the correct model structure.

Table 12.2 *Posterior means of regression coefficients and estimated posterior inclusion probabilities from separate models for the simulation example. Effects with inclusion probabilities larger than 0.5 are in bold.*

Effect	Binary component $\widehat{\beta}$	Binary component $P(\delta_j = 1)$	Normal component $\widehat{\beta}$	Normal component $P(\delta_j = 1)$
Intercept	−0.68	—	0.51	—
x_1	**0.33**	**1.00**	**0.05**	**1.00**
x_2	0.06	0.36	−0.01	0.20
x_3	−0.01	0.12	**0.04**	**0.98**
x_4	0.00	0.07	−0.00	0.11

12.4   Modeling material deprivation and household income

The European statistics on income and living conditions (EU-SILC) provides data on income, living conditions, and social exclusion, which are comparable for the EU countries. The data are collected in a survey among households, which focuses on income and living, but also includes questions about socio-demographic attributes. For policy and social research purposes, several key indicators are derived from these data. One line of investigation focuses on income and indicators related to the income, whereas a second line of research looks at material or financial deprivation. Although there is an obvious relationship between income and deprivation, analyses of this topic usually concentrate on one aspect only. Our study presents a broadening of these common approaches, as we include both aspects: income and material deprivation. According to European guidelines, a person is hit by material deprivation if at least four out of nine criteria are fulfilled. These nine criteria are

1. arrears on mortgage or rent payments, utility bills, hire purchase instalments, or other loan payments;

2. household cannot afford paying for one week's annual holiday away from home;

3. household cannot afford a meal with meat, chicken, fish (or vegetarian equivalent) every second day;

4. household cannot bear unexpected financial expenses of an amount, which varies for different countries, and is about 900 euros for Austria;

5. household cannot afford a telephone (including mobile phone);

6. household cannot afford a color TV;

7. household cannot afford a washing machine;

8. household cannot afford a car; and

9. household is not able to pay for keeping its home adequately warm.

Our data set contains 3,704 households from the EU-SILC 2009 survey in Austria (BMASK, 2011). Following Fusco *et al.* (2010), we consider only those data sets, where the main-income-earner of the household, i.e., the person with the highest income, is not retired, and at least one adult person is younger than 60 years.

We combine the indicator of material deprivation and the logarithm of the household income in a logit-normal model, and include covariates which are supposed to have an influence on these two responses. One type of covariates corresponds to attributes of the main-income earner, whereas the second type pertains to attributes of the household. The variables of the main-income-earner are gender, age, activity status, education, and migration background. The variable activity-status has categories full-time work, part-time work, unemployed, and out-of-labor-force (note that households with a retired main-income-earner are excluded from the data), whereas the variable education has categories lower education, medium education, higher education, and university. A person has migration background if he or she either now has or once had a non-EU/EFTA citizenship. The remaining covariates are the type of household, the type of building, and the population density. The variable type of household has categories single, two adults/no children, single-parent household, two adults/one or two children, two adults/more than two children, and other household. The variable type of building is categorized as single-family house, house with two families, multi-family house with three to nine households, multifamily house with more than nine households, and other. For population density, we distinguish between high, medium, and low.

We run 25,000 iterations of our MCMC scheme and skip the first 5,000 as burn-in. The first 1,000 iterations are drawn from the unrestricted model, and variable selection is carried out for the rest of the sampling steps. The mean parameter β follows a spike and slab prior, where the slab has mean $\mathbf{0}$ and covariance matrix $\mathbf{B}_0 = 5\mathbf{I}$. The prior for θ is standard normal, and the normal prior $N(0,1)$ for ρ is restricted to $(-1,1)$.

In Table 12.3, we give estimates of the regression coefficients and the corresponding inclusion probabilities for the two responses. It turned out that the gender variable does not help to explain

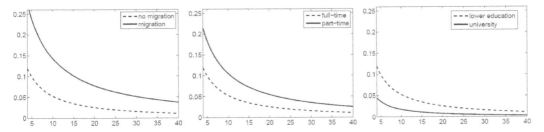

Figure 12.2 *Probability of material deprivation conditional on income (in 1,000) for different households.*

the two responses, as its inclusion probability is below 0.5 for both. Although this is in contrast to studies about gender inequalities, it is in line with analyses in the household context. Since the household income is derived by adding the incomes of all household members, individual differences between males and females are hardly relevant. In social studies, the age effect is known to be non-linear, and it is usually modeled as quadratic. However, in our case, the data are restricted to households with at least one adult member being younger than 60 years, and this yields a quadratic effect, which is restricted to zero. As expected, the activity status plays an important role in our model. Households with the main-income earner working only on a part-time basis have less income and higher risk for material deprivation. Naturally, these effects are stronger for households with a main-income-earner who is unemployed or out-of-labour force. It is well-known that education has an important influence on income; this is also confirmed by our study. The higher the level of education, the bigger is the estimated effect for income and the smaller for the material deprivation response. The migration effect is included in the model with probability 1 for both responses. Households with a main-income-earner who currently has or once had a non-EU/EFTA citizenship have less income and a higher risk for material deprivation. The household type helps to explain the income variable, but it is excluded for the material deprivation response during most of the MCMC iterations. The income of households with two adults and no children is higher, and the income of households with more than two children is smaller than the income of the baseline category single-household. For the building variable, the effects for many categories are restricted to 0 during most of the iterations. However, we find that households living in big buildings with many flats have less income and are more likely to be in a situation of material deprivation. Households living in an area with low population density are not so often hit by material deprivation as households located in areas with high population density.

The correlation parameter ρ is equal to -0.25, and indicates that households with higher income are less likely to be in a situation of material deprivation and vice versa. In Figure 12.2, we analyze this correlation for different compositions of households in more detail. We depict the conditional probability of material deprivation as a function of the log-income, which is given by

$$p(y_i^b = 1 | y_i^n) \quad = \quad \sum \Phi \left(\frac{m_{i,r}}{s_{i,r}} \right) \pi_r,$$

see (12.21), and use the posterior means for the parameters ρ, σ and $\boldsymbol{\beta}$.

In each plot, we compare households with all covariates being equal to the baseline categories, to households differing in only one covariate value. In the left plot, households with the main-income-earner having no migration background are compared to households where such a migration background exists. We find that for a fixed income level, households with no migration background are less likely to be in a situation of material deprivation. For the median household income of 21, 196 euros, the estimated probability for material deprivation is 2.4% for households with no migration background, and 7.3% for those with migration background. Ten percent of the households have an income below 11, 513 euros. For these households, the effect of migration is even bigger with a probability of 4.6% without, and 12.8% with migration background, respectively. In the second plot,

Table 12.3 *Posterior means and estimated posterior inclusion probabilities from the model for material depri-vation and household income. Effects with inclusion probabilities larger than 0.5 are in bold.*

Effect	Material deprivation		log(Earnings)	
	$\widehat{\beta}$	$P(\delta_j = 1)$	$\widehat{\beta}$	$P(\delta_j = 1)$
Intercept	−3.43	—	9.82	—
Gender	0.11	0.34	−0.00	0.05
(Base: Male)				
Age	−0.00	0.03	**0.01**	**1.00**
(Centered at median 43 yrs.)				
Age^2	−0.00	0.00	0.00	0.00
Activity status				
(Base: Full-time)				
Part-time	**1.02**	**1.00**	**−0.26**	**1.00**
Unemployed	**2.29**	**1.00**	**−0.45**	**1.00**
Out-of-labor	**1.78**	**1.00**	**−0.69**	**1.00**
Education				
(Base: Lower)				
Medium	**−0.40**	**0.78**	**0.15**	**1.00**
Higher	**−1.55**	**1.00**	**0.29**	**1.00**
University	**−1.73**	**1.00**	**0.45**	**1.00**
Migration	**1.40**	**1.00**	**−0.28**	**1.00**
(Base: No migration)				
Type of household				
(Base: Single)				
2 adults/no children	−0.17	0.42	**0.19**	**1.00**
Single-parent	0.19	0.45	−0.02	0.29
2 adults/1 or 2 children	−0.03	0.18	0.00	0.04
2 adults/+3 children	−0.01	0.21	**−0.20**	**1.00**
Other	−0.08	0.26	**0.11**	**1.00**
Type of building				
(Base: Single-family)				
2 families	0.02	0.20	−0.00	0.02
3 to 9 families	**0.69**	**0.92**	−0.01	0.14
+10 families	**0.94**	**0.99**	**−0.09**	**1.00**
Other	−0.12	0.41	−0.01	0.09
Population density				
(Base: High)				
Medium	−0.05	0.22	0.00	0.04
Low	**−0.40**	**0.75**	**−0.03**	**0.54**

we see the difference between households with full-time employed, and households with part-time employed main-income-earner. For the median income, we find that 5.1% of the households with part-time employed, and 2.4% of the households with full-time employed main-income-earner, face material deprivation. On the right-hand side of Figure 12.2, we depict the probability for material deprivation, conditional on fixed incomes for baseline households with lower education and with a university degree. Naturally, those with a university degree are less likely to be in a situation of material deprivation. For the median income, the rate of households in material deprivation equals

0.7%, if the main-income earner has a university degree, and 2.4%, if he or she has only a lower education.

12.5 Estimating consumer behavior from panel data

In this section, we analyze a data set of monthly customer data from an apparel retailer, who is interested to see whether his marketing actions have an effect on the buying behavior of his customers; see also Wagner and Tüchler (2010) for more details. In our model, we include one continuous response — the customers' monthly profitability contributions — and one count response — the number of different items purchased in the respective time period. The three covariates measuring marketing activities are the fraction of spending the customer made on weekends (*weekend*), the fraction of shopping trips the customer made with coupon redemption (*coupon*), and the number of mailings the customer received in the time period (*mail*). The data set comprises monthly data over five years for 2, 157 customers.

The dependence structure of the panel data is captured via a random effects specification for the augmented model; see (12.10)–(12.11). We use a spike and slab prior, with fractional slab for $\boldsymbol{\beta}_{\boldsymbol{\delta}}$ and $\mathbf{C}_{\boldsymbol{\gamma}}$, and $1/f = \sum_i T_i$. For the model error variance of the normal component, we use the improper inverted gamma $\mathscr{G}^{-1}(0,0)$-prior. We run 50, 000 iterations, and skip a burn-in of 10, 000 iterations. The first 1, 000 iterations are drawn from the unrestricted model. We check convergence by starting the chain from different starting values.

Posterior estimates and inclusion probabilities for the mean parameter $\boldsymbol{\beta}$ are given in Table 12.4. We find that the marketing action *coupon* has a considerable effect on both responses, whereas *weekend* and *mail* have only a rather small effect on profitability, and for the count response, they are only included with a probability of 0.55 and 0.22, respectively.

Table 12.4 *Posterior means and estimated posterior inclusion probabilities from the model for consumer data. Effects with inclusion probabilities larger than 0.5 are in bold.*

Effect	Profitability		Number of items	
	$\widehat{\boldsymbol{\beta}}$	$P(\delta_j = 1)$	$\widehat{\boldsymbol{\beta}}$	$P(\delta_j = 1)$
Intercept	4.65	—	0.50	—
weekend	**0.06**	**1.00**	**0.01**	**0.55**
coupon	**0.71**	**1.00**	**0.37**	**1.00**
mail	**0.05**	**0.96**	0.00	0.22

Table 12.5 shows posterior means and inclusion probabilities for the elements of the covariance matrix $\mathbf{Q}$ which are easily obtained at each MCMC step from the indicators $\boldsymbol{\gamma}$ of the Cholesky factors $\mathbf{C}$ and the identity $\mathbf{Q} = \mathbf{C}\mathbf{C}^{\top}$. Including both responses jointly in one model enables the data to be interpreted in a new way. Furthermore, we benefit from modeling the dependence between the two responses via the joint covariance matrix, and we find additional structure by doing this in a sparse way.

In our example, we find high pairwise correlations of almost 1 between the random effects of the count and normal parts, for the *intercept*, the *weekend*, the *coupon*, and *mail* effects. This implies that the profit value increases as the number of items purchased rises. The positive correlation between *weekend* and *mail* shows that people who go shopping on weekends also tend to have a positive reaction to mailings. Interestingly, both these variables are negatively correlated with the *intercepts*. Thus, high mailing frequencies might lead to diminishing returns on profit values, and might reduce the number of items purchased for high-profit-customers, whereas an increased mailing activity might stimulate low-profit-customers' interest. Similar conclusions might be drawn for the effect of the fractions of spendings made on weekends. These results are very important for the marketing

Table 12.5 *Posterior means (upper row) and estimated posterior inclusion probabilities (lower row, in italics) for the elements of* **Q** *from the model for consumer data. Effects with inclusion probabilities larger than 0.5 are in bold.*

	Profitability				Number of items			
	Intercept	weekend	coupon	mail	Intercept	weekend	coupon	mail
Profitability								
Intercept	**0.11**							
	1.00							
weekend	**−0.04**	**0.09**						
	1.00	*1.00*						
coupon	−0.00	0.00	**0.08**					
	0.09	*0.12*	*1.00*					
mail	**−0.05**	**0.01**	0.00	**0.08**				
	1.00	*1.00*	*0.10*	*1.00*				
Number of items								
Intercept	**0.07**	**−0.03**	−0.00	**−0.03**	**0.04**			
	1.00	*1.00*	*0.13*	*1.00*	*1.00*			
weekend	**−0.03**	**0.06**	0.00	**0.01**	**−0.02**	**0.04**		
	1.00	*1.00*	*0.14*	*1.00*	*1.00*	*1.00*		
coupon	−0.00	0.00	**0.06**	0.00	−0.00	0.00	**0.04**	
	0.02	*0.06*	*1.00*	*0.05*	*0.07*	*0.08*	*1.00*	
mail	**−0.03**	**0.01**	0.00	**0.06**	**−0.02**	**0.01**	−0.00	**0.04**
	1.00	*1.00*	*0.12*	*1.00*	*1.00*	*1.00*	*0.06*	*1.00*

strategy, but may only be derived from the mixed data approach, as separate modeling led to different estimates. See Wagner and Tüchler (2010) for results on separate models.

12.6   Discussion

We proposed a class of regression models for multidimensional responses of mixed types which relies on data augmentation and allows a representation as a linear Gaussian model. Dependence between response components can be modeled via the correlations of either the error terms or the random effects.

A great advantage of these models is that Bayesian estimation by MCMC is straightforward and allows Gibbs sampling for most of the parameters. Bayesian variable selection and model averaging can easily be incorporated by using spike and slab priors for regression effects. This is an important issue, as it allows to specify a rather general model with all covariates at hand. By performing MCMC, we learn from the data which of these covariates actually has an effect on the different response components.

Whereas our models for cross-sectional data consider mixtures of Gaussian copulas, other copula functions can also be used to join the marginal models. In this case, MCMC would be more involved, as the vector of regression coefficients can no longer be drawn in a Gibbs step. As marginal likelihoods are no longer available in closed form, spike and slab priors with a continuous spike should be used to perform variable selection.

Our method applies to a wide range of data from applied sciences, like economics, social science, and others. With our two examples, we showed that the mixed data approach yields deeper insights and opens up new perspectives. The incorporation of the dependence structure gives results that could not be obtained by separate modeling of the different data types.

Chapter 13

Bayesian methods for the analysis of mixed categorical and continuous (incomplete) data

Michael J. Daniels and Jeremy T. Gaskins

13.1 Introduction

The literature on Bayesian methods for the analysis of mixed discrete and continuous data is not extensive. The classes of models used mirror, to a large extent, those used for likelihood-based inference. However, there are important issues involving priors and computational techniques.

In the following, we review the key literature for Bayesian methods. The recent non-Bayesian reviews of de Leon and Carrière Chough (2010) and Teixeira–Pinto and Normand (2009) are general introductions to the various methods available. de Leon and Carriére Chough (2010) pay particular attention to the drawbacks of each method, while Teixeira–Pinto and Normand (2009) use an extensive simulation study and a number of examples for their comparison of various methods.

The two most commonly used approaches for constructing joint models for mixed continuous and categorical outcomes involve either factorizing the joint distribution (in a sensible way), or introducing latent variables to tie the responses together. The former is sometimes termed the "direct" approach while the latter is the "indirect" method. See Chapter 1 for a brief discussion.

For specifying a factorizable joint distribution, general location models (GLOMs) are a common and convenient approach. These were first introduced by Olkin and Tate (1961) and extended and adapted to a variety of applications, including developmental toxicity studies (Fitzmaurice and Laird, 1995) and incomplete data settings (Fitzmaurice and Laird, 1997; Little and Schluchter, 1985). The idea is to factorize the joint distribution of the continuous and categorical responses such that the marginal distribution of the categorical responses is some (restricted) multinomial distribution, and the continuous responses, conditional on the categorical responses, follow a multivariate normal distribution with mean (and sometimes covariance matrix) depending on the categorical responses. Liu and Rubin (1998) discuss extensions of this model, including replacing the multivariate normal with a multivariate t-distribution and allowing the conditional covariance matrix of the continuous responses to depend (in some way) on the categorical responses. They also discuss Bayesian estimation using the Gibbs sampler. Barnard *et al.* (2000) further the extensions in Liu and Rubin (1998) by allowing more flexibility in the conditional covariance matrix of the continuous responses. In particular, they allow the variances to differ across levels of the categorical responses, but the correlation matrix remains constant (Liu and Rubin, 1998, propose proportional covariance matrices). They propose priors and outline Bayesian inference. Related Bayesian models using different decompositions of the covariance matrix that can be applied in these settings are given in Daniels (2006). de Leon and Carrière (2007) further extend the GLOM by incorporating latent variables corresponding to ordinal responses with the continuous outcomes. A joint distribution for the continuous variables and ordinal latent variables is specified conditionally on the nominal categorical variables.

In incomplete data settings, Fitzmaurice and Laird (1997) consider a factorization model in the context of a multivariate binary response and a multivariate continuous response. They model the correlation among the binary responses using a log-linear model that allows for direct specification of the marginal mean regressions (Fitzmaurice and Laird, 1995) and correlate these with the continuous responses via a conditional mean structure that allows interpretation of the parameters as marginal mean regressions as well. The full likelihood based formulation allows for missing data that is missing at random (MAR), or ignorable (in this case), and parameter estimates are obtained using an Expectation-Maximization (EM) algorithm. An extension to Bayesian inference is straightforward. Fitzmaurice and Laird (1997) apply their technique to a data example from the Harvard Six Cities Study on air pollution. A second example uses a portion of the St. Louis Risk Research Project data, previously analyzed by Little and Schluchter (1985) (see Section 13.2, Example III, for more details about this data). A logistic model is considered for the presence of psychiatric symptoms in the child, given the level of risk of psychiatric symptoms for the parents. Using the parental group and the residuals of the binary model, continuous models for the child's reading and verbal scores are constructed.

Fitzmaurice and Laird's (1997) model differs from the GLOM of Little and Schluchter (1985) in that the focus is on obtaining a joint model that maintains marginal regressions. By building the continuous piece of the joint model using the residuals from the binary model as a covariate, the marginal model is obtained by dropping these binary residuals from the regressions. In contrast, the marginal distribution of the continuous responses from the GLOM is a mixture over all combinations of the categorical variables. A good discussion of ignorable missingness in GLOMs can be found in Schafer (1997).

An alternative modeling approach is to introduce latent variables to induce a joint distribution for the entire vector of responses. Dunson (2000) proposes a general Bayesian framework based on latent variables for clustered mixed outcome data. Multiple responses per subject are modeled with independent generalized linear models (GLMs), conditional on latent variables. The subject and unit level latent variables are assumed to be multivariate normal or given a more general formulation. Bayesian inference using Markov chain Monte Carlo (MCMC) is proposed. The latent variable method is demonstrated on a rat reproductive toxicity example. The responses of interest, litter size and a binary indicator for time to birth (i.e., long vs. normal), are measured on up to five litters (i.e., subunits) per dam (i.e, unit). Modeling the responses is accomplished by assuming underlying Poisson variables with mean functions dependent on unit and subunit latent variables.

Landrum et al. (2003) builds a similar model for health care provider profiling in a cross-sectional setting with both categorical and continuous outcomes. The motivating application consisted of both binary and continuous variables for subjects (i.e., patients) within units (i.e., hospitals). The mixed responses are at the patient level, so inference is performed on the unit-level latent variables.

In the setting of longitudinal data, Dunson (2003) proposes separate GLMs for each response at each time, with subject-time latent variables and subject-response errors. The subject-time latent variable is modeled through a linear transition (i.e., Markov) model to induce dependence across time. Daniels and Normand (2006) extend the work in Landrum et al. (2003) to the longitudinal setting and formulate the model differently from Dunson (2003), as they directly model the longitudinal correlation via latent factors that follow a multivariate normal distribution. These models are used for profiling providers over time based on multiple patient outcomes. We provide further details on this example in Section 13.2 (see Example I). Liu et al. (2009) introduce latent variables to correlate longitudinal continuous and binary responses. We present this model as Example II in Section 13.2. See Chapter 8 for details on latent variable models for longitudinal data in a frequentist setting.

In these latent variable models, one issue is the interpretation of regression coefficients. They are interpreted conditional on fixed values for the latent variables. The corresponding unconditional regression coefficient is attenuated (Diggle et al., 2002). Marginal (unconditional) regression coefficients can be modeled directly in the presence of latent variables using marginalized models (Lee and Daniels, 2008; Heagerty, 1999).

Beyond the factorization and latent variable methods, another approach to creating a categorical-continuous joint distribution is to use a Gaussian copula. One can view copula methods as a latent variable formulation, since an underlying latent variable is generally introduced for the categorical responses. Briefly, Gaussian copula regression (Pitt *et al.*, 2006) involves specifying a marginal regression model for each component of the response vector. The marginal regressions are used to obtain the quantiles of the response, which are then transformed to the quantiles of a standard normal distribution. Dependence between responses is induced by assigning this vector of normal quantiles a mean-zero multivariate normal distribution, with covariance matrix set to be a correlation matrix for identifiability. An example of continuous-binary data modeling with copulas is given by de Leon and Wu (2011). They choose normal and robit regressions for the marginal models to jointly model burn area and the probability of death for burn injury patients. Details on this approach are discussed in Chapter 10.

In this chapter, we also investigate appropriate methodology when data is incomplete. A default Bayesian analysis implicitly assumes the missingness is ignorable (Little and Rubin, 2002; Schafer, 1997). We provide details on this assumption and alternative assumptions in Section 13.5. In the context of latent variable modeling, most of the Bayesian methods for nonignorable missingness in the literature introduce additional latent variables to create a shared parameter model (Wu and Carroll, 1988); some Bayesian references include Rizopoulos and Ghosh (2011), Dunson and Perreault (2001), and Cowles *et al.* (1996). However, these are often for just a univariate (longitudinal) response.

Another important issue is parsimoniously characterizing the dependence of the mixed outcomes, potentially over time (or even space) in both latent variable models and GLOMs, as discussed earlier. We explore this in the context of several examples in the chapter.

The chapter is structured as follows. Section 13.2 presents three examples that we use to clarify concepts throughout the chapter. Section 13.3 discusses characterizing dependence in models for mixed outcomes. The importance of Bayesian inference via the introduction of informative priors is explored in Section 13.4. Section 13.5 introduces some key concepts for missing data and explains corresponding issues in the models considered here. General computational issues for posterior inference is discussed in Section 13.6. Two of the data examples from Section 13.2 are analyzed in Section 13.7. Section 13.8 contains some concluding remarks.

13.2 Examples

We outline three motivating examples of mixed-outcome (longitudinal) data, possibly with missingness.

Example I: Health care provider profiling

The U. S. Veteran's Health Administration (VHA) provides health care to more than 4 million veterans annually and is the largest provider of behavioral health services in the United States. In the late 1990s (1995–2001), a study found that the shift from inpatient to outpatient mental health care in the VHA resulted in a 21% decrease in inpatient spending and a 63% increase in outpatient spending (Shuo *et al.*, 2003). As a result of these drastic changes of where care was being administered, concerns were raised about corresponding changes in the quality of the care. The data here are patient-level, and the patients are grouped into 22 regional networks. The main inferential focus is the performance of the regional networks based on the patient-level outcomes.

We describe a model to examine network performance based on mixed patient data for 1995. First, we introduce some notation. Let Y_{kij} denote response $j = 1, \cdots, J$, for subject $i = 1, \cdots, n_k$, within unit (network) $k = 1, \cdots, K$. Subjects are clustered within networks. The $(J = 4)$ patient-level measures are a binary measure of whether the patient was readmitted within 180 days, and three continuous measures, namely, number of readmissions, number of bed days and days to first readmission. These continuous responses are defined only if the patient is readmitted, i.e., if $Y_{ki1} = 1$.

The patient-level model specifies the distribution for the four responses, conditional on a network-level latent variable $\boldsymbol{\eta}_k = (\eta_{k1}, \eta_{k2}, \eta_{k3}, \eta_{k4})^\top$. For the binary response, we introduce a latent variable $Z_{ki1} \sim N(\xi_{ki1}, 1)$, such that $Y_{ki1} = \mathrm{I}\{Z_{ki1} > 0\}$, where

$$\xi_{ki1} = \beta_{0j} + \beta_{1j}x_{1ki} + \cdots + \beta_{Pj}x_{Pki} + \eta_{k1}. \tag{13.1}$$

After a log-transformation the continuous variables, Y_{kij}, $j = 2, 3, 4$, are assumed to follow a trivariate normal distribution, conditional on a positive value for the corresponding binary random variable (i.e., having an inpatient event),

$$
\begin{aligned}
(Y_{ki2}, Y_{ki3}, Y_{ki4})^\top | Y_{ki1} = 1 &\sim N_3(\boldsymbol{\xi}_{ki}, \boldsymbol{\Sigma}), \\
\xi_{kij} &= \beta_{0j} + \beta_{1j}x_{1ki} + \cdots + \beta_{Pj}x_{Pki} + \eta_{kj};
\end{aligned}
\tag{13.2}
$$

$\boldsymbol{\Sigma}$ is the covariance matrix of the trivariate continuous responses, and x_{pki}, $p = 1, \cdots, P$, are the patient-specific covariates.

The parameter $\boldsymbol{\eta}_k$ represents the effect that the provider network has on each of the responses. These J network effects, η_{kj}, are assumed to be a linear combination of L underlying *a priori* independent latent factors $\theta_{k1}, \cdots, \theta_{kL}$:

$$
\begin{aligned}
\eta_{kj} | \boldsymbol{\theta}_k, \boldsymbol{\lambda}_j, \psi_j^2 &\sim N(\lambda_{1j}\theta_{k1} + \cdots + \lambda_{Lj}\theta_{kL}, \psi_j^2), \\
\theta_{k\ell} &\sim N(0, 1), \quad \ell = 1, \cdots, L,
\end{aligned}
\tag{13.3}
$$

where $\boldsymbol{\lambda}_j = (\lambda_{1j}, \cdots, \lambda_{Lj})^\top$ is a $L \times 1$ vector of discrimination parameters. The latent factors, $\theta_{k\ell}$, represent unobserved traits of each network. The variance ψ_j^2 of the network effects represents heterogeneity among network effects not explained by the L latent factors. The four dimensional patient-level responses are correlated through the η_{kj}, as are the patient responses within a network.

For Bayesian inference, we need to specify priors for the parameters, which is very important in latent variable models. We provide details on this in Section 13.4. We also outline how to sample from the posterior distribution of the parameters in Section 13.6. Finally, we further explore this application and extend the model to the longitudinal setting in Section 13.7, which provides an interesting application of the issues with the dependence structure discussed in Section 13.3.

Example II: Mixed longitudinal behavioral processes

In some longitudinal studies, several time-varying mixed outcomes may be measured, and how they are related is of interest. Examples in smoking cessation studies include smoking status and alcohol use, as well smoking status and weight change. In these studies, the association between the processes can lead to an improved understanding of the mechanism of behavior change. For example, treatments for smoking cessation that include an exercise regimen do so to ideally weaken the dependence between weight gain and relapse back to smoking.

An example of such a study is Commit to Quit II (CTQ II) (Marcus *et al.*, 2005; Marcus *et al.*, 2003), a 4-year randomized trial undertaken to test the efficacy of moderate-intensity physical activity as an aid for smoking cessation among women. The two treatments were a moderate-intensity exercise condition (denoted as exercise) and a contact condition (denoted as wellness). The relevant outcomes for understanding the mechanism of behavior change were quit status (a longitudinal binary outcome) and weight change (a longitudinal continuous outcome). The goal of the modeling was to test whether the association between smoking status and weight change was weakened by exercise.

To assess this question, we can construct a joint model for the longitudinal binary and continuous outcomes using latent variables. More specifically, we introduce normal latent variables underlying the binary responses, which results in a multivariate probit model for the vector of longitudinal binary responses (the weekly quit process). We then assume the joint distribution of these latent variables and the vector of continuous responses (the weekly weight change process) follow a multivariate normal distribution. We provide further details on the notation and the model next.

We let Q_{it} denote (binary) weekly smoking cessation status, W_{it} denote weekly weight change for subject i, and a_i denotes the indicator of whether subject i was randomized to the exercise treatment. The corresponding vectors of responses for binary and continuous outcomes are given by $\mathbf{Q}_i = (Q_{i1}, \cdots, Q_{iT})^\top$ and $\mathbf{W}_i = (W_{i1}, \cdots, W_{iT})^\top$, respectively. The vector of (normal) latent variables underlying the binary vector $\mathbf{Q}_i$ are given by $\mathbf{Z}_i = (Z_{i1}, \cdots, Z_{iT})^\top$. Finally, we combine the latent variables representing the binary responses with the continuous responses to form a vector of the joint processes $\mathbf{Y}_i = (\mathbf{Z}_i^\top, \mathbf{W}_i^\top)^\top$.

The joint distribution of binary and continuous variables over time follows a multivariate normal specification

$$\mathbf{Y}_i \;\sim\; N_{2T}(\mathbf{X}_i\boldsymbol{\beta}, \boldsymbol{\Sigma}(a_i)), \tag{13.4}$$

where $\mathbf{X}_i$ is the design matrix for the mean effects, and $\boldsymbol{\beta}$ is the vector of regression coefficients; $\mathbf{X}_i$ is specified to have $2T$ components, giving $\mathbf{Y}_i$ a distinct mean for each treatment and each time.

An important issue here is the dimension of the covariance matrix $\boldsymbol{\Sigma}$. In this study, $T = 8$, so each covariance matrix is 16×16. In addition, due to the multivariate probit formulation for the vector of binary responses, the $T \times T$ upper block of $\boldsymbol{\Sigma}$ is a correlation matrix (for identifiability). We discuss how to address the complication of this relatively large, constrained covariance matrix, and how to ensure its estimate is positive definite in Section 13.3.

Example III: Mental illness and effect on academic development

Little and Schluchter (1985) analyze data from the St. Louis Risk Research Project using GLOMs. The goal of this study was to analyze the effect of parental mental illness on the children. The ordinal random variable G_i takes values 1 to 3 for normal, moderate, and high risk, respectively, identifying the severity of (any) psychological disorders of the parents in the ith family. Each household in the study had two children, and the variables D_{ij}, R_{ij}, and V_{ij} are recorded for child $j = 1, 2$, where D_{ij} is a binary response indicating the number of symptoms of psychological disorders for the jth child, either 1 for low or 2 for high. The continuous responses R_{ij} and V_{ij} give measures of reading and verbal skills, respectively.

We now describe modeling this data using a GLOM. We first consider a model for the categorical components G and D, followed by specification of the distribution of R and V given the categorical values.

Let $\mathbf{W}_i = (G_i, D_{i1}, D_{i2})^\top$ denote the set of categorical variables for household i. The vector $\mathbf{W}_i$ represents an element within a $C = 3 \times 2 \times 2$ contingency table. For $\mathbf{w} = (j, k, \ell)$, $j \in \{1, 2, 3\}$, $k, \ell \in \{1, 2\}$, we specify the contingency cell probability for $\mathbf{W}_i$ by $\pi_\mathbf{w} = P(\mathbf{W}_i = \mathbf{w}) = P(G_i = j, D_{i1} = k, D_{i2} = \ell)$. We may leave the set of cell probabilities to be fully saturated, or employ a more parsimonious log-linear model to reduce the dimension of the cell probabilities.

The distribution of the continuous response vector $\mathbf{Y}_i = (R_{i1}, V_{i1}, R_{i2}, V_{i2})^\top$ is defined conditional on the categorical responses. We assume that given $\mathbf{W}_i = \mathbf{w}$, $\mathbf{Y}_i \sim N_4(\boldsymbol{\mu}_\mathbf{w}, \boldsymbol{\Sigma}_\mathbf{w})$. Here, the mean vector $\boldsymbol{\mu}_\mathbf{w}$ depends on categorical responses, and we have written a general form whereby the covariance matrix also depends on categorical responses. For the latter, it is often assumed that $\boldsymbol{\Sigma}_\mathbf{w} = \boldsymbol{\Sigma}$, as mentioned in the introduction; we discuss parsimonious ways to allow the covariance matrix to depend on the categorical responses in Sections 13.3 and 13.4. Similar to utilizing the log-linear model for $\pi_\mathbf{w}$, it is often beneficial to estimate the C mean vectors in a lower dimensional space. This can be accomplished by writing $\boldsymbol{\mu}_\mathbf{w} = \mathbf{B}^\top \mathbf{x}_\mathbf{w}$, where $\mathbf{x}_\mathbf{w}$ is a cell-specific design vector, and $\mathbf{B}$ serves as a matrix of regression coefficients. Care must be taken in the construction of the set of design vectors, but a potential model might be

$$\boldsymbol{\mu}_\mathbf{w} \;=\; \boldsymbol{\beta}_0 + \boldsymbol{\beta}_1 \mathrm{I}\{j = 1\} + \boldsymbol{\beta}_2 \mathrm{I}\{j = 2\} + \boldsymbol{\beta}_3 \mathrm{I}\{k = 1\} + \boldsymbol{\beta}_4 \mathrm{I}\{l = 1\},$$

where each $\boldsymbol{\beta}$ is a four-dimensional column vector from $\mathbf{B}$. It is clear how one might add higher order interactions into the model for $\boldsymbol{\mu}_\mathbf{w}$.

One important component to the analysis of this data is that many of the responses are missing. The parental risk level G is always observed, but many of the child-specific responses are absent. If missingness is assumed to be ignorable (see Section 13.5), then parameters may be estimated using an EM algorithm as in the original analysis. When operating under a Bayesian framework, it is possible to facilitate computations for posterior inference by sampling the missing values (data augmentation). We provide details in Section 13.6.

13.3   Characterizing dependence

A key issue for models for mixed outcomes is specifying the dependence between the outcomes in both flexible and parsimonious ways. For categorical and continuous responses, this is often done with latent variables (as in Example I). In some mixed data models, the resulting covariance matrix can be relatively high-dimensional (as in Example II). For modeling based on GLOMs (as in Example III), the covariance structure among the continuous responses, given the categorical responses can be problematic; this is often remedied by assuming a constant covariance matrix across all values of the categorical responses. We explore all these issues in the context of the three examples introduced in Section 13.2.

Example I: Incorporating longitudinal dependence

Latent variables are introduced to induce dependence among the patient-level responses of a common network. An extension of this model to account for longitudinal dependence is given in Section 13.7. This model extends the univariate latent factors $\theta_{k\ell}$ to a multivariate factor with components $\theta_{k\ell t}$ giving the effect for year t. The longitudinal dependence is structured through a correlation matrix on $(\theta_{k\ell 1}, \cdots, \theta_{k\ell T})$. In fact, we explore flexible and parsimonious correlation structures and choice of priors for the correlation matrix of the latent longitudinal (network) factors by re-parameterizing it using partial autocorrelations (Wang and Daniels, 2012b; Daniels and Pourahmadi, 2009).

Example II: Dependence structure in longitudinal smoking cessation and weight change data

For simplicity, we remove the treatment dependence of the covariance matrix, so that $\Sigma(a) = \Sigma$, for all treatments a. To address the high-dimensional covariance matrix, we first partition the covariance matrix Σ as follows:

$$\Sigma \;=\; \begin{pmatrix} \Sigma_{11} & \Sigma_{12} \\ \Sigma_{21} & \Sigma_{22} \end{pmatrix}.$$

Using this partition, we factor the joint distribution of $\mathbf{Y}_i$ into two components: a marginal model for $\mathbf{Z}_i$ and a correlated regression model for $\mathbf{W}_i$, given $\mathbf{Z}_i$, by extending the ideas from Fitzmaurice and Laird (1995) and Gueorguieva and Agresti (2001). Let

$$\mathbf{X}_i \;=\; \begin{pmatrix} \mathbf{X}_{1i} & \mathbf{0} \\ \mathbf{0} & \mathbf{X}_{2i} \end{pmatrix} \quad \text{and} \quad \boldsymbol{\beta} \;=\; \begin{pmatrix} \boldsymbol{\beta}_1 \\ \boldsymbol{\beta}_2 \end{pmatrix},$$

then the new models can be expressed as

$$\mathbf{Z}_i \;=\; \mathbf{X}_{1i}\boldsymbol{\beta}_1 + \boldsymbol{\varepsilon}_{1i} \tag{13.5}$$

$$\mathbf{W}_i \;=\; \mathbf{X}_{2i}\boldsymbol{\beta}_2 + \mathbf{B}(\mathbf{Z}_i - \mathbf{X}_{1i}\boldsymbol{\beta}_1) + \boldsymbol{\varepsilon}_{2i}, \tag{13.6}$$

where $\mathbf{B} = \Sigma_{21}\Sigma_{11}^{-1}$ is the matrix that reflects association between $\mathbf{Z}_i$ and $\mathbf{W}_i$, $\boldsymbol{\varepsilon}_{1i} \sim N_T(\mathbf{0}, \Sigma_{11})$, and $\boldsymbol{\varepsilon}_{2i} \sim N_T(\mathbf{0}, \Sigma_{22}^*)$, with $\Sigma_{22}^* = \Sigma_{22} - \Sigma_{21}\Sigma_{11}^{-1}\Sigma_{12}$. The reparameterization of Σ to $(\Sigma_{11}, \mathbf{B}, \Sigma_{22}^*)$ is known in the literature as the Bartlett decomposition of a covariance matrix (Bartlett, 1933).

It is easy to see that (13.5) is a correlated probit model and (13.6) is a standard correlated regression model, conditional on the latent variable $\mathbf{Z}_i$. For identifiability, $\mathbf{\Sigma}_{11}$ is a correlation matrix $\mathbf{R}$ (Chib and Greenberg, 1998). In addition, this factorization provides a convenient parameterization to examine the association between $\mathbf{W}_i$ and $\mathbf{Z}_i$ (the latent variables for the categorical values $\mathbf{Q}_i$), since the components of the $T \times T$ $\mathbf{B}$ matrix are unconstrained. These components also have an intuitive interpretation, which suggests that this matrix might be sparse in practice. The tth row of $\mathbf{B}$ characterizes the association of the continuous process at week $t = 1, \cdots, T$, with the binary process at all weeks. In particular, it corresponds to the regression model

$$W_{it}|\mathbf{Z}_i \quad = \quad \mathbf{x}_{2it}^\top \boldsymbol{\beta}_2 + \mathbf{b}_t^\top (\mathbf{Z}_i - \mathbf{X}_{1i}\boldsymbol{\beta}_1) + \boldsymbol{\varepsilon}_{2it},$$

where $\mathbf{b}_t = (b_{t1}, \cdots, b_{tT})^\top$ is the tth row of $\mathbf{B}$. Since the covariates associated with $\mathbf{b}_t$, $\mathbf{Z}_i - \mathbf{X}_{1i}\boldsymbol{\beta}_1$, are centered with variance one (recall the marginal covariance matrix of $\mathbf{Z}_i$ is a correlation matrix), the components of $\mathbf{B}$ are standardized regression coefficients. This property of the components of $\mathbf{B}$ facilitates between-component comparisons, and motivate ideas for modeling it parsimoniously. In particular, we expect this matrix to be sparse via conditional independence arguments.

So far, we have only suggested a re-parameterization, but not reduced the dimension. In Section 13.4, we discuss a prior that can be used to generate parsimony in a data-dependent way using this re-parameterization.

Example III: Continuous and discrete dependence structures for a GLOM

For GLOMs, a major issue is how to allow the covariance matrices of the continuous responses to vary across different levels/combinations of the categorical variables. Approaches that allow the marginal variances to differ, but with a constant correlation structure, and proportionality across the levels of the categorical values have been proposed in Barnard *et al.* (2000) and Liu and Rubin (1998), respectively. However, more flexible approaches can be used. For example, allowing components of the covariance matrix to depend on levels of the categorical variables could be accomplished by building regressions on the matrix-logarithm (Chiu *et al.*, 1996) or the modified Cholesky parameters (Daniels, 2006; Pourahmadi, 1999), the latter if the responses are time ordered. For example, let $\mathbf{\Sigma}_\mathbf{w}$ be the covariance matrix corresponding to the categorical response pattern $\mathbf{w}$, with generalized autoregressive parameters $\phi_{\mathbf{w},tj}$ and innovation variances $\gamma_{\mathbf{w},k}$, $t = 1, \cdots, p-1$, $j = 2, \cdots, t-1, k = 1, \cdots, p$. Models for these covariance parameters are specified by $\phi_{\mathbf{w},tj} = \mathbf{g}_{\mathbf{w},tj}^\top \boldsymbol{\omega}$ and $\log \gamma_{\mathbf{w},k} = \mathbf{h}_{\mathbf{w},k}^\top \boldsymbol{\beta}$, where $\mathbf{g}_{\mathbf{w},tj}$ and $\mathbf{h}_{\mathbf{w},k}$ are design vectors, and $\boldsymbol{\omega}$ and $\boldsymbol{\beta}$ are global regression parameters. For details on this parameterization and corresponding priors, see Daniels and Pourahmadi (2002).

Dependence among the categorical variables is controlled by the specification of the log-linear model for $\pi_\mathbf{w}$. For example, we might consider the model for $\mathbf{w} = (j, k, l)$ given by

$$\log \pi_{jk\ell} \quad = \quad \alpha^{(0)} + \alpha_j^{(1)} + \alpha_k^{(2)} + \alpha_\ell^{(3)} + \alpha_{jk}^{(12)} + \alpha_{j\ell}^{(13)},$$

where we place appropriate identifiability constraints on the set of α parameters. Note that this two-way interaction model excludes a term between the second and third discrete responses. This implies conditional independence between D_{i1} and D_{i2}, given G_i, that is, independence between the mental health status of the two children, conditional on the health status of the parents. Marginal, conditional, and/or mutual independence relationships can be specified by the modeling choice (Agresti, 2002, Chapter 8). We also discuss more flexible and robust modeling using shrinkage priors in Section 13.4.

13.4   (Informative) Priors

An integral component of Bayesian inference is the specification of priors for the model parameters. Diffuse or improper priors are often used for regression parameters and variances (Robert, 2001), but informative priors can facilitate model identifiability, bring external information into the problem, "shrink" complex regression models, and reduce the dimension of a covariance matrix.

In latent variable models, informative priors are needed for identifiability, for example, in specifying a prior for the factor loading matrix. This can be done structurally, using point mass priors (Ghosh and Dunson, 2009; Lopes and West, 2004) or by non-degenerate priors (e.g., Daniels and Normand, 2006), which can often facilitate interpretation of the latent variables (more details in Example I below). For the former, where interpretation of the latent factors is not of interest, the number of latent factors can be assumed to be unknown, and reversible jump MCMC approaches can be used (Green, 1995). In addition, in the context of certain types of missing data, informative priors are essential; for details, see Section 13.5.

Another relevant situation is sparseness in GLOMs. This can happen in the marginal distribution of the categorical responses and/or in the multivariate distributions assumed for the continuous responses, conditioned on the categorical responses. We provide some remedies via priors in Example III below.

Example I: Priors for parameters for VA data

For the remainder of the chapter, we assume the presence of a single latent variable θ_k representing the quality of inpatient care. In the notation from Section 13.2, $L = 1$, so we drop the ℓ subscript. For this single latent variable model, (13.3) becomes

$$\eta_{kj}|\theta_k,\lambda_j,\psi_j^2 \ \sim \ N(\lambda_j\theta_k,\psi_j^2), \tag{13.7}$$
$$\theta_k \ \sim \ N(0,1).$$

We specify an informative prior on the discrimination parameters $\boldsymbol{\lambda} = (\lambda_1,\cdots,\lambda_J)^\top \sim N_J(\boldsymbol{\pi},c^2\mathbf{I}_J)$, where

$$\boldsymbol{\pi} \ = \ \pi^* \left(\ -1 \quad -1 \quad -1 \quad 1 \ \right)^\top, \tag{13.8}$$

for $\pi^* > 0$. The 1s and -1s in the prior specification correspond to that response's (of the J) contribution to the latent variable. For example, favorable characteristics include many days to first readmission (response $j = 4$), but fewer number of bed days ($j = 3$); in (13.8), these get the multipliers 1 and -1, respectively. Here, the components of the prior mean for $\boldsymbol{\lambda}$ are set to equally weight the outcomes; obviously, there is subjectivity in this choice, and such choices are made in the context of the problem. For the analysis in Section 13.7, we ultimately set $\pi^* = 0.25$ and $c^2 = 0.01$. Identifiability issues arose if the ratio π^*/c is less than two. Here, we want to guide the factor loadings via this prior, but not determine them via the prior.

For the other parameters, diffuse (proper) priors can be specified. In particular, for the regression coefficients β_{pj}, we specify normal priors with mean 0 and variance 10^3, and for the variance components, gamma priors on their inverses, $1/\psi_j^2$, with parameters $(0.1,0.1)$. A Wishart prior is placed on $\boldsymbol{\Sigma}^{-1}$ with degrees of freedom equal to the dimension of $\boldsymbol{\Sigma}$ and scale matrix equal to the identity matrix. The scale matrix is chosen as the identity, since the continuous responses are standardized and the degrees of freedom is set to the dimension of $\boldsymbol{\Sigma}$ to be as uninformative as possible but still have the prior be proper. The Wishart prior is computationally tractable, but it is restrictive (see, e.g., Daniels and Kass, 1999).

Example II: Priors for mixed longitudinal behavioral processes

Based on the re-parameterization of $\boldsymbol{\Sigma}$ in Section 13.3, we discuss specification of priors for the individual components (as well as the mean parameters). In the following, we denote by $\mathbf{b}$ the column vector obtained by stringing together the rows of $\mathbf{B}$; that is, $\mathbf{b} = (b_{11},\cdots,b_{1T},\cdots,b_{TT})^\top$. Except for $\mathbf{b}$, we use diffuse priors for the parameters under the factorization

$$p(\boldsymbol{\beta},\mathbf{R},\mathbf{b},\boldsymbol{\Sigma}_{22}^*) \ = \ p(\boldsymbol{\beta})p(\mathbf{R})p(\boldsymbol{\Sigma}_{22}^*)p(\mathbf{b}|\boldsymbol{\beta},\boldsymbol{\Sigma}_{22}^*).$$

The above specification implies that $\boldsymbol{\beta}$, $\mathbf{b}$, and $\boldsymbol{\Sigma}_{22}^*$ are *a priori* jointly independent of $\mathbf{R}$ and that marginally, $\boldsymbol{\beta}$ and $\boldsymbol{\Sigma}_{22}^*$ are *a priori* independent. Since we have little prior information for $\boldsymbol{\beta}$, $\boldsymbol{\Sigma}_{22}^*$ and $\mathbf{R}$, we specify flat priors for $\boldsymbol{\beta}$ and $\boldsymbol{\Sigma}_{22}^*$, and a joint uniform prior (Barnard *et al.*, 2000) for $\mathbf{R}$:

$$p(\boldsymbol{\beta}) \quad \propto \quad 1,$$
$$p(\boldsymbol{\Sigma}_{22}^*) \quad \propto \quad \mathrm{I}\{\boldsymbol{\Sigma}_{22}^* \in (-\infty, +\infty)^{T(T+1)/2}, \text{ and } \boldsymbol{\Sigma}_{22}^* \text{ is positive definite}\},$$
$$p(\mathbf{R}) \quad \propto \quad \mathrm{I}\{r_{jk}\colon r_{jk} = 1 \text{ for } j = k, |r_{jk}| < 1 \text{ for } j \neq k, \text{ and } \mathbf{R} \text{ is positive definite}\},$$

where r_{jk}, $j \neq k$, $j, k = 1, \cdots, T$, is the off-diagonal element of the jth row and kth column in $\mathbf{R}$. Alternative default priors could be specified for $\mathbf{R}$ that recognize it as being a longitudinal correlation matrix (Wang and Daniels, 2012b; Daniels and Pourahmadi, 2009). We provide further details on these priors in the data example in Section 13.7.

We now provide some details on the prior $p(\mathbf{b}|\boldsymbol{\beta}, \boldsymbol{\Sigma}_{22}^*)$, which facilitates estimation of the relatively high-dimensional covariance matrix by encouraging sparsity. Since the components of $\mathbf{b}$ are the regression coefficients of $\mathbf{W}_i$ on $\mathbf{Z}_i$, we expect many of these to be zeros, based on conditional independence (Markov type) arguments for longitudinal data. For example, consider the regression for W_{it}. Once we condition on $\{Z_{ik} : t - 1 \leq k \leq t + 1\}$ (i.e., current value and one lag value forward and backward), we might expect W_{it} to be independent of $\{Z_{ik} : k > t + 1, k < t - 1\}$ (i.e., values more than one lag away). To incorporate these features into the model, we specify a hierarchical prior distribution that essentially allows components of $\mathbf{b}$ to be zeros, borrowing ideas from Smith and Kohn (2002), with higher lag coefficients more likely to be zero. This facilitates reducing the large number of dependence parameters in a data-dependent way that avoids searching through the space of many possible models for the best one. We refer the reader to Liu *et al.* (2009), for specific details on this prior. Related priors to reduce the dimension of a correlation matrix in Gaussian copulas can be found in Pitt *et al.* (2006).

Example III: Priors for a GLOM

The priors chosen for a GLOM depend on the modeling decisions for $\pi_{\mathbf{w}}$, $\boldsymbol{\mu}_{\mathbf{w}}$, and $\boldsymbol{\Sigma}_{\mathbf{w}}$. Rather than specifying a lower-dimensional model, one could use hierarchical priors that encourage shrinkage toward an overall center. For instance, let $\boldsymbol{\mu}_{\mathbf{w}} \sim N_4(\boldsymbol{\mu}_0, \boldsymbol{\Sigma}_0)$. The covariance matrix $\boldsymbol{\Sigma}_0$ determines the extent to which the cell-specific means $\boldsymbol{\mu}_{\mathbf{w}}$ deviate from the global mean $\boldsymbol{\mu}_0$. Without knowledge of this overall center, one will generally place a diffuse prior on $\boldsymbol{\mu}_0$, such as $N_4(\boldsymbol{\mu}^*, \boldsymbol{\Sigma}^*)$. One can choose a similar hierarchy for $\boldsymbol{\Sigma}_{\mathbf{w}}$ with inverse Wishart distributions.

Log-linear models, like the one proposed in Section 13.3, for $\mathbf{w} = (j, k, l)$

$$\log \pi_{jkl} \quad = \quad \alpha^{(0)} + \alpha_j^{(1)} + \alpha_k^{(2)} + \alpha_\ell^{(3)} + \alpha_{jk}^{(12)} + \alpha_{j\ell}^{(13)}, \tag{13.9}$$

can be specified more flexibly via shrinkage priors. For example, the two-way interactions can be shrunk toward zero using normal priors, e.g., $\alpha_{jk}^{(ab)} \sim N(0, \tau^{(ab)})$, for unknown $\tau^{(ab)}$. The data, along with the prior for $\tau^{(ab)}$, can then determine the importance of this interaction, and model selection is not necessary. One must still deal with the issue of model identifiability under these prior choices; a common fix is to set the initial category parameters to zero, i.e., $0 = \alpha_1^{(a)} = \alpha_{1i}^{(ab)} = \alpha_{i1}^{(ab)}$, for all a, b, i. More computationally efficient priors could be used by modeling $\mathbf{W}$ as a multinomial, with a Dirichlet prior centered on a parsimonious log-linear model as in (13.9), similar to the approach in Daniels *et al.* (2012b) in a related setting and to that for binary data using beta priors in Wang *et al.* (2010). We illustrate these shrinkage approaches in the data example in Section 13.7.

Alternative approaches to prior construction can be found in the ideas of Bayesian nonparametrics. For an introduction to Bayesian nonparametrics, see Gershman and Blei (2012) and accompanying references. Most of these methods impose a flexible clustering upon the observations (in this case, categorical cells $\mathbf{w}$), so that the mean vectors of certain sets of $\mathbf{w}$ can be equal. In many cases, this clustering may be useful for inference and does not require parametric models for mean vectors. Through posterior inference, such quantities as the posterior probability that the mean vectors for cells $\mathbf{w}$ and $\mathbf{w}'$ being equal are obtained. Dunson and Xing (2009) propose a nonparametric prior for the setting of multivariate categorical data, which is applicable for the categorical components of a GLOM. Additionally, Gaskins and Daniels (2012b) provide a nonparametric prior for the set of covariance matrices that would be appropriate for the continuous components of a GLOM in a longitudinal setting.

13.5   Incomplete responses

It is not uncommon to have some missingness in settings with multiple responses per observational unit (in particular, for longitudinal data). A standard Bayesian analysis using the observed data likelihood (for the responses) and priors implicitly makes an assumption that the missingness is ignorable. We introduce some notation and define ignorable missingness next.

Let R_{ij} be an indicator that response Y_{ij} is observed, i.e., $R_{ij} = \mathrm{I}\{Y_{ij}$ is observed$\}$. Define the full data response model as the probability model $p(\mathbf{y}|\boldsymbol{\theta})$ for $\mathbf{Y}$, parameterized by $\boldsymbol{\theta}$. Define the missing data mechanism as $p(\mathbf{r}|\mathbf{y},\boldsymbol{\phi})$, parameterized by $\boldsymbol{\phi}$. Finally, define $\mathbf{y}_{obs}$ as the set of $\mathbf{Y}$s that are observed; correspondingly, $\mathbf{y}_{mis}$ as the set of $\mathbf{Y}$s that are missing. The missing data mechanism is ignorable, if the following three assumptions hold:

1. $p(\mathbf{r}|\mathbf{y},\boldsymbol{\phi}) = p(\mathbf{r}|\mathbf{y}_{obs},\boldsymbol{\phi})$, i.e., the missingness only depends on the observed responses, so it is missing at random (MAR).

2. The parameters of the full data response model, $\boldsymbol{\theta}$, and the missing data mechanism, $\boldsymbol{\phi}$, are distinct.

3. The joint prior distribution for the parameters, $p(\boldsymbol{\theta},\boldsymbol{\phi})$, can be factored as $p(\boldsymbol{\theta})p(\boldsymbol{\phi})$.

Note that the validity of inferences under ignorability relies on correct model specification, including the covariance structure. In fact, misspecification of the dependence structure typically leads to biased estimates of regression parameters (see, e.g., Chapter 6 in Daniels and Hogan, 2008). We provide an example of ignorable missingness in the context of Example III below.

When the three conditions above are not satisfied, the missingness is called nonignorable, most often due to condition 1 not holding, so the missingness is missing not at random (MNAR). In a Bayesian framework, the joint distribution of the full data response $\mathbf{Y}$ and missingness indicator $\mathbf{R}$ now needs to be specified. Sensitivity analysis and/or specification of informative priors is essential. This can be seen by factorizing the full data model as follows:

$$p(\mathbf{y},\mathbf{r}|\boldsymbol{\omega}) \quad = \quad p(\mathbf{y}_{mis}|\mathbf{y}_{obs},\mathbf{r},\boldsymbol{\omega}_E)p(\mathbf{y}_{obs},\mathbf{r}|\boldsymbol{\omega}_O). \qquad (13.10)$$

This factorization has been called the extrapolation factorization (Daniels and Hogan, 2008). The second factor in (13.10) is the distribution of the observed data; clearly, $\boldsymbol{\omega}_O$ is identified. However, the data never provide any information for inference on the first factor, the extrapolation factor and its corresponding vector of parameters $\boldsymbol{\omega}_E$. Thus, to account for uncertainty regarding this first factor, informative priors are essential. Both are based on finding sensitivity parameters. Such parameters are not identified by the observed data (so a function of $\boldsymbol{\omega}_E$) and do not change the value of the observed data likelihood when they are varied (i.e., they do not affect the fit of the model to the observed data). When these parameters are fixed, all the parameters in the model are identified; for a formal definition, see Chapter 8 in Daniels and Hogan (2008).

Many parametric approaches for nonignorable missingness in the literature are specified such that all model parameters are identified and sensitivity parameters do not exist. Three examples include the models in Dunson and Perreault (2001) and Cowles et al. (1996), which are often called shared parameter models, and the model in Chapter 5, which is called a (parametric) selection model. Pattern mixture models (PMM), which factorize the joint distribution of $p(\mathbf{y},\mathbf{r})$ as $p(\mathbf{r})p(\mathbf{y}|\mathbf{r})$, provide an alternative and convenient factorization to deal with nonignorable missingness. We provide an illustration in the context of Example II next.

Example II: Nonignorable missingness in the smoking cessation data

The CTQ II study had considerable missingness in the form of dropout. This was addressed under an assumption of ignorable missingness in Liu *et al.* (2009). Here, we propose a mixture model approach to address nonignorable missingness.

Consider the data and the model described in Section 13.2. The missingness was monotone for each response (i.e., quit status and weight change), and if one response was missing at time t, so was the other. Define D as the dropout time from the study with $D = T + 1$ for those with observed values at all T time points. Given that missingness was monotone, this random variable contains the same information as the set of missingness indicators, $\mathbf{R}$. We can specify the conditional distribution of the full data response $\mathbf{Y}$, given D, as a multivariate normal as in (13.4), but with the mean depending on D (for now, we assume the covariance matrix is constant as a function of D). Given the unstructured mean specified in Section 13.2, the (mean) parameters identified and not identified, given D, are obvious. All of these unidentified mean parameters could be viewed as sensitivity parameters. However, we can reduce the number of sensitivity parameters by using a non-future dependence restriction (Kenward *et al.*, 2003).

Let $\mathbf{U}_{it} = (Z_{it}, W_{it})^{\top}$ be the quit status latent variable and weight-change pair at time t. Consider a time-ordered rearrangement $\mathbf{U}_i = (Z_{i1}, W_{i1}, Z_{i2}, W_{i2}, \cdots, Z_{iT}, W_{iT})^{\top}$ of the components of $\mathbf{Y}_i$. There exists an orthogonal permutation matrix $\boldsymbol{\Gamma}$ such that $\mathbf{U}_i = \boldsymbol{\Gamma}\mathbf{Y}_i$, implying that $\mathbf{U}_i \sim N_{2T}(\boldsymbol{\Gamma}\mathbf{X}_i\boldsymbol{\beta}, \boldsymbol{\Gamma}\boldsymbol{\Sigma}\boldsymbol{\Gamma}^{\top})$. Recall that $\mathbf{X}_i\boldsymbol{\beta}$ is simply specified by a distinct mean for each component at each treatment level, so $\boldsymbol{\Gamma}\mathbf{X}_i\boldsymbol{\beta}$ just provides the appropriate rearrangement. Define $\boldsymbol{\mu} = \boldsymbol{\Gamma}\mathbf{X}_i\boldsymbol{\beta}$ and $\boldsymbol{\Omega} = \boldsymbol{\Gamma}\boldsymbol{\Sigma}\boldsymbol{\Gamma}^{\top}$. For the PMM, we assume the mean structure $\boldsymbol{\beta}_d$ depends on pattern but not the covariance structure. Hence, we have a distinct mean $\boldsymbol{\mu}_d$ by pattern and common dependence $\boldsymbol{\Omega}$ under the rearrangement.

The historical data at week t is denoted by $\overline{\mathbf{U}}_{it} = (\mathbf{u}_{i1}^{\top}, \cdots, \mathbf{u}_{i,t-1}^{\top})^{\top}$, where $\mathbf{u}_{it}$ is the observed value of $\mathbf{U}_{it}$. We construct the distributions for each pattern d through the factorization

$$f_d(\mathbf{u}_i) = f_d^1(\mathbf{u}_{i1}) \prod_{t=2}^{T} f_d^t(\mathbf{u}_{it} | \overline{\mathbf{U}}_{it}),$$

where $f_d^t(\cdot | \overline{\mathbf{U}}_{it})$ is the density for the pair of measurements $\mathbf{U}_{it}$ at time t, for subjects who leave the study at time d, conditional on the earlier measurements $\overline{\mathbf{U}}_{it}$. Only those distributions $f_d^t(\cdot)$ with $t < d$, are identified by the observed data. The remaining distributions, which taken together form the extrapolation distribution in (13.10), need to be specified through a combination of sensitivity parameters and modeling choices.

In specifying these extrapolation distributions, we assume each identifiable $f_d^t(\cdot)$ is bivariate normal. We also want the unidentifiable distributions to differ from their identified counterparts by a sensitivity parameter such that MAR is a special case (when the sensitivity parameter is set to zero). For this to be the case, we have that $f_d^t(\cdot)$ is the same across pattern d for $1 < t < d$ (Wang and Daniels, 2011, Corollary 4; see correction in Wang and Daniels, 2012a); we denote this common distribution by $f_{t+}^t(\cdot)$. We do not need to similarly restrict any of the week one distributions $f_2^1(\cdot), \cdots, f_{T+1}^1(\cdot)$.

The week one distributions $f_d^1(\cdot)$ have the form $N_2(\boldsymbol{\zeta}_d^1, \boldsymbol{\Omega}_1)$, $d = 2, \cdots, T+1$; the distribution $f_{t+}^t(\cdot | \overline{\mathbf{U}}_{it})$ is $N_2(\boldsymbol{\zeta}^t + \boldsymbol{\Phi}_t^{\top} \overline{\mathbf{U}}_{it}, \boldsymbol{\Omega}_t)$. The $2(t-1) \times 2$ matrix $\boldsymbol{\Phi}_t$ contains regression coefficients for patient's history up to time t, controlling the dependence on the earlier observations; $\boldsymbol{\Omega}_t$ is the covariance matrix for this conditional regression. These two matrices are determined by the time-ordered covariance matrix $\boldsymbol{\Omega}$. After marginalizing over the history $\overline{\mathbf{U}}_{it}$, the marginal mean at time t is obtained iteratively from $\boldsymbol{\mu}_d^1 = \boldsymbol{\zeta}_d^1$ and $\boldsymbol{\mu}_d^t = \boldsymbol{\zeta}^t - \boldsymbol{\Phi}_t^{\top}(\{\boldsymbol{\mu}_d^1\}^{\top}, \cdots, \{\boldsymbol{\mu}_d^{t-1}\}^{\top})^{\top}$.

We now specify the unidentified distributions $f_d^t(\cdot)$, $d \leq t$, under a non-future dependence assumption. The distribution for the first missing value of each pattern, $t = d$, is identified by incorporating a sensitivity parameter $\boldsymbol{\Delta}^t$. Under the non-future dependence assumption, there is only one unidentified distribution in each pattern, $f_t^t(\cdot)$, which we specify as $N_2(\boldsymbol{\zeta}^t + \boldsymbol{\Delta}^t + \boldsymbol{\Phi}_t^{\top} \overline{\mathbf{U}}_{it}, \boldsymbol{\Omega}_t)$, with one sensitivity parameter $\mathbf{\Delta}^t$, and the distributions $f_d^t(\cdot)$, $d < t$, are all the same, with the form of a mixture of the observed distribution and the first missing distribution

$$f_{t-}^t(\mathbf{u}_{it}|\overline{\mathbf{U}}_{it}) \;=\; \frac{\sum_{d=t+1}^{T+1} P(D_i = d) f_d^1(\mathbf{u}_{i1})}{\sum_{d=t}^{T+1} P(D_i = d) f_d^1(\mathbf{u}_{i1})} f_{t+}^t(\mathbf{u}_{it}) + \frac{P(D_i = t) f_t^1(\mathbf{u}_{i1})}{\sum_{d=t}^{T+1} P(D_i = d) f_d^1(\mathbf{u}_{i1})} f_t^t(\mathbf{u}_{it}),$$

where $f_{t-}^t(\cdot)$ and $f_d^t(\cdot)$ are the same, for all $d < t$.

Under this modeling framework, the contribution of individual i with full data $\{D_i = d, \mathbf{U}_i = \mathbf{u}_i\}$ to the overall likelihood is

$$\left\{ P(D_i = d) f_d^1(\mathbf{u}_{i1}) \prod_{1 < t < d} f_{t+}^t(\mathbf{u}_{it}|\overline{\mathbf{U}}_{it}) \right\} f_d^d(\mathbf{u}_{id}|\overline{\mathbf{U}}_{id}) \prod_{d < t} f_{t-}^t(\mathbf{u}_{it}|\overline{\mathbf{U}}_{it}),$$

where the terms in the brackets are those that are identified by the observed data (and their fit can be checked).

We have specified a PMM for the CTQ II data that allows separate mean parameters for each of the T patterns. The model has only $2(T-1)$ sensitivity parameters and includes MAR as a special case when $\mathbf{\Delta}^1 = \cdots = \mathbf{\Delta}^{T-1} = \mathbf{0}$. Each of the identified distributions and the first unidentified distribution have a bivariate normal form, with the distributions $f_{t-}^t(\cdot)$ being a mixture over these two normals. Under the MAR model, $f_{t-}^t(\cdot)$, $f_t^t(\cdot)$, and $f_{t+}^t(\cdot)$ are all the same.

What remains is to specify/elicit a range and/or a prior for the sensitivity parameters. To do this, we first recognize the $\mathbf{\Delta}$'s as differences in means, conditional on a fixed history of responses. We make a simplifying assumption (not checkable from the data) that the $\mathbf{\Delta}$s do not depend on time. Specifically, $\mathbf{\Delta} = (\Delta_1, \Delta_2)^\top$, where

$$\Delta_1 \;=\; E(Z_t|\overline{\mathbf{U}}_t = \overline{\mathbf{u}}_t, D = t) - E(Z_t|\overline{\mathbf{U}}_t = \overline{\mathbf{u}}_t, D > t),$$
$$\Delta_2 \;=\; E(W_t|\overline{\mathbf{U}}_t = \overline{\mathbf{u}}_t, D = t) - E(W_t|\overline{\mathbf{U}}_t = \overline{\mathbf{u}}_t, D > t),$$

for all times t. Prior elicitation for Δ_2 is straightforward as it represents the difference in weight change for those who dropout and those who remain in the study. We refer the reader to Chapter 10 in Daniels and Hogan (2008), for strategies and ideas for eliciting Δ_2.

Eliciting Δ_1 is less obvious, since it is the difference on the scale of the latent variable for quit status (not its binary realization). To facilitate the elicitation, we first point out that the standard normal distribution function can be approximated by a scaled logistic distribution $F(x) = \{1 + \exp(-kx)\}^{-1}$, at $k = 1.749$ (Savalei, 2006). As such, the change in the log-odds of $Q_t = 1$ for those who drop out at t and those who remain in the study is approximately $k\Delta_1$, i.e.,

$$\log\left\{ \frac{P(Q_t = 1|\overline{\mathbf{U}}_t = \overline{\mathbf{u}}_t, D = t)}{P(Q_t = 0|\overline{\mathbf{U}}_t = \overline{\mathbf{u}}_t, D = t)} \right\} - \log\left\{ \frac{P(Q_t = 1|\overline{\mathbf{U}}_t = \overline{\mathbf{u}}_t, D > t)}{P(Q_t = 0|\overline{\mathbf{U}}_t = \overline{\mathbf{u}}_t, D > t)} \right\} \;\approx\; k\Delta_1.$$

We specify a prior for Δ_1 in terms of our prior beliefs about this change in log-odds, scaled by k^{-1}. We refer the reader to the OASIS Case Study in Chapter 10 of Daniels and Hogan (2008), and to Wang et al. (2010) for suggestions on eliciting a log-odds ratio such as Δ_1.

For the full analysis of this example, see Gaskins and Daniels (2012a).

Example III: Ignorable missingness in the GLOM

For the analysis of this data in Little and Schluchter (1985), the missingness is assumed to be ignorable. This assumption avoids the need to specify a joint model for $\mathbf{Y}$ and $\mathbf{R}$. As such, it is only necessary to specify a GLOM for the full data response, and then use data augmentation to fill in the missing data (for details, see Example III in Section 13.6); for details on the full data analysis, see Example III in Section 13.7.

An MAR analysis in a semiparametric setting with mixed longitudinal data (using generalized estimating equations) can be found in Chapter 5; in this approach, the authors build a separate imputation model to fill in the missing data.

13.6 General computational issues

All the models that have been discussed, and most models for mixed categorical and continuous data, do not have a posterior distribution of the parameters that is available in closed form. Typically, the posterior distribution of the parameters for these models must be computed by using a sampling-based Markov chain Monte Carlo (MCMC) algorithm (Robert and Casella, 2001).

We briefly describe the the Gibbs sampler, the most widely used MCMC sampling method. Assume that we have a distribution $f(\theta_1, \cdots, \theta_p)$ from which we wish to sample, where each θ_i lives on support Θ_i, possibly of dimension greater than one. In the context of Bayesian data analysis, $f(\cdot)$ is generally the posterior distribution, and each θ_i represents a set of parameters from the model, such as regression coefficients or variance parameters. We can not directly sample from $f(\cdot)$, but we are able to sample from each of the conditionals $f_i(\theta_i | \boldsymbol{\theta}_{-i})$, where $\boldsymbol{\theta}_{-i}$ represents the vector $\boldsymbol{\theta}$ without the ith component. After specifying a starting point $\boldsymbol{\theta}^{(0)} = (\theta_1^{(0)}, \cdots, \theta_p^{(0)})$, the sampler proceeds by performing the following p steps at each iteration $t+1$:

$$
\begin{aligned}
\theta_1^{(t+1)} &\sim & f_1(\theta_1 | \theta_2^{(t)}, \theta_3^{(t)}, \cdots, \theta_p^{(t)}) \\
\theta_2^{(t+1)} &\sim & f_2(\theta_2 | \theta_1^{(t+1)}, \theta_3^{(t)}, \cdots, \theta_p^{(t)}) \\
&\vdots& \\
\theta_p^{(t+1)} &\sim & f_p(\theta_p | \theta_1^{(t+1)}, \cdots, \theta_{p-1}^{(t+1)}).
\end{aligned}
$$

Under suitable regularity conditions and t sufficiently large, the vector $\boldsymbol{\theta}^{(t)}$ approximately follows the joint distribution $f(\cdot)$. After running the Gibbs sampler for many iterations, we consider the set of $\boldsymbol{\theta}^{(t)}$s as an approximate sample from $f(\cdot)$. Usually, we exclude the first t_0 iterations as a burn-in to minimize the effect of the choice of starting point $\boldsymbol{\theta}^{(0)}$ and thin the sample by using only every kth iteration to reduce the autocorrelation between iterations, for some appropriate choice of t_0 and k. Diagnostic measures such as those reviewed in Cowles et al. (1996), can be used to assess whether the chain can be considered as an approximate posterior sample.

In some cases, it may be possible to improve the mixing of the MCMC chain by sampling the variables in "blocks." Instead of sampling each θ_i conditional on the others, we combine sets of variables (without loss of generality to two blocks) so that $\theta_1^* = (\theta_1, \cdots, \theta_\ell)$ and $\theta_2^* = (\theta_{\ell+1}, \cdots, \theta_p)$. The Gibbs sampler now alternates sampling from θ_1^* given θ_2^*, and θ_2^* given θ_1^*.

Computations can often be simplified (in terms of tractable full conditional distributions) by sampling the latent variables (e.g., in (multivariate) probit regressions in Examples I and II; details below) and by sampling the missing responses (under ignorability or non-ignorability) using a data augmentation step (Hobert, 2011). Multivariate probit regressions and Gaussian copulas introduce correlation matrices which can be difficult to sample. Such problems can sometimes be lessened by using parameter expansion-type approaches (Liu, 2001) recently proposed in the literature (Liu and Daniels, 2006; Zhang et al., 2006). General computational strategies for mixed continuous and categorical outcomes using Gaussian copulas can be found in Pitt et al. (2006).

Example I: Gibbs sampling scheme for VA data

We list the conditional distributions which are sampled in a Gibbs sampler from the model and priors, as specified in Sections 13.2 and 13.4. Note that we sample the patient-level latent variables Z_{ki1}, as well as the network-level latent variables η_{kj} and θ_k, instead of integrating them out to obtain simpler forms for the full conditional distributions for all the parameters.

Step 1. The conditional distribution of the latent variable Z_{ki1}, corresponding to hospital readmittance is the truncated version of the $N(\xi_{ki1}, 1)$ distribution, where ξ_{ki1} is given by (13.1). The truncation regions are $(-\infty, 0]$, if $y_{ki1} = 0$, and $(0, +\infty)$, otherwise.

Step 2. The regression parameters $\boldsymbol{\beta}_1 = (\beta_{01}, \cdots, \beta_{P1})^\top$ for the categorical component have conditional distribution $N_{P+1}((\mathbf{XX}^\top + \boldsymbol{\Omega}^{-1})^{-1}\mathbf{X}(\mathbf{Z} - \mathbf{H}_1), (\mathbf{XX}^\top + \boldsymbol{\Omega}^{-1})^{-1})$, when the prior for $\boldsymbol{\beta}_1$ is $N_{P+1}(\mathbf{0}, \boldsymbol{\Omega})$ (the form of $\boldsymbol{\Omega}$ is given in Section 13.4). Here, $N = \sum_k n_k$ is the total number of patients across all networks, $\mathbf{X}$ is the $(P+1) \times N$ matrix with columns $\mathbf{x}_{ki} = (1, x_{1ki}, \cdots, x_{Pki})^\top$, $\mathbf{Z}$ is the vector of Z_{ki1}s, and $\mathbf{H}_1$ is the vector with η_{k1} in the location corresponding to Z_{ki1} in $\mathbf{Z}$. The regression parameters $\boldsymbol{\beta}_2, \boldsymbol{\beta}_3, \boldsymbol{\beta}_4$, for the continuous responses also follow a normal distribution calculated in a similar fashion.

Step 3. The full conditional distribution of the inverse covariance matrix $\boldsymbol{\Sigma}^{-1}$ follows a Wishart distribution with shape $(\mathbf{B} + \mathbf{E}^\top \mathbf{E})^{-1}$ and degrees of freedom $m + \sum_{k,i} Y_{ki1}$. The prior for $\boldsymbol{\Sigma}^{-1}$ is Wishart$(\mathbf{B}, m)$ (hyperparameters specified in Section 13.4), and the $(\sum_{k,i} Y_{ki1}) \times 3$ matrix $\mathbf{E}$ contains the residuals from the continuous response regression. An element in column j corresponds to $Y_{kij} - \xi_{kij}$, for some k, i, such that $y_{ki1} = 1$.

Step 4. The full conditional distribution of the network-level latent variable for the categorical response η_{k1} follows from the distribution in (13.7). Conditional on all other parameters, η_{k1} is normally distributed with mean $(n_k + \psi_1^{-2})^{-1}\{\psi_1^{-2}\lambda_1\theta_k + \sum_i(Z_{ki1} - \mathbf{x}_{ki}^\top\boldsymbol{\beta}_1)\}$ and variance $(n_k + \psi_1^{-2})^{-1}$. Note that $\{\eta_{k2}, \eta_{k3}, \eta_{k4}\}$ should be drawn jointly because they correspond to the continuous responses, which are correlated through $\boldsymbol{\Sigma}$. The conditional distribution is a three-dimensional normal distribution of similar form to that of η_{k1}.

Step 5. The full conditional distribution for the latent factors θ_k is a normal distribution with mean $(\boldsymbol{\lambda}^\top\boldsymbol{\Psi}^{-1}\boldsymbol{\lambda} + 1)^{-1}\boldsymbol{\lambda}^\top\boldsymbol{\Psi}^{-1}\boldsymbol{\eta}_i$ and variance $(\boldsymbol{\lambda}^\top\boldsymbol{\Psi}^{-1}\boldsymbol{\lambda} + 1)^{-1}$, where $\boldsymbol{\Psi} = diag(\psi_1^2, \cdots, \psi_J^2)$.

Step 6. The conditional distribution for the factor loadings $\boldsymbol{\lambda}$ follows a J-dimensional normal distribution with mean vector $(\boldsymbol{\Theta}^\top\boldsymbol{\Psi}_2^{-1}\boldsymbol{\Theta} + c^{-2}\mathbf{I}_J)^{-1}(c^{-2}\boldsymbol{\pi} + \boldsymbol{\Theta}^\top\boldsymbol{\Psi}_2^{-1}\mathbf{H})$ and covariance matrix $(\boldsymbol{\Theta}^\top\boldsymbol{\Psi}_2^{-1}\boldsymbol{\Theta} + c^{-2}\mathbf{I}_J)^{-1}$. Here $\mathbf{H} = (\boldsymbol{\eta}_1^\top, \cdots, \boldsymbol{\eta}_K^\top)^\top$, $\boldsymbol{\Psi}_2$ is block diagonal with K blocks of $\boldsymbol{\Psi}$, and $\boldsymbol{\Theta}$ is a $JK \times J$ block matrix with blocks $\theta_k\mathbf{I}_J$.

Step 7. Under an inverse gamma prior with shape a and scale b, the sampling conditional distribution for ψ_j^2 is inverse gamma, with shape $a + I/2$ and scale $b + \sum_k(\eta_{kj} - \lambda_j\theta_k)^2$, where a and b are specified in Section 13.4.

The fact that simple full conditionals can be obtained by using normal latent variables underlying a probit regression, often leads to their use in these settings. However, similar approximate computations can be used with a logistic regression by taking advantage of the fact that the logistic cumulative distribution function can be approximated by the cumulative distribution of a t-distribution with 7 degrees of freedom (O'Brien and Dunson, 2004). The t-distribution can be written as a gamma mixture of normals (Albert and Chib, 1993); thus, for $X \sim t_\nu(\mu, \sigma^2)$, we can equivalently state that $X|\gamma \sim N(\mu, \gamma^{-1}\sigma^2)$, with $\gamma \sim Gamma(\nu/2, \nu/2)$. The benefit to this mixture of normals form is that conditional conjugacy is maintained for μ and σ^2, if they have normal and inverse gamma priors as above, if we condition on γ; the full conditional distribution for γ can be shown to follow a gamma distribution.

Example II: Data augmentation steps for CTQ II data

In the CTQ II analysis, there are two elements to the data augmentation: the unobserved (missing) values of $\mathbf{W}_i$ and $\mathbf{Q}_i$, and the values of the latent variables $\mathbf{Z}_i$. Adapting the notation from Section 13.5, let $\mathbf{W}_{i,obs}$ and $\mathbf{Q}_{i,obs}$ denote the observed values for weight change and quit status, respectively. We also define $\mathbf{Z}_{i,obs}$ as the values of the latent variables $\mathbf{Z}_i$, which correspond to $\mathbf{Q}_{i,obs}$; these values are not actually observed but are constrained by the observed $\mathbf{Q}_{i,obs}$. The vectors of the missing values are defined similarly.

Under the ignorable missingness assumption, data augmentation proceeds with the following steps.

Step 1. To sample $\mathbf{W}_{i,mis}$, recall from (13.6) that $\mathbf{W}_i$ follows a normal distribution with mean

$\mathbf{X}_{2i}\boldsymbol{\beta}_2 + \mathbf{B}(\mathbf{Z}_i - \mathbf{X}_{i1}\boldsymbol{\beta}_1)$ and covariance matrix $\boldsymbol{\Sigma}_{22}^*$. The full conditional distribution of $\mathbf{W}_{i,mis}$ can be derived from this normal distribution after conditioning on $\mathbf{W}_{i,obs}$.

Step 2. To sample the latent variables $\mathbf{Z}_i$, we first define the sets $\mathscr{Z}_{it}$, $1 \leq t \leq T$, as the subsets of the real line designating the support of Z_{it}. If Q_{it} is observed to be 1, then $\mathscr{Z}_{it} = (0, +\infty)$, and if Q_{it} is observed to be 0, then $\mathscr{Z}_{it} = (-\infty, 0]$. If Q_{it} is unobserved, then there is no restriction on Z_{it}, and $\mathscr{Z}_{it}$ is the entire real line. Based on the multivariate normal distribution in (13.4), the conditional distribution of $\mathbf{Z}_i$, given $\mathbf{W}_i$, is $N_T(\boldsymbol{\mu}^*(\mathbf{X}_i, \mathbf{W}_i), \boldsymbol{\Sigma}_{11}^*)$, where $\boldsymbol{\mu}^*(\mathbf{X}_i, \mathbf{W}_i) = \mathbf{X}_{i1}\boldsymbol{\beta}_1 + \boldsymbol{\Sigma}_{12}\boldsymbol{\Sigma}_{22}^{-1}(\mathbf{W}_i - \mathbf{X}_{i2}\boldsymbol{\beta}_2)$ and $\boldsymbol{\Sigma}_{11}^* = \boldsymbol{\Sigma}_{11} - \boldsymbol{\Sigma}_{12}\boldsymbol{\Sigma}_{22}^{-1}\boldsymbol{\Sigma}_{12}^\top$. The matrices $\boldsymbol{\Sigma}_{11}, \boldsymbol{\Sigma}_{12}, \boldsymbol{\Sigma}_{22}$, refer to the blocks of $\boldsymbol{\Sigma}$ obtained by reversing the Bartlett decomposition of $(\mathbf{R}, \mathbf{B}, \boldsymbol{\Sigma}_{22}^*)$. The latent variables $\mathbf{Z}_i$ are drawn from the restricted normal distribution

$$N_T(\boldsymbol{\mu}^*(\mathbf{X}_i, \mathbf{W}_i), \boldsymbol{\Sigma}_{11}^*)\prod_{t=1}^{T}\mathrm{I}\{Z_{it} \in \mathscr{Z}_{it}\}. \tag{13.11}$$

Sampling from the distribution in (13.11) is generally nontrivial. The elements of $\mathbf{Z}_i$ are typically sampled one at a time. Sampling one Z_{it} given the remaining elements of $\mathbf{Z}_i$, can be accomplished quickly (Robert, 1995), but this sampling strategy can lead to slow mixing and high autocorrelation between iterations. A way to lessen this problem is to "orthogonalize" and sample according to the algorithm suggested by Proposition 1 in Liu *et al.* (2009).

Under the pattern mixture model in Section 13.5, it is still advantageous to sample the missing responses to provide simple full conditionals for the mean and dependence parameters. For the specifics of the sampling algorithm under this non-future PMM, see Gaskins and Daniels (2012a).

Example III: Data augmentation for a GLOM

We briefly outline the data augmentation steps for analyzing the GLOM under ignorable missingness. We sample the missing values for the discrete and continuous components separately due to the conditional factorization of the continuous responses. Sampling steps for the parameters associated with the log-linear probabilities $\pi_\mathbf{w}$ and the normal mean and covariance terms $\boldsymbol{\mu}_\mathbf{w}, \boldsymbol{\Sigma}_\mathbf{w}$, depend on their respective modeling choices but are generally straightforward.

Step 1. To sample the missing categorical responses conditional on the continuous response $\mathbf{Y}_i$ and the current parameter values, we draw the missing components of $\mathbf{W}_i$ from a multinomial distribution with probabilities $P(\mathbf{W}_i = \mathbf{w}) \propto \pi_\mathbf{w} f(\mathbf{y}_i|\boldsymbol{\mu}_\mathbf{w}, \boldsymbol{\Sigma}_\mathbf{w})$, where $f(\cdot|\boldsymbol{\mu}, \boldsymbol{\Sigma})$ denotes the multivariate normal density at $\mathbf{y}$, with mean $\boldsymbol{\mu}$ and covariance $\boldsymbol{\Sigma}$.

Step 2. Given the discrete response $\mathbf{W}_i = \mathbf{w}$, the missing continuous responses follow a normal distribution derived from the $N_4(\boldsymbol{\mu}_\mathbf{w}, \boldsymbol{\Sigma}_\mathbf{w})$ distribution, conditional on the observed values of $\mathbf{Y}_i$.

We discuss the analysis of a GLOM for the St. Louis Risk Research Project data (i.e., Example III) in Section 13.7.

13.7 Analysis of examples

We now consider the analysis of two of the examples we have considered here. We describe the analysis of Example I in slightly more detail, drawing together many of the issues described throughout the chapter. In particular, we extend the single-year model to one that considers the performance of the health care networks over a six-year period. We also introduce a new model for the St. Louis Risk Research Project (Example III) that takes advantage of shrinkage priors as discussed in Section 13.4.

Example I: Analysis of longitudinal VA data

The extension to a longitudinal setting now adds a time subscript to each patient level response, and there are now J network effects η_{ktj}, for each time t, represented by L latent variables that are *a*

Table 13.1 *Model comparisons for the longitudinal dependence structure of the network factors.*

Prior for $\mathbf{R}$	DIC	p_D
Uniform on PACs	−576.5	8.2
Triangular prior	−568.5	8.8
Uniform on $\mathbf{R}$	−558.0	9.9
AR(1) structure	−552.0	4.0
Independence	−495.2	5.1

priori independent:

$$\eta_{ktj}|\boldsymbol{\theta}_{kt}, \boldsymbol{\lambda}_j, \psi_j^2 \sim N(\lambda_{1j}\theta_{k1t} + \cdots + \lambda_{Lj}\theta_{kLt}, \psi_j^2),$$

$$\boldsymbol{\theta}_{k\ell} = (\theta_{k\ell 1}, \cdots, \theta_{k\ell T})^\top \sim N_T(\mathbf{0}, \mathbf{R}_\ell), \quad \ell = 1, \cdots, L,$$

where $\boldsymbol{\lambda}_j = (\lambda_{1j}, \cdots, \lambda_{Lj})^\top$ is a $L \times 1$ vector of discrimination parameters, assumed constant over time. The variance ψ_j^2 of the unit random effects represents heterogeneity among outcomes in a given year not explained by the latent variable. $\mathbf{R}_\ell$ denotes a $T \times T$ correlation matrix that permits serial dependence in the L latent variables.

Analogous to the single year case, we assume one (i.e., $L = 1$) latent variable θ_{kt} for each time, which then represents the overall inpatient quality of network k in year t. We drop the subscript on ℓ and define the vectors $\boldsymbol{\theta}_k = (\boldsymbol{\theta}_{k1}, \cdots, \boldsymbol{\theta}_{kT})^\top$ and $\boldsymbol{\lambda} = (\lambda_1, \cdots, \lambda_J)^\top$. We comment on the consequences of the choice of the number of latent factors in the discussion of Section 13.8.

We discuss different approaches to handle the longitudinal (temporal) correlation induced through the correlation matrix $\mathbf{R}$, both through structure and priors. If no structure is placed on the correlation matrix $\mathbf{R}$, there are several priors that can be considered as mentioned in Section 13.4. Barnard *et al.* (2000) specify a uniform prior over the compact subspace of the $T(T-1)/2$-dimensional cubic $[-1, 1]^{T(T-1)/2}$, such that $\mathbf{R}$ is positive definite. However, recent papers by Daniels and Pourahmadi (2009) and Wang and Daniels (2012b) suggest instead independent uniform and triangular priors, respectively, on the partial autocorrelations (PAC) as discussed in Section 13.4.

We compare inferences about the network level latent factors under these three priors on the correlation matrix, in addition to a simple autoregressive model and an independence model. Note for the simple first order autoregressive model, we specify a uniform prior on the interval $(-1, 1)$ for the correlation parameter ρ.

For model comparison, we employ the deviance information criteria (DIC) (Spiegelhalter *et al.*, 2002). This quantity is formed as the sum of a term measuring model fit, the expected deviance, and p_D, which measures model complexity, given by the difference between the expected deviance and the deviance measured at the parameter estimate. Because our interest is in model selection on the network model, not the patient model, we desire a network-level likelihood function. As in Daniels and Normand (2006), we obtain the maximum likelihood estimate $\hat{\boldsymbol{\eta}}_k$ of $\boldsymbol{\eta}_k = (\eta_{k11}, \cdots, \eta_{kTJ})^\top$, as well as its covariance matrix $\mathbf{V}_k$, and assume a normal approximation $\hat{\boldsymbol{\eta}}_k | \boldsymbol{\eta}_k \sim N_{TJ}(\boldsymbol{\eta}_k, \mathbf{V}_k)$. Using the previously defined model $\boldsymbol{\eta}_k | \boldsymbol{\theta}_k \sim N_{TJ}(\boldsymbol{\Lambda}\boldsymbol{\theta}_k, \boldsymbol{\Psi}_2)$ and $\boldsymbol{\theta}_k \sim N_T(\mathbf{0}, \mathbf{R})$, where $\boldsymbol{\Lambda} = \mathbf{I}_T \otimes \boldsymbol{\lambda}$, with "$\otimes$" denoting the Kronecker product operator, the 'observed' data $\hat{\boldsymbol{\eta}}_k$ have (after integrating out $\boldsymbol{\eta}_k$ and $\boldsymbol{\theta}_k$) a multivariate normal distribution with mean $\mathbf{0}$ and covariance $\mathbf{V}_k + \boldsymbol{\Psi}_2 + \boldsymbol{\Lambda}\mathbf{R}\boldsymbol{\Lambda}^\top$. We use this likelihood for the "observations" $\hat{\boldsymbol{\eta}}_k$ in the calculation of the DIC. The parameterization for the plug-in we use for the DIC is the inverse of the posterior expectation of $(\mathbf{V}_k + \boldsymbol{\Psi}_2 + \boldsymbol{\Lambda}\mathbf{R}\boldsymbol{\Lambda}^\top)^{-1}$. For further details and justification, see Daniels and Normand (2006).

Table 13.1 provides the DIC values for the five correlation structures under consideration. The prior that is uniform on the PACs performs the best (i.e., smallest DIC), followed by the one that places a triangular distribution on the PACs. The parsimonious models (i.e., independence and autoregressive) provide poorer model fit. In particular, the prior that assumes the latent factors θ_{kt} are

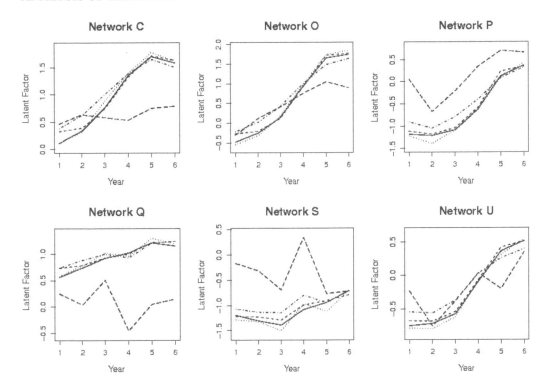

Figure 13.1 *Paths of latent factors $\boldsymbol{\theta}_k$ by correlation structure for six selected networks. The path of the latent factor under the uniform PAC prior is depicted by solid lines, triangular by short dashed lines, uniform on* **R** *by dotted lines, AR(1) by dot-dash lines, independence by long dashed lines.*

independent across year, is unsatisfactory, implying the presence of independence in the level of patient care across years. We note that alternative parameterizations of the plug-in in the DIC yield the same ordering; see discussion in the following section.

Another approach to handle the correlation matrix would be to reduce the dimension in a data-dependent way using priors, similar to those described in Example II in Section 13.4; for more details in general, see Wong *et al.* (2003) and Liu *et al.* (2009). For a way to do this via partial autocorrelations, as discussed above, see Gaskins *et al.* (2012).

We conclude the discussion of the VA example by describing how one can compare the 22 networks using the posterior distributions of the latent factors $\boldsymbol{\theta}_k$. Recall from Section 13.4 that the prior for $\boldsymbol{\lambda}$ is constructed such that positive values of θ_{kt} indicate networks that perform better than average, while negative values represent poorly performing networks.

Figure 13.1 shows the posterior means of the longitudinal paths $\boldsymbol{\theta}_k$ for six representative networks under each of the correlation matrix priors. Overall, we see that networks tend to be improving over the six-year study, as indicated by the upward trajectories of the factors. However, some networks, such as network S, remain consistently poor performers across all years, as evidenced by the consistently negative values of θ_{kt}. In fact, we see worsening patient care in this network for the first three years. Characterizing the effectiveness of the networks through the single longitudinal parameter $\boldsymbol{\theta}_k$ provides an immediate measure of the high and low performing health systems across years, as well as those with improving inpatient care. Posterior credible intervals can also be placed on these plots to characterize the uncertainty in the trajectories (Daniels and Normand, 2006).

Comparing the plots across the different priors for **R**, there are only small differences between the factor paths under the uniform prior on the PACs, the uniform prior on **R**, and the triangular prior on the PACs. As these priors imply similar structures, we are not surprised by their relatively

Table 13.2 *Posterior probability $P(\widehat{\beta}(\boldsymbol{\theta}_k) > 0|\mathbf{y})$, that is, the probability that the slope of the latent factor is positive (increasing) for selected networks.*

Prior	Network					
	C	O	P	Q	S	U
Uniform on PACs	0.99	1.00	1.00	0.85	0.81	0.98
Triangular prior	0.99	1.00	1.00	0.81	0.79	0.99
Uniform on **R**	0.99	1.00	1.00	0.83	0.82	0.97
AR(1) structure	0.98	1.00	1.00	0.81	0.76	0.96
Independence	0.62	0.91	0.88	0.40	0.33	0.76

close agreement. The autoregressive prior yield posterior paths that tend to move more smoothly than those produced by the other priors. Conversely, the independence prior yields fairly jagged movements, which do not coincide with our intuition. We do not expect to see extreme differences in the system performance across years, such as the large jumps between years for networks Q and S.

We may also consider the general trajectory of the networks as either increasing or decreasing by fitting least squares lines to each posterior sample of the $\boldsymbol{\theta}_k$s, which we denote by $\widehat{\beta}(\boldsymbol{\theta}_k)$. Table 13.2 contains the posterior probabilities that this slope is positive, i.e., that the quality of patient care in the network is improving. Under the uniform PAC prior, all six of the above networks have $P(\widehat{\beta}(\boldsymbol{\theta}_k) > 0|\mathbf{y})$ greater than 0.8. The triangular, uniform on **R**, and AR(1) priors, additionally, all yield very high posterior probabilities of increasing latent factors across all groups. In addition, networks C, O, P, and U, all have very strong evidence of improvement with posterior probabilities of positive slopes greater than .95. As is clear from Figure 13.1, the factor paths tend not to increase as quickly under the independence correlation structure, an observation further supported by Table 13.2. For example, network C has positive slope with posterior probability 0.62, compared to values that are nearly one for the other priors. Additionally, inference under the independence assumption leads to the conclusion that networks Q and S have declining patient care, since these factors have positive slope with probabilities less than a half.

Example III: Analysis of GLOM model

We now describe in depth an analysis using a GLOM for the St. Louis Risk Research Project. As we have previously described general modeling strategies for the GLOM without limiting ourselves to a single structure, we narrow our focus to models that depend on shrinkage-type priors.

For the categorical component of the model, we introduce a Dirichlet prior for the contingency cell probabilities that shrinks toward an independence model. To that end, let $\boldsymbol{\alpha}^{(1)} = (\alpha_1^{(1)}, \alpha_2^{(1)}, \alpha_3^{(1)})^{\top} \sim Dir(3, (1/3)\mathbf{1}_3)$, and $\boldsymbol{\alpha}^{(2)} = (\alpha_1^{(2)}, \alpha_2^{(2)})^{\top}$ and $\boldsymbol{\alpha}^{(3)} = (\alpha_1^{(3)}, \alpha_2^{(3)})^{\top}$ be independently drawn from $Dir(2, (1/2)\mathbf{1}_2)$, where $\mathbf{1}_p$ is a p-dimensional vector of ones. Here, we parameterize the Dirichlet distribution by a scale factor c and a probability vector $\mathbf{v}$, so that the mean and variance of the ith element are v_i and $v_i(1 - v_i)/(c+1)$. These three vectors give the marginal probabilities of the three discrete components G, D_1, D_2, respectively. We can form the prior cell probabilities from these marginal probabilities by $p_{jkl} = \alpha_j^{(1)} \alpha_k^{(2)} \alpha_l^{(3)}$. The distribution for the contingency cell probabilities is the conjugate Dirichlet distribution, with shrinkage parameter γ, i.e., $\{\pi_{\mathbf{w}}\} \sim Dir(\gamma, \{p_{\mathbf{w}}\})$.

To specify the prior on γ, we use a uniform shrinkage prior (Daniels, 1999). We note that under the Dirichlet prior, the conditional posterior expectation for $\pi_{\mathbf{w}}$ is the linear combination

$$E(\pi_{\mathbf{w}}|n_{\mathbf{w}}, \gamma, p_{\mathbf{w}}) = \left(\frac{N}{N+\gamma}\right)\left(\frac{n_{\mathbf{w}}}{N}\right) + \left(1 - \frac{N}{N+\gamma}\right)p_{\mathbf{w}},$$

Table 13.3 *Model comparisons for the St. Louis Risk Research Project.*

Prior for $\pi_{\mathbf{w}}$	Prior for $\mathbf{B}$	Prior for Σ	DIC	p_D
Shrinkage	Shrinkage	Full	1,951.5	32.6
Shrinkage	Shrinkage	Block diagonal	1,953.0	28.9
Shrinkage	No 2-way terms	Block diagonal	1,953.6	29.0
Shrinkage	No 2-way terms	Full	1,954.5	32.5
Conjugate	Shrinkage	Full	1,956.5	36.3
Conjugate	No 2-way terms	Block diagonal	1,958.3	32.6
Conjugate	Shrinkage	Block diagonal	1,959.3	33.3
Conjugate	No 2-way terms	Full	1,959.6	36.5
Shrinkage	No shrinkage	Full	1,974.7	46.1
Shrinkage	No shrinkage	Block diagonal	1,975.7	42.1
Conjugate	No shrinkage	Block diagonal	1,979.7	45.5
Conjugate	No shrinkage	Full	1,982.1	50.3

where $n_{\mathbf{w}}$ is the number of observations in cell $\mathbf{w}$, and N is the total sample size. It is clear from this that γ controls how close $\pi_{\mathbf{w}}$ is to $p_{\mathbf{w}}$, with large values of γ yielding $\pi_{\mathbf{w}}$ close to the prior probability $p_{\mathbf{w}}$. Smaller values of γ (relative to N) lead to less shrinkage of $\pi_{\mathbf{w}}$ (so closer the observed proportion of observations in cell $\mathbf{w}$). The uniform shrinkage prior for γ is $p(\gamma) = N/(N+\gamma)^2$, for $\gamma > 0$; this is derived by placing a $U(0,1)$ distribution on the shrinkage factor $N/(N+\gamma)$ from the (conditional) posterior expectation.

The distribution of the continuous component $\mathbf{Y}$ in cell $\mathbf{w}$ is $N_4(\boldsymbol{\mu}_{\mathbf{w}}, \Sigma)$. As introduced in Section 13.2, the cell means can be written in a regression form with $\boldsymbol{\mu}_{\mathbf{w}} = \mathbf{B}^\top \mathbf{x}_{\mathbf{w}}$. The q-dimensional design vector $\mathbf{x}_{\mathbf{w}}$ depends only on the values of the contingency cell $\mathbf{w}$, and the matrix of regression coefficients $\mathbf{B}$ has dimension $q \times 4$. For cell $\mathbf{w} = (j,k,\ell)$, the covariate vector is $\mathbf{x}_{\mathbf{w}} = (1, \mathrm{I}\{j=1\}, \mathrm{I}\{j=2\}, \mathrm{I}\{k=1\}, \mathrm{I}\{\ell=1\}, \mathrm{I}\{j=1,k=1\}, \mathrm{I}\{j=2,k=1\}, \mathrm{I}\{j=1,\ell=1\}, \mathrm{I}\{j=2,\ell=1\}, \mathrm{I}\{k=1,\ell=1\})^\top$. The prior for $\mathbf{B} = (b_{ij})$ is specified element-wise. The prior for b_{ij} is a dispersed $N(0,10^3)$, for $i \leq 5$, i.e., those regression coefficients corresponding to the global mean or the one-way interaction term between Y_j and one of the discrete components. For $i > 5$, $b_{ij} \sim N(0, \tau^2)$, which gives the prior for the two-way interactions between discrete components. The variance τ^2 shrinks these higher-order terms toward zero, that is, shrinking toward a mean structure with independent additive effects for the discrete components. The prior for τ^2 is $p(\tau^2) = c/(c+\tau^2)^2$, where the parameter c is the median. To provide a substantial amount of shrinkage, we use $c = 1$.

Note that we assume a constant covariance matrix Σ as a function of $\mathbf{w}$, and we consider two possible structures for this matrix. First, we specify a uniform prior for Σ over the space of 4×4 positive-definite matrices. We additionally consider a block diagonal specification. This covariance structure implies that (R_{i1}, V_{i1}), the reading and verbal measures for child 1, are independent of (R_{i2}, V_{i2}), the academic markers for child 2, conditional on the categorical responses $\mathbf{W}_i$. The prior is formed by independently drawing the two blocks from a uniform prior on the space of 2×2 positive-definite matrices. We denote these as the full and block diagonal priors for Σ, respectively.

We have introduced a shrinkage prior for the categorical component of the model that centers the cell probabilities around an independence model. We compare this choice with the simple conjugate $\mathrm{Dir}(12, (1/12)\mathbf{1}_{12})$ prior; we label this prior as conjugate. For the continuous component of the model, in addition to the shrinkage prior on the interaction terms, we consider a "no shrinkage" prior that provides little structure by giving all elements of $\mathbf{B}$ independent $N(0,10^3)$ prior. We finally consider a model with no two-way interaction terms, i.e., $b_{ij} = 0$, for $i > 5$. All the combinations of the priors on $\{\pi_{\mathbf{w}}\}, \mathbf{B}$, and Σ, give us twelve models to consider.

As in the previous data analysis, we use the DIC for model comparison. Since there is missingness in the data, we use the observed data likelihood in the DIC calculations (Wang and

Daniels, 2011). The DIC values and model complexity statistics p_D are displayed in Table 13.3. We first notice that the four models that left **B** unconstrained, produce the worst model fits. Of the eight remaining models, the shrinkage prior on $\{\pi_w\}$ outperforms the conjugate prior. We gained strength by shrinking the cell probabilities toward an independence structure. There seems to be little difference between the full and block diagonal structures for the covariance matrix; comparing the priors on Σ for each pair of $\{\pi_w\}$ and **B** priors, the models with the full prior have a larger DIC than the block diagonal prior in three instances, and a smaller DIC in the other three. The shrinkage prior on the mean vector seems to be slightly better than the one that excludes the two-way interaction terms. We conclude that the best model (as determined by the DIC), out of those considered, is the one formed by using shrinkage priors on the cell probabilities and two-way regression coefficients, along with a flat prior on the full covariance matrix.

13.8 Discussion

We laid out in this chapter the basic issues with modeling mixed-outcome data using Bayesian methods, including prior specification and computations. We also discussed Bayesian approaches to address the complex (potentially high-dimensional) covariance structure in such data, along with approaches to deal with incomplete (missing) data.

In the analysis of the VA data, we assumed a single latent factor. This is, of course, a nontrivial assumption and can have important consequences for our analysis. The choice of the number of factors should be guided by the desire for model adequacy and interpretation. We choose a single factor, in large part, for its interpretation as the overall quality of the network's inpatient care. Additional analysis might include fitting the model with $L = 2$ and comparing DIC values, or stochastically mixing over models with different numbers of latent factors through reversible jump steps (Green, 1995).

We did not discuss Bayesian model fit and checking in detail, though we used the DIC to choose the best correlation model in the VA data analysis of Section 13.7, as well as the model comparison for the St. Louis data. The DIC is typically a simple criterion to compute for choosing between models. However, one must be careful in defining the likelihood given the differences in the models being compared. For example, should one use the likelihood conditional on random effects/latent variables, or the marginalized likelihood? In Example I of Section 13.7, we were interested in comparing models different in parameters at the network-level (not the patient-level). Thus, we focused on the network-level likelihood. A nice discussion of these issues in the context of latent variable/random effects models can be found in Trevisani and Gelfand (2003) and Celeux *et al.* (2006). In the situation with Example III, where the data is not fully observed, we used the observed data likelihood for computing the DIC; see Wang and Daniels (2011) and Daniels *et al.* (2012*a*) for further discussion.

Posterior predictive checks (Gelman *et al.*, 2003) can be used to assess specified aspects of model fit. A nice discussion of these checks in the context of latent variable models can be found in Gelman *et al.* (2005).

Bibliography

Aas, K., C. Czado, A. Frigessi and H. Bakken (2009). Pair-copula constructions of multiple dependence. *Insurance: Mathematics and Statistics* **44**, 182–198.

Abramowitz, M. and I. Stegun (1972). *Handbook of Mathematical Functions*. Dover.

Aburatani, S. (2011). Application of structural equation modeling for inferring a serial transcriptional regulation in yeast. *Gene Regulation and Systems Biology* **5**, 75–88.

Aerts, M., H. Geys, G. Molenberghs and L. M. Ryan (Eds.) (2002). *Topics in Modelling of Clustered Data*. Chapman & Hall/CRC Press.

Afifi, A. A. and R. M. Elashoff (1969). Multivariate two sample tests with dichotomous and continuous variables. I. The location model. *Annals of Mathematical Statistics* **40**, 290–298.

Agresti, A. (2002). *Categorical Data Analysis*. 2nd ed.. Wiley & Sons, Inc.

Albert, J. H. and S. Chib (1993). Bayesian analysis of binary and polychotomous response data. *Journal of the American Statistical Association* **88**, 669–679.

Amos, C. I., W. V. Chen, M. F. Seldin, E. F. Remmers, K. E. Taylor, L. A. Criswell, A. T. Lee, R. M. Plenge, D. L. Kastner and P. K. Gregersen (2009). Data for Genetic Analysis Workshop 16 Problem 1, association analysis of rheumatoid arthritis data. *BMC Proceedings* **3**(Suppl 7), S2.

An, Q., D. Bowman and E. O. George (2012). Joint modeling of clustered multivariate continuous and discrete endpoints. In preparation.

An, X. and P. M. Bentler (2011). Nesting Monte Carlo EM for high-dimensional item factor analysis. *Journal of Statistical Computation and Simulation*. In press.

An, X. and P. M. Bentler (2012). Efficient direct sampling MCEM algorithm for latent variable models with binary responses. *Computational Statistics & Data Analysis* **56**, 231–244.

Anderson, J. A. and J. D. Pemberton (1985). The grouped continuous model for multivariate ordered categorical variables and covariate adjustment. *Biometrics* **41**, 875–885.

Anderson, J. A. and P. R. Philips (1981). Regression, discrimination and measurement models for ordered categorical variables. *Applied Statistics* **30**, 22–31.

Anderson, J. C. and D. W. Gerbing (1984). The effects of sampling error on convergence, improper solutions and goodness-of-fit indices for maximum likelihood confirmatory factor analysis. *Psychometrika* **49**, 155–173.

Anderson, T. W. (1984). *An Introduction to Multivariate Statistical Analysis*. 2nd ed.. Wiley & Sons, Inc.

Arnold, B. C. and D. Strauss (1991). Pseudolikelihood estimation: Some examples. *Sankhya B* **53**, 233–243.

Asparouhov, T. and B. Muthèn (2009). Exploratory structural equation modeling. *Structural Equation Modeling* **16**, 397–438.

Bahadur, R. R. (1961). A representation of the joint distribution of responses to *n* dichotomous items. In: *Studies in Item Analysis and Prediction* (H. Solomon, Ed.). pp. 158–168. Stanford University Press.

Baker, F. B. and S. H. Kim (2004). *Item Response Theory: Parameter Estimation Techniques*. 2nd ed.. Marcel Dekker.

Balakrishnan, N. and C.−D. Lai (2009). *Continuous Bivariate Distributions*. 2nd ed.. Springer.

Bandalos, D. L. (2002). The effects of item parceling on goodness-of-fit and parameter estimate bias in structural equation modeling. *Structural Equation Modeling* **9**, 78–102.

Bandalos, D. L. and S. J. Finney (2001). Item parceling issues in structural equation modeling. In: *New Developments and Techniques in Structural Equation Modeling* (G. A. Marcoulides and R. E. Schumacker, Eds.). pp. 269–296. Erlbaum.

Bar−Hen, A. and J. J. Daudin (1995). Generalization of the Mahalanobis distance in the mixed case. *Journal of Multivariate Analysis* **53**, 332–342.

Barcena, M. J. and F. Tusell (2004). Fitting multivariate responses using scalar trees. *Statistics & Probability Letters* **69**, 253–259.

Barnard, J., R. McCulloch and X.−L. Meng (2000). Modeling covariance matrices in terms of standard deviations and correlations, with application to shrinkage. *Statistica Sinica* **10**, 1281–1311.

Bartels, R. and D. G. Fiebig (1991). A simple characterization of seemingly unrelated regressions models in which OLS is BLUE. *American Statistician* **45**, 137–140.

Bartholomew, D. J. (2007). Three faces of factor analysis. In: *Factor Analysis at 100: Historical Developments and Future Directions* (R. Cudeck and R. MacCallum, Eds.). pp. 9–21. Erlbaum.

Bartholomew, D. J., M. Knott and I. Moustaki (2011). *Latent Variable Models and Factor Analysis: A Unified Approach*. 3rd ed.. Wiley & Sons, Inc.

Bartlett, M. S. (1933). On the theory of statistical regression. *Proceedings of the Royal Society of Edinburgh* **53**, 260–283.

Bauer, D.J. and P. J. Curran (2004). The integration of continuous and discrete latent variable models: Potential problems and promising opportunities. *Psychological Methods* **9**, 3–29.

Bedrick, E. J., J. Lapidus and J. F. Powell (2000). Estimating the Mahalanobis distance from mixed continuous and discrete data. *Biometrics* **56**, 394–401.

Bello, N. M., J. P. Steibel and R. J. Tempelman (2012). Hierarchical Bayesian modeling of heterogeneous cluster- and subject-level associations between continuous and binary outcomes in dairy production. *Biometrical Journal* **54**, 230–248.

Bentler, P. M. (1983). Some contributions to efficient statistics in structural equation models: Specification and estimation of moment structure. *Psychometrika* **48**, 493–517.

Bentler, P. M. and D. G. Weeks (1980). Linear structural equations with latent variables. *Psychometrika* **45**, 289–308.

Bentler, P. M. and J. de Leeuw (2011). Factor analysis via components analysis. *Psychometrika* **76**, 461–470.

Berridge, D. M. and R. Crouchley (2011). *Multivariate Generalized Linear Mixed Models in R*. Chapman & Hall/CRC Press.

Betensky, R. A. and A. S. Whittemore (1996). An analysis of correlated multivariate binary data: Application to familial cancers of the ovary and breast. *Applied Statistics* **45**, 411–429.

Bhat, C. (2001). Quasi-random maximum simulated likelihood estimation of the mixed multinomial logit model. *Transportation Research: Part B* **35**, 677–693.

Bhat, C., C. Varin and N. Ferdous (2010). A comparison of the maximum simulated likelihood and composite marginal likelihood estimation approach in the context of the multivariate ordered-response model system. In: *Advances in Econometrics: Maximum Simulated Likelihood Methods and Applications* (R. Carter Hill and W. Green, Eds.). Vol. 26. pp. 65–106. Emerald.

Biau, G., L. Devroye and G. Lugosi (2008). Consistency of random forests and other averaging classifiers. *Journal of Machine Learning Research* **9**, 2015–2033.

Bishop, Y. M. M., S. E. Fienberg and P. W. Holland (1975). *Discrete Multivariate Analysis: Theory and Practice*. MIT Press.

BMASK (2011). Armutsgefährdung und Lebensbedingungen in Österreich, Ergebnisse aus EU-SILC 2009. In: *Studie der Statistik Austria im Auftrag des BMASK*. Bd5 ed. Wien: Sozialpolitische Studienreihe.

Bock, R. D. (1972). Estimating item parameters and latent ability when responses are scored in two or more nominal categories. *Psychometrika* **37**, 29–51.

Bock, R. D. and M. Aitkin (1981). Marginal maximum likelihood estimation of item parameters: Application of an EM algorithm. *Psychometrika* **46**, 443–459.

Bodnar, O., T. Bodnar and A. K. Gupta (2010). Estimation and inference for dependence in multivariate data. *Journal of Multivariate Analysis* **101**, 869–881.

Bollen, K. A. (2002). Latent variables in psychology and the social sciences. *Annual Review of Psychology* **46**, 605–634.

Bollen, K. A. and P. J. Curran (2006). *Latent Curve Models: A Structural Equation Approach*. Wiley & Sons, Inc.

Bollen, K. A. and S. Bauldry (2011). Three Cs in measurement models: Causal indicators, composite indicators, and covariates. *Psychological Methods* **16**, 265–284.

Booth, J. G. and J. P. Hobert (1999). Maximizing generalized linear mixed model likelihoods with an automated Monte Carlo EM algorithm. *Journal of the Royal Statistical Society–B* **61**, 265–285.

Bou–Hamad, I., D. Larocque and H. Ben–Ameur (2011). A review of survival trees. *Statistics Surveys* **5**, 44–71.

Bowman, D. and E. O. George (1995). A saturated model for analyzing exchangeable binary data: Applications to clinical and developmental toxicity studies. *Journal of the American Statistical Association,* **90**, 871–879.

Breiman, L. (2001). Random forests. *Machine Learning* **45**, 5–32.

Breiman, L. and J. H. Friedman (1997). Predicting multivariate responses in multiple linear regression. *Journal of the Royal Statistical Society–B* **59**, 3–54.

Breiman, L., J. Friedman, R. Olshen and C. Stone (1984). *Classification and Regression Trees*. Wadsworth International Group.

Breslow, N. E. and D. G. Clayton (1993). Approximate inference in generalized linear mixed models. *Journal of the American Statistical Association* **88**, 9–25.

Broyden, C. G. (1970). The convergence of a class of double-rank minimization algorithms. *Journal of the Institute of Mathematics and Its Applications* **6**, 76–90.

Buntin, M. and A. Zaslavsky (2004). Too much ado about two-part models and transformation? Comparing methods of modeling medicare expenditures. *Journal of Health Economics* **23**, 525–542.

Burzykowski, T., G. Molenberghs, M. Buyse, D. Renard and H. Geys (2001). Validation of surrogate endpoints in multiple randomized clinical trials with failure-time endpoints. *Applied Statistics* **50**, 405–422.

Caflisch, R. (1998). Monte Carlo and quasi-Monte Carlo methods. *Acta Numerica* **7**, 1–49.

Cai, L. (2010). High-dimensional exploratory item factor analysis by a Metropolis–Hastings Robbins–Monro algorithm. *Psychometrika* **75**, 33–57.

Cameron, A. and P. Trivedi (1986). Econometric models based on count data: Comparisons and applications of some estimators. *Journal of Applied Econometrics* **1**, 29–53.

Catalano, P. J. (1997). Bivariate modelling of clustered continuous and ordered categorical outcomes. *Statistics in Medicine* **16**, 883–900.

Catalano, P. J. and L. M. Ryan (1992). Bivariate latent variable models for clustered discrete and continuous outcomes. *Journal of the American Statistical Association* **87**, 651–658.

Cattell, R. B. and C. A. Burdsall (1975). The radial parcel double factor design: A solution to the item-versus-parcel controversy. *Multivariate Behavioral Research* **10**, 165–179.

Celeux, G., F. Forbes, C. Robert and D. Titterington (2006). Deviance information criteria for missing data models. *Bayesian Analysis* **1**, 651–674.

Chandra, A., J. Gruber and R. McKnight (2007). Patient cost sharing, hospitalization offsets, and the design of optimal health insurance for the elderly. *Working Paper* 12972. National Bureau of Economic Research.

Chang, P. C. and A. A. Afifi (1974). Classification based on dichotomous and continuous variables. *Journal of the American Statistical Association* **69**, 336–339.

Chen, F., K. A. Bollen, P. Paxton, P. J. Curran and J. B. Kirby (2001). Improper solutions in structural equation models: Causes, consequences, and strategies. *Sociological Methods & Research* **29**, 468–508.

Chen, J. and R. Kodell (1989). Quantitative risk assessment for teratological effects. *Journal of the American Statistical Association* **84**, 966–971.

Chen, L., M. Zhong, W. V. Chen, C. I. Amos and R. Fan (2009). A genomewide association scan for rheumatoid arthritis data by Hotelling's T^2 tests. *BMC Proceedings* **3**(Suppl 7), S6.

Chib, S. (2007). Analysis of treatment response data without the joint distribution of potential outcomes. *Journal of Econometrics* **140**, 401–412.

Chib, S. and E. Greenberg (1998). Analysis of multivariate probit models. *Biometrika* **85**, 347–361.

Chiu, T. Y. M., T. Leonard and K.–W. Tsui (1996). The matrix-logarithmic covariance model. *Journal of the American Statistical Association* **91**, 198–210.

Ciampi, A., R. du Berger, H. G. Taylor and J. Thiffault (1991). RECPAM: A computer program for recursive partition and amalgamation for survival data and other situations frequently occurring in biostatistics. III. Classification according to a multivariate construct. Application to data on Haemophilus influenzae type b meningitis. *Computer Methods and Programs in Biomedicine* **36**, 51–61.

Clayton, D. (1978). A model for association in bivariate life tables and its application in epidemiological studies of familial tendency in chronic disease incidence. *Biometrika* **65**, 141–151.

Cowles, M. K., B. P. Carlin and J. E. Connett (1996). Bayesian tobit modeling of longitudinal ordinal clinical trial compliance data with nonignorable missingness. *Journal of the American Statistical Association* **91**, 86–98.

Cox, D. R. (1972). The analysis of multivariate binary data. *Applied Statistics* **21**, 113–120.

Cox, D. R. and N. Wermuth (1992). Response models for mixed binary and quantitative variables. *Biometrika* **79**, 441–461.

Cox, D. R. and N. Wermuth (1994). A note on the quadratic exponential binary distribution. *Biometrika* **81**, 403–408.

Cox, D. R. and N. Wermuth (1996). *Multivariate Dependencies: Models, Analysis and Interpretations*. Chapman & Hall/CRC Press.

Cox, N. R. (1974). Estimation of the correlation between a continuous and a discrete variable. *Biometrics* **30**, 171–178.

Craiu, V. R. and A. Sabeti (2012). In mixed company: Bayesian inference for bivariate conditional copula models with discrete and continuous outcomes. *Journal of Multivariate Analysis* **110**, 106–120.

Cuadras, C. M. (1992). Some examples of distance based discrimination. *Biometrical Letters* **29**, 3–20.

Cuadras, C. M., J. Fortiana and F. Oliva (1997). The proximity of an individual to a population with applications in discriminant analysis. *Journal of Classification* **14**, 117–136.

Czado, C., R. Kastenmeier, E. Brechmann and A. Min (2011). A mixed copula model for insurance claims and claim sizes. *Scandinavian Actuarial Journal*. In press.

Czado, C., V. Erhardt, A. Min and S. Wagner (2007). Zero-inflated generalized Poisson models with regression effects on the mean, dispersion and zero-inflation level applied to patent outsourcing rates. *Statistical Modelling* **7**, 125–153.

Daniels, M. J. (1999). A prior for the variance in hierarchical models. *Canadian Journal of Statistics* **27**, 567–578.

Daniels, M. J. (2006). Bayesian modeling of several covariance matrices and some results on the propriety of the posterior for linear regression with correlated and/or heterogeneous errors. *Journal of Multivariate Analysis* **97**, 1185–1207.

Daniels, M. J., A. Chatterjee and C. Wang (2012*a*). Bayesian model selection for incomplete data using the posterior predictive distribution. *Biometrics*. In press.

Daniels, M. J. and J. W. Hogan (2008). *Missing Data in Longitudinal Studies: Strategies for Bayesian Modeling and Sensitivity Analysis*. Chapman & Hall/CRC Press.

Daniels, M. J. and M. Pourahmadi (2002). Bayesian analysis of covariance matrices and dynamic models for longitudinal data. *Biometrika* **89**, 553–566.

Daniels, M. J. and M. Pourahmadi (2009). Modeling covariance matrices via partial autocorrelations. *Journal of Multivariate Analysis* **100**, 2352–2363.

Daniels, M. J. and R. E. Kass (1999). Nonconjugate Bayesian estimation of covariance matrices and its use in hierarchical models. *Journal of the American Statistical Association* **94**, 1254–1263.

Daniels, M. J. and S.–L. T. Normand (2006). Longitudinal profiling of health care units based on continuous and discrete patient outcomes. *Biostatistics* **7**, 1–15.

Daniels, M. J., C. Wang and B. H. Marcus (2012*b*). Fully Bayesian inference under ignorable missingness in the presence of auxiliary covariates. Submitted.

Darroch, J. N. (1981). The Mantel–Haenszel test and tests of marginal symmetry; fixed-effects and mixed models for a categorical response. *International Statistical Review* **49**, 285–307.

de Boeck, P. and M. Wilson (2004). *Explanatory Item Response Models: A Generalized Linear and Nonlinear Approach*. Springer–Verlag.

de Finetti, B. (1974). *Theory of Probability*. Wiley & Sons, Inc.

de Leon, A. R. (2005). Pairwise likelihood approach to grouped continuous model and its extension. *Statistics & Probability Letters* **75**, 49–57.

de Leon, A. R. (2007). One-sample likelihood ratio tests for mixed data. *Communications in Statistics–Theory & Methods* **36**, 129–141.

de Leon, A. R., A. Soo and T. Williamson (2011). Classification with discrete and continuous variables via general mixed-data models. *Journal of Applied Statistics* **38**, 1021–1032.

de Leon, A. R. and B. Wu (2011). Copula-based regression models for a bivariate mixed discrete and continuous outcome. *Statistics in Medicine* **30**, 175–185.

de Leon, A. R. and K. C. Carrière (2000). On the one-sample location hypothesis for mixed bivariate data. *Communications in Statistics–Theory & Methods* **29**, 2573–2581.

de Leon, A. R. and K. C. Carrière (2005). A generalized Mahalanobis distance for mixed data.

Journal of Multivariate Analysis **92**, 174–185.

de Leon, A. R. and K. C. Carrière (2007). General mixed-data model: extension of general location and grouped continuous models. *Canadian Journal of Statistics* **35**, 533–548.

de Leon, A. R. and K. C. Carrière Chough (2010). Mixed-outcome data. In: *Encyclopedia of Biopharmaceutical Statistics* (S. C. Chow, Ed.). 3rd ed.. pp. 817–822. Informa Healthcare.

de Leon, A. R. and Y. Zhu (2008). ANOVA extensions for mixed discrete and continuous data. *Computational Statistics & Data Analysis* **52**, 2218–2227.

de Leon, A. R., B. Wu and N. Withanage (2012). Gaussian copula-generated distributions for correlated discrete and continuous variables. Submitted.

Deb, P. and P. Trivedi (1997). Demand for medical care among the elderly: A finite mixture approach. *Journal of Applied Econometrics* **12**, 313–326.

Deb, P. and P. Trivedi (2006). Specification and simulated likelihood estimation of a non-normal treatment-outcome model with selection: Application to health care utilization. *Econometrics Journal* **9**, 307–331.

Deb, P., P. Trivedi and D. Zimmer (2009). Dynamic cost-offsets of prescription drug expenditures: Panel data analysis using a copula-based hurdle model. *Working Paper* 15191. National Bureau of Economic Research.

Delyon, B., M. Lavielle and E. Moulines (1999). Convergence of a stochastic approximation version of the EM algorithm. *Annals of Statistics* **27**, 94–128.

Demidenko, E. (2004). *Mixed Models: Theory and Applications*. Wiley & Sons, Inc.

Denuit, M. and P. Lambert (2005). Constraints on concordance measures in bivariate discrete data. *Journal of Multivariate Analysis* **93**, 40–57.

Diamanti–Kandarakis, E., J. Bourguignon, L. C. Giudice, R. Hauser, G. S. Prins, A. M. Soto, R. T. Zoeller and A. C. Gore (2009). Endocrine-disrupting chemicals: An Endocrine Society scientifc statement. *Endocrine Reviews* **30(4)**, 293–342.

Diggle, P. J., K.–Y. Liang and S. L. Zeger (1994). *Analysis of Longitudinal Data*. Clarendon Press.

Diggle, P. J., P. J. Heagerty, K.–Y. Liang and S. L. Zeger (2002). *Analysis of Longitudinal Data*. 2nd ed.. Clarendon Press.

Dine, A., D. Larocque and F. Bellavance (2009). Multivariate trees for mixed outcomes. *Computational Statistics & Data Analysis* **53**, 3795–3804.

Dobra, A. and A. Lenkoski (2011). Copula Gaussian graphical models and their application to modeling functional disability data. *Annals of Applied Statistics* **5**, 969–993.

Doornik, J. A. (2007). *Object-Oriented Matrix Programming using Ox*. 3rd ed.. Timberlake Consultants Press/Oxford.

Doull, J., R. Cattley, C. Elcombe, B. G. Lake, J. Swenberg, C. Wilkinson, G. Williams and M. vanGemert (1999). A cancer risk assessment of Di(2-ethylhexyl) phthalate: Application of the new U. S. EPA Risk Assessment Guidelines. *Regulatory Toxicology and Pharmacology* **29**, 327–357.

Draper, D., J. S. Hodges, C. L. Mallows and D. Pregibon (1993). Exchangeability and data analysis. *Journal of the Royal Statistical Society–A* **156**, 9–37.

Dunson, D. B. (2000). Bayesian latent variable models for clustered mixed outcomes. *Journal of the Royal Statistical Society–B* **62**, 355–366.

Dunson, D. B. (2003). Dynamic latent trait models for multidimensional longitudinal data. *Journal of the American Statistical Association* **98**, 555–563.

Dunson, D. B. (2006). Bayesian dynamic modeling of latent trait distributions. *Biostatistics* **7**, 551–568.

Dunson, D. B. and C. Xing (2009). Nonparametric Bayes modeling of multivariate categorical data. *Journal of the American Statistical Association* **104**, 1042–1051.

Dunson, D. B. and S. D. Perreault (2001). Factor analytic models of clustered multivariate data with informative censoring. *Biometrics* **57**, 302–308.

Dunson, D., B. Chen and J. Harry (2003). A Bayesian approach for joint modeling of cluster size and subunit-specific outcomes. *Biometrics* **59**, 521–530.

Edwards, D. (1995). *Introduction to Graphical Modelling*. Springer.

Embretson, S. E. and S. P. Reise (2000). *Item Response Theory for Psychologists*. Erlbaum.

Engelhardt, G. and J. Gruber (2010). Medicare Part D and the financial protection of the elderly. *Working Paper* 16155. National Bureau of Economic Research.

Erhardt, V. and C. Czado (2012). Modeling dependent yearly claim totals including zero-claims in private health insurance. *Scandinavian Actuarial Journal* **2012**, 106–129.

Faes, C., H. Geys and P. Catalano (2009). Joint models for continuous and discrete longitudinal data. In: *Longitudinal Data Analysis* (G. Fitzmaurice, M. Davidian, G. Verbeke and G. Molenberghs, Eds.). pp. 327–348. Chapman & Hall/CRC Press.

Faes, C., H. Geys, M. Aerts and G. Molenberghs (2006). A hierarchical modeling approach for risk assessment in developmental toxicity studies. *Computational Statistics & Data Analysis* **51**, 1848–1861.

Faes, C., H. Geys, M. Aerts, G. Molenberghs and P. Catalano (2004). Modeling combined continuous and ordinal outcomes in a clustered setting. *Journal of Agricultural, Biological and Environmental Statistics* **9**, 515–530.

Faes, C., M. Aerts, G. Molenberghs, H. Geys, G. Teuns and L. Bijnens (2008). A high-dimensional joint model for longitudinal outcomes of different nature. *Statistics in Medicine* **27**, 4408–4427.

Fahrmeir, L. and G. Tutz (2001). *Multivariate Statistical Modelling Based on Generalized Linear Models*. Springer–Verlag.

Fedorov, V., Y. Wu and R. Zhang (2012). Optimal dose-finding designs with correlated continuous and discrete responses. *Statistics in Medicine* **31**, 217–234.

Feller, W. (1971). *An Introduction to Probability Theory and Its Applications*. Vol. 2. 2nd ed.. Wiley & Sons, Inc.

Fieuws, S. and G. Verbeke (2006). Pairwise fitting of mixed models for the joint modeling of multivariate longitudinal profiles. *Biometrics* **62**, 424–431.

Fitzmaurice, G. M. and N. M. Laird (1995). Regression models for a bivariate discrete and continuous outcome with clustering. *Journal of the American Statistical Association* **90**, 845–852.

Fitzmaurice, G. M. and N. M. Laird (1997). Regression models for mixed discrete and continuous responses with potentially missing values. *Biometrics* **53**, 110–122.

Fitzmaurice, G., M. Davidian, G. Verbeke and G. Molenberghs (Eds.) (2009). *Longitudinal Data Analysis*. Chapman & Hall/CRC Press.

Fletcher, R. (1970). A new approach to variable metric algorithms. *Computer Journal* **13**, 317–322.

Fletcher, R. (1987). *Practical Methods of Optimization*. 2nd ed.. Wiley & Sons, Inc.

Freidlin, B., G. Zheng, Z. Li and J. L. Gastwirth (2002). Trend tests for case-control studies of genetic markers: power, sample size and robustness. *Human Heredity* **53**, 146–152.

Frühwirth–Schnatter, S. and H. Wagner (2006). Auxiliary mixture sampling for parameter-driven models of time series of small counts with applications to state space modelling. *Biometrika* **93**, 827–841.

Frühwirth–Schnatter, S. and R. Frühwirth (2010). Data augmentation and MCMC for binary and multinomial logit models. In: *Statistical Modelling and Regression Structures – Festschrift in Honour of Ludwig Fahrmeir* (T. Kneib and G. Tutz, Eds.). pp. 111–132. Physica–Verlag.

Frühwirth–Schnatter, S. and R. Tüchler (2008). Bayesian parsimonious covariance estimation for hierarchical linear mixed models. *Statistics and Computing* 18, 1–13.

Frühwirth–Schnatter, S., R. Frühwirth, L. Held and H. Rue (2009). Improved auxiliary mixture sampling for hierarchical models of non-Gaussian data. *Statistics and Computing* 19, 479–492.

Fusco, A., A.–C. Guio and E. Marlier (2010). Income poverty and material deprivation in European countries. *Working Paper*. Eurostat-Methodologies.

Gaskins, J. T. and M. J. Daniels (2012*a*). Modeling a joint binary-continuous process with a pattern mixture model under nonignorable missingness. *Working Paper*. University of Florida.

Gaskins, J. T. and M. J. Daniels (2012*b*). A nonparametric prior for simultaneous covariance estimation. Submitted.

Gaskins, J. T., M. J. Daniels and B. H. Marcus (2012). A prior for partial autocorrelation selection. Submitted.

Gass, C. S. (1996). MMPI-2 variables in attention and memory test performance. *Psychological Assessment* 8, 135–138.

Gates, K. M., P. C. M. Molenaar, F. G. Hillary and S. Slobounov (2011). Extended unified SEM approach for modeling event-related fMRI data. *Neuroimage* 54, 1151–1158.

Gaylor, D. W. (1994). Biostatistical approaches to low level exposures. In: *Biological Effects of Low Level Exposure* (E. J. Calabrese, Ed.). Chap. 6. CRC Press.

Gelman, A., I. Van Mechelen, G. Verbeke, D. F. Heitjan and M. Meulders (2005). Multiple imputation for model checking: Completed-data plots with missing and latent data. *Biometrics* 61, 74–85.

Gelman, A., J. B. Carlin, H. S. Stern and D. B. Rubin (2003). *Bayesian Data Analysis*. 2nd ed.. Cambridge University Press.

Genest, C. and J. Nešlehovà (2007). A primer on copula for count data. *ASTIN Bulletin* 37, 475–515.

George, E. and R. McCulloch (1997). Approaches for Bayesian variable selection. *Statistica Sinica* 7, 339–373.

George, E. O. and D. Bowman (1995). A full likelihood procedure for analysing exchangeable binary data. *Biometrics* 51, 512–523.

George, E. O., D. Armstrong, P. J. Catalano and D. K. Srivastava (2007). Regression models for analyzing clustered binary and continuous outcomes under an assumption of exchangeability. *Journal of Statistical Planning and Inference* 137, 3462–3474.

Gershman, S. J. and D. M. Blei (2012). A tutorial on Bayesian nonparametric models. *Journal of Mathematical Psychology* 56, 1–12.

Geys, H., G. Molenberghs and L. M. Ryan (1999). Pseudo-likelihood modeling of multivariate outcomes in developmental toxicology. *Journal of the American Statistical Association* 94, 734–745.

Geys, H., M. M. Regan, P. J. Catalano and G. Molenberghs (2001). Two latent variable risk assessment approaches for mixed continuous and discrete outcomes from developmental toxicity. *Journal of Agricultural, Biological and Environmental Statistics* 6, 340–355.

Ghosh, J. and D. B. Dunson (2009). Default prior distributions and efficient posterior computation in Bayesian factor analysis. *Journal of Computational and Graphical Statistics* 18, 306–320.

Gibbons, R. D. (2000). Mixed-effects models for mental health services research. *Health Services and Outcomes Research Methodology* 1, 91–129.

Goldfarb, D. (1970). A family of variable metric updates derived by variational means. *Mathematics of Computation* **24**, 23–26.

Goldman, D., G. Joyce and Y. Zheng (2007). Prescription drug cost sharing: Associations with medication and medical utilization and spending and health. *Journal of the American Medical Association* **298**, 61–69.

Goldstein, H. (2011). *Multilevel Statistical Models*. 4th ed.. Wiley & Sons, Inc.

Gong, G. and F. J. Samaniego (1981). Pseudo maximum likelihood estimation: Theory and applications. *Annals of Statistics* **9**, 861–869.

Gourieroux, C. and A. Monfort (1996). *Simulation Based Econometric Methods*. Oxford University Press.

Grace, J. B. (2006). *Structural Equation Modeling and Natural Systems*. Cambridge.

Gradstein, M. (1986). Maximal correlation between normal and dichotomous variables. *Journal of Educational and Behavioral Statistics* **11**, 259–261.

Green, P. J. (1995). Reversible jump Markov Chain Monte Carlo computation and Bayesian model determination. *Biometrika* **82(4)**, 711–732.

Grossman, M. (1972). On the concept of health capital and the demand for health. *Journal of Political Economy* **80**, 223–255.

Gueorguieva, R. and A. Agresti (2001). A correlated probit model for joint modeling of clustered binary and continuous responses. *Journal of the American Statistical Association* **96**, 1102–1112.

Gueorguieva, R. and G. Sanacora (2006). Joint analysis of repeatedly observed continuous and ordinal measures of disease severity. *Statistics in Medicine* **25**, 1307–1322.

Gueorguieva, R. V. (2001). A multivariate generalized linear mixed model for joint modelling of clustered outcomes in the exponential family. *Statistical Modelling* **1**, 177–194.

Gueorguieva, R. V. (2005). Comments about joint modeling of cluster size and binary and continuous subunit-specific outcomes. *Biometrics* **61**, 862–867.

Gueorguieva, R. V. (2006). Correlated probit model. In: *Encyclopedia of Biopharmaceutical Statistics* (S. C. Chow, Ed.). 3rd ed.. pp. 355–362. Informa Healthcare.

Gumbel, E. (1960). Distributions des valeurs extremes en plusieurs dimensions. *Publications de l'Institut de Statistique de l'Université de Paris* **9**, 171–173.

Hamza, M. and D. Larocque (2005). An empirical comparison of ensemble methods based on classification trees. *Journal of Statistical Computation and Simulation* **75**, 629–643.

Hannan, J. F and R. F. Tate (1965). Estimation of the parameters for a multivariate normal distribution when one variable is dichotomized. *Biometrika* **52**, 664–668.

Hardin, A. M., J. C. Chang and M. A. Fuller (2011). Formative measurement and academic research: In search of measurement theory. *Educational and Psychological Measurement* **71**, 270–284.

Hastie, T. J. and R. J. Tibshirani (1990). *Generalized Additive Models*. Chapman & Hall/CRC Press.

Hays, R. D., D. Revicki and K. Coyne (2005). Application of structural equation modeling to health outcomes research. *Evaluation and the Health Professions* **28**, 295–309.

Heagerty, P. J. (1999). Marginally specified logistic-normal models for longitudinal binary data. *Biometrics* **55**, 688–698.

Hirakawa, A. (2012). An adaptive dose-finding approach for correlated bivariate binary and continuous outcomes in phase I oncology trials. *Statistics in Medicine* **31**, 516–532.

Hobert, J. P. (2011). The data augmentation algorithm: Theory and methodology. In: *Handbook of Markov Chain Monte Carlo* (S. Brooks, A. Gelman, G. Jones and X.–L. Meng, Eds.). Chapman & Hall/CRC Press.

Hoff, P. D. (2007). Extending the rank likelihood for semiparametric copula estimation. *Annals of Applied Statistics* **1**, 265–283.

Holmes, D. R., M. B. Leon, J. W. Moses, J. J. Popma, D. Cutlip, P. J. Fitzgerald, C. Brown, T. Fischell, S. C. Wong, M. Midei, D. Snead and R. E. Kuntz (2004). Analysis of 1-year clinical outcomes in the SIRIUS trial: a randomized trial of a sirolimus-eluting stent versus a standard stent in patients at high risk for coronary restenosis. *Circulation* **109**, 634–640.

Hoshino, T. and K. Shigemasu (2008). Standard errors of estimated latent variables with estimated structural parameters. *Applied Psychological Measurement* **32**, 181–189.

Hoshino, T. and P. M. Bentler (2012). A note on factor score regression. Under review.

Huizinga, T.W., C. I. Amos, A. H. van der Helm–van Mil, W. Chen, F. A. van Gaalen, D. Jawaheer, G. M. Schreuder, M. Wener, F.C. Breedveld, N. Ahmad, R. F. Lum, R. R. de Vries, P. F. Gregersen, R. E. Toes and L. A. Criswell (2005). Refining the complex rheumatoid arthritis phenotype based on specificity of the HLA-DRB1 shared epitope for antibodies to citrullinated proteins. *Arthritis and Rheumatism* **52**, 3433–3438.

Hunt, D. and D. Bowman (2006). Modeling developmental data using U-shaped threshold dose-response curves. *Journal of Applied Statistics* **33**, 35–47.

Husler, J. and R. Reiss (1989). Maxima of normal random vectors: Between independence and complete dependence. *Statistics & Probability Letters* **7**, 283–286.

Hutchinson, T. and C. Lai (1990). *Continuous Bivariate Distributions, Emphasizing Applications*. Rumsby.

Ibrahim, J. G., M. H. Chen and S. R. Lipsitz (2001). Missing responses in generalised linear mixed models when the missing data mechanism is nonignorable. *Biometrika* **88**, 551–564.

Ibrahim, J. G., M. H. Chen, S. R. Lipsitz and A. H. Herring (2005). Missing-data methods for generalized linear models: A comparative review. *Journal of the American Statistical Association* **100**, 332–345.

Ishwaran, H. and S. J. Rao (2005). Spike and slab variable selection: Frequentist and Bayesian strategies. *Annals of Statistics* **33**, 730–773.

Joe, H. (1990). Families of mini-stable multivariate exponential and multivariate extreme value distributions. *Statistics & Probability Letters* **9**, 75–81.

Joe, H. (1993). Parametric family of multivariate distributions with given margins. *Journal of Multivariate Analysis* **46**, 262–282.

Joe, H. (1994). Multivariate extreme value distributions with applications to environmental data. *Canadian Journal of Statistics* **22**, 47–64.

Joe, H. (1997). *Multivariate Models and Dependence Concepts*. Wiley & Sons, Inc.

Joe, H. and J. J. Xu (1996). The estimation method of inference functions for margins for multivariate models. *Technical Report*. University of British Columbia.

Johnson, R. A. and D. W. Wichern (2002). *Applied Multivariate Statistical Analysis*. 5th ed.. Prentice Hall.

Jöreskog, K. G. (1967). Some contributions to maximum likelihood factor analysis. *Psychometrika* **32**, 443–482.

Jöreskog, K. G. (1970). A general method for the analysis of covariance structure. *Biometrika* **57**, 239–251.

Jöreskog, K. G. (1977). Structural equation models in the social sciences: Specification, estimation and testing. In: *Applications of Statistics* (P. R. Krishnaiah, Ed.). pp. 265–287. North–Holland.

Kano, Y. (1983). Consistency of estimators in factor analysis. *Journal of the Japan Statistical Society* **13**, 137–144.

Karlis, D. (2003). An EM algorithm for multivariate Poisson distribution and related models. *Journal of Applied Statistics* **30**, 63–77.

Karlis, D. and L. Meligkotsidou (2005). Multivariate Poisson regression with covariance structure. *Statistics and Computing* **15**, 255–265.

Kenward, M., G. Molenberghs and H. Thijs (2003). Pattern-mixture models with proper time dependence. *Biometrika* **90**, 53–71.

Khan, N. and R. Kaestner (2009). Effect of prescription drug coverage on the elderly's use of prescription drugs. *Inquiry* **46**, 35–45.

Kiefer, J. and J. Wolfowitz (1956). Consistency of the maximum likelihood estimator in the presence of infinitely many incidental parameters. *Annals of Mathematical Statistics* **27**, 887–906.

Kimmel, C. A. and D. W. Gaylor (1988). Issues in qualitative and quantitative risk analysis for developmental toxicology. *Risk Analysis* **8**, 15–20.

Kirkemo, A., M. Peabody, A. C. Diokno, A. Afanasyev, L. M. Nyberg, J. R. Landis, Y. L. M. Cook and L. J. Simon (1997). Associations among urodynamic findings and symptoms in women enrolled in the Interstitial Cystitis Data Base (ICDB) Study. *Urology* **49**, 76–80.

Kishton, J. M. and K. F. Widaman (1994). Unidimensional versus domain representative parceling of questionnaire items: An empirical example. *Educational & Psychological Measurement* **54**, 757–765.

Klaassen, C. A. J. and J. A. Wellner (1997). Efficient estimation in the bivariate normal copula model: normal margins are least favourable. *Bernoulli* **3**, 55–77.

Koch, I. and K. Naito (2010). Prediction of multivariate responses with a selected number of principal components. *Computational Statistics & Data Analysis* **54**, 1791–1807.

Komárek, A. and L. Komárková (2012). Clustering for multivariate continuous and discrete data. *Annals of Applied Statistics*. In press.

Koop, G. and D. J. Poirier (1997). Learning about the across-regime correlation in switching regression models. *Journal of Econometrics* **78**, 217–227.

Kotz, S. and S. Nadarajah (2004). *Multivariate t Distributions and Their Applications*. Cambridge.

Krzanowski, W. J. (1975). Discrimination and classification using both binary and continuous variables. *Journal of the American Statistical Association* **70**, 782–790.

Krzanowski, W. J. (1976). Canonical representation of the location model for discrimination or classification. *Journal of the American Statistical Association* **71**, 845–848.

Krzanowski, W. J. (1983). Distance between populations using mixed continuous and categorical variables. *Biometrika* **70**, 235–243.

Krzanowski, W. J. (1984). On the null distribution of distance between two groups, using mixed continuous and categorical variables. *Journal of Classification* **1**, 243–253.

Krzanowski, W. J. (1988). *Principles of Multivariate Analysis*. Clarendon Press.

Krzanowski, W. J. (1993). The location model for mixtures of categorical and continuous variables. *Journal of Classification* **10**, 25–49.

Kugiumtzis, D. and E. Bora–Senta (2010). Normal correlation coefficient of non-normal variables using piecewise linear approximation. *Computational Statistics* **25**, 645–662.

Kuk, A. Y. C. (2004). A litter-based approach to risk assessment in developmental toxicity studies via a power family of completely monotone functions. *Applied Statistics* **53**, 369–386.

Kullback, S. (1968). *Information Theory and Statistics*. 2nd ed.. Dover.

Laird, N. M. (1995). Longitudinal panel data: An overview of current methodology. In: *Likelihood Time Series With Econometric and Other Applications* (D. R. Cox, D. V. Hinkley and O. E. Barndoff-Nielsen, Eds.). Chap. 4, pp. 143–175. Chapman & Hall/CRC Press.

Landrum, M. B., S.–L. T. Normand and R. A. Rosenheck (2003). Selection of related multivariate means: Monitoring psychiatric care in the Department of Veterans Affairs. *Journal of the American Statistical Association* **98**, 7–16.

Lauritzen, S. L. and N. Wermuth (1989). Graphical models for association between variables, some of which are qualitative and some quantitative. *Annals of Statistics* **17**, 31–54.

Lee, K. and M. J. Daniels (2008). Marginalized models for longitudinal ordinal data with application to quality of life studies. *Statistics in Medicine* **27**, 4359–4380.

Lee, S. K. (2005). On generalized multivariate decision tree by using GEE. *Computational Statistics & Data Analysis* **49**, 1105–1119.

Lee, S.–Y. and W.–Y. Poon (1986). Maximum likelihood estimation of polyserial correlations. *Psychometrika* **51**, 113–121.

Lee, S.–Y. (Ed.) (2007). *Handbook of Latent Variable and Related Models.* North–Holland.

Lee, S.–Y., W.–Y. Poon and P. M. Bentler (1989). Simultaneous analysis of multivariate polytomous variates in several groups. *Psychometrika* **54**, 63–73.

Lee, S.–Y., W.–Y. Poon and P. M. Bentler (1990). A three-stage estimation procedure for structural equation models with polytomous variables. *Psychometrika* **55**, 45–51.

Lee, S.–Y., W.–Y. Poon and P. M. Bentler (1995). A two-stage estimation of structural equation models with continuous and polytomous variables. *British Journal of Mathematical and Statistical Psychology* **48**, 339–358.

Lesaffre, E. and G. Molenberghs (1991). Multivariate probit analysis: A neglected procedure in medical statistics. *Statistics in Medicine* **10**, 1391–1403.

Levy, H. and D. Weir (2010). Take-up of Medicare Part D: Results from the Health and Retirement Study. *Journal of Gerontology: Psychological Sciences* **65**, 492–501.

Li, J. and W. K. Wong (2011). Two-dimensional toxic dose and multivariate logistic regression, with application to decompression sickness. *Biostatistics* **12**, 143–155.

Liang, K.–Y. and S. L. Zeger (1986). Longitudinal data analysis using generalized linear models. *Biometrika* **73**, 13–22.

Liang, K.–Y., S. L. Zeger and B. Qaqish (1992). Multivariate regression analyses for categorical data. *Journal of the Royal Statistical Society–B* **54**, 3–24.

Lin, D. Y. and D. Zeng (2009). Proper analysis of secondary phenotype data in case-control association studies. *Genetic Epidemiology* **33**, 256–265.

Lin, L., D. Bandyopadhyay, S. R. Lipsitz and D. Sinha (2010). Association models for clustered data with binary and continuous responses. *Biometrics* **66**, 287–293.

Lin, X. (1997). Variance component testing in generalized linear models with random effects. *Biometrika* **84**, 309–326.

Lin, X., L. Ryan, M. Sammel, D. Zhang, C. Padungtod and X. Xu (2000). A scaled linear mixed model for multiple outcomes. *Biometrics* **56**, 593–601.

Lindsay, B., C. C. Clogg and J. Grego (1991). Semiparametric estimation in the Rasch model and related exponential response models, including a simple latent class model for item analysis. *Journal of the American Statistical Association* **86**, 96–107.

Lindsey, J. K. (1993). *Models for Repeated Measurements.* Clarendon Press.

Little, R. J. and D. B. Rubin (2002). *Statistical Analysis with Missing Data.* 2nd ed.. Wiley & Sons, Inc.

Little, R. J. and M. D. Schluchter (1985). Maximum likelihood estimation for mixed continuous and categorical data with missing values. *Biometrika* **72**, 496–512.

Liu, C. (2001). Comment on "The art of data augmentation," by D. A. van Dyk and X.–L. Meng.

Journal of Computational and Graphical Statistics **10**, 7–81.

Liu, C. (2004). Robit regression: A simple robust alternative to logistic and probit regression. In: *Applied Bayesian Modeling and Causal Inference from Incomplete Data Perspectives* (Gelman A. and X.–L. Meng, Eds.). pp. 227–238. Wiley & Sons, Inc.

Liu, C. and D. B. Rubin (1998). Ellipsoidally symmetric extensions of the general location model for mixed categorical and continuous data. *Biometrika* **85**, 673–688.

Liu, C., D. B. Rubin and Y. N. Wu (1998). Parameter expansion to accelerate EM: The PX-EM algorithm. *Biometrika* **85**, 755–770.

Liu, J. (2007). Multivariate ordinal data analysis with pairwise likelihood and its extension to SEM. PhD thesis. University of California–Los Angeles.

Liu, X. and M. J. Daniels (2006). A new algorithm for simulating a correlation matrix based on parameter expansion and re-parameterization. *Journal of Computational and Graphical Statistics* **15**, 897–914.

Liu, X., M. J. Daniels and B. Marcus (2009). Joint models for the association of longitudinal binary and continuous processes with application to a smoking cessation trial. *Journal of the American Statistical Association* **104**, 429–438.

Lopes, H. F. and M. West (2004). Bayesian model assessment in factor analysis. *Statistica Sinica* **14**, 41–67.

Lord, F. M. and M. R. Novick (1968). *Statistical Theories of Mental Test Scores*. Addison–Wesley.

Lubke, G. H. (2010). Latent variable mixture modeling. In: *The Reviewer's Guide to Quantitative Methods in the Social Sciences* (G. R. Hancock and R. O. Mueller, Eds.). pp. 209–220. Routledge.

MacCallum, R. C. and J. T. Austin (2000). Applications of structural equation modeling in psychological research. *Annual Review of Psychology* **51**, 201–226.

MacCallum, R. C., K. F. Widaman, S. Zhang and S. Hong (1999). Sample size in factor analysis. *Psychological Methods* **4**, 84–99.

Magnus, J.R. and H. Neudecker (1999). *Matrix Differential Cauculus with Applications in Statistics and Econometrics*. 2nd ed.. Wiley & Sons, Inc.

Malsiner–Walli, G. and H. Wagner (2011). Comparing spike and slab priors for Bayesian variable selection. *Austrian Journal of Statistics* **40**, 241–264.

Manning, W., J. Newhouse and J. Ware (1982). The status of health in demand estimation: Beyond excellent, good, fair, poor. In: *Economic Aspects of Health* (V. Fuchs, Ed.). pp. 141–184. University of Chicago Press.

Marcus, B., B. Lewis, T. King, A. Albrecht, J. Hogan, B. Bock, A. Parisi and D. Abrams (2003). Rationale, design, and baseline data for Commit to Quit II: An evaluation of the efficacy of moderate-intensity physical activity as an aid to smoking cessation in women. *Preventive Medicine* **36**, 479–492.

Marcus, B. H., B. Lewis, J. Hogan, T. K. King, A. Albrecht, B. Bock and A. Parisi (2005). The efficacy of moderate-intensity exercise as an aid for smoking cessation in women: A randomized controlled trial. *Nicotine and Tobacco Research* **7**, 871–880.

Mardia, K. V., J. T. Kent and J. M. Bibby (1979). *Multivariate Analysis*. Academic Press.

Marsh, H. W., K. T. Hau, J. R. Balla and D. Grayson (1998). Is more ever too much? The number of indicators per factors in confirmatory factor analysis. *Multivariate Behavioral Research* **33**, 181–222.

Marshall, A. and I. Olkin (1988). Families of multivariate distributions. *Journal of the American Statistical Association* **83**, 834–841.

Matsuyama, Y. and Y. Ohashi (1997). Mixed models for bivariate response repeated measures data using Gibbs sampling. *Statistics in Medicine* **16**, 1587–1601.

Matusita, K. (1956). Decision rule, based on distance, for the classification problem. *Annals of the Institute of Statistical Mathematics* **16**, 305–315.

McCullagh, P. (1980). Regression models for ordinal data. *Journal of the Royal Statistical Society–B* **42**, 109–142.

McCulloch, C. (2008). Joint modeling of mixed outcome types using latent variables. *Statistical Methods in Medical Research* **17**, 53–73.

McCulloch, C. E., S. R Searle and J. M. Neuhaus (2008). *Generalized, Linear, and Mixed Models.* 2nd ed.. Wiley & Sons, Inc.

McFadden, D. (1974). Conditional logit analysis of qualitative choice behaviour. In: *Frontiers of Econometrics* (P. Zarembka, Ed.). pp. 105–142. Academic Press.

McLachlan, G. J. and T. Krishnan (1997). *The EM Algorithm and Extensions.* Wiley & Sons, Inc.

Meester, S. and J. MacKay (1994). A parametric model for cluster correlated categorical data. *Biometrics* **50**, 954–963.

Meng, X.–L. and D. A. van Dyk (1998). Fast EM-type implementations for mixed effects models. *Journal of the Royal Statistical Society–B* **60**, 559–578.

Meng, X.–L. and D. B. Rubin (1993). Maximum likelihood estimation via the ECM algorithm: A general framework. *Biometrika* **80**, 267–278.

Mesfioui, M. and A. Tajar (2005). On the properties of some nonparametric concordance measures in the discrete case. *Journal of Nonparametric Statistics* **17**, 541–554.

Miglioretti, D. L. (2003). Latent transition regression for mixed outcomes. *Biometrics* **59**, 710–720.

Mitchell, T. and J. J. Beauchamp (1988). Bayesian variable selection in linear regression. *Journal of the American Statistical Association* **404**, 1023–1032.

Molenberghs, G. and G. Verbeke (2005). *Models for Discrete Longitudinal Data.* Springer.

Molenberghs, G. and H. Geys (2001). Multivariate clustered data analysis in developmental toxicity studies. *Statistica Neerlandica* **55**, 319–345.

Molenberghs, G. and L. M. Ryan (1999). Likelihood inference for clustered multivariate binary data. *Environmetrics* **10**, 279–300.

Molenberghs, G., H. Geys and M. Buyse (2001). Evaluation of surrogate endpoints in randomized experiments with mixed discrete and continuous outcomes. *Statistics in Medicine* **20**, 3023–3038.

Monsees, G. M., R. M. Tamimi and P. Kraft (2009). Genome-wide association scans for secondary traits using case-control samples. *Genetic Epidemiology* **33**, 717–728.

Morales, D., L. Pardo and K. Zografos (1998). Informational distances and related statistics in mixed continuous and categorical variables. *Journal of Statistical Planning and Inference* **75**, 47–63.

Moustafa, M. D. (1957). Tests of hypotheses on a multivariate population, some of the variables being continuous and the rest categorical. *Institute of Statistics Mimeograph Series* 179. University of North Carolina–Chapel Hill.

Moustaki, I. and M. Knott (2000). Generalized latent trait models. *Psychometrika* **65**, 391–411.

Moustaki, I., K. G. Jöreskog and D. Mavridis (2004). Factor models for ordinal variables with covariate effects on the manifest and latent variables: A comparison of LISREL and IRT approaches. *Structural Equation Modeling* **11**, 487–513.

Mulaik, S. A. (2009). *Linear Causal Modeling with Structural Equations.* Chapman & Hall/CRC Press.

Munkin, M. and P. Trivedi (1999). Simulated maximum likelihood estimation of multivariate mixed-Poisson regression models, with application. *Econometrics Journal* **2**, 29–48.

Murteira, J. and O. Lourenco (2011). Health care utilization and self-assessed health: Specification of bivariate models using copulas. *Empirical Economics* **41**, 447–472.

Muthèn, B. and K. Shedden (1999). Finite mixture modeling with mixture outcomes using the EM algorithm. *Biometrics* **55**, 463–469.

Muthèn, B. O. (1984). A general structural equation model with dichotomous, ordered categorical, and continuous latent variable indicators. *Psychometrika* **49**, 115–132.

Najita, J. S., Y. Li and P. J. Catalano (2009). A novel application of a bivariate regression model for binary and continuous outcomes to studies of fetal toxicity. *Journal of the Royal Statistical Society–C* **58**, 555–573.

Nakanishi, H. (1996). Distance between populations in a mixture of categorical and continuous variables. *Journal of the Japan Statistical Society* **26**, 221–230.

Narotsky, M. G. (1995). Nonadditive developmental toxicity in mixtures of trichloroethylene, Di(2-ethylhexyl)phthalate, and heptachlor in a $5 \times 5 \times 5$ design. *Fundamental and Applied Toxicology* **27**, 203–216.

Nelsen, R. (1999). *An Introduction to Copulas*. Springer.

Nešlehovà, J. (2007). On rank correlation measures for non-continuous random variables. *Journal of Multivariate Analysis* **95**, 544–567.

Neyman, J. and E. L. Scott (1948). Consistent estimation from partially consistent observations. *Econometrica* **16**, 1–32.

Nikoloulopoulos, A. K. (2012). Comment on "Two-dimensional toxic dose and multivariate logistic regression, with application to decompression sickness," by J. Li and W. K. Wong. *Biostatistics* **13**, 1–3.

Nikoloulopoulos, A. K. and D. Karlis (2008). Multivariate logit copula model with an application to dental data. *Statistics in Medicine* **27**, 6393–6406.

Nikoloulopoulos, A. K. and D. Karlis (2009). Finite normal mixture copulas for multivariate discrete data modeling. *Journal of Statistical Planning and Inference* **139**, 3878–3890.

Nikoloulopoulos, A. K. and D. Karlis (2010). Modeling multivariate count data using copulas. *Communications in Statistics–Simulations & Computation* **39**, 172–187.

Nuñez, M., A. Villaroya and J. M. Oller (2003). Minimum distance probability discriminant analysis for mixed variables. *Biometrics* **59**, 248–253.

O'Brien, S. and D. Dunson (2004). Bayesian multivariate logistic regression. *Biometrics* **60**, 739–746.

OECD Health Data (2005). Organisation for Economic Co-operation and Development.

Ogasawara, H. (1998). Standard errors for rotation matrices with an application to the promax solution. *British Journal of Mathematical and Statistical Psychology* **51**, 163–178.

Ogawa, J., M. D. Moustafa and S. N. Roy (1957). On the asymptotic distribution of the likelihood ratio in some problems on mixed-variate populations. *Institute of Statistics Mimeograph Series* 180. University of North Carolina–Chapel Hill.

Olkin, I. and R. F. Tate (1961). Multivariate correlation models with mixed discrete and continuous variables. *Annals of Mathematical Statistics* **32**, 448–465. (with correction in **36**, 343–344).

Olsson, U., F. Drasgow and N. J. Dorans (1982). The polyserial correlation coefficient. *Psychometrika* **47**, 337–347.

Oman, S. D. (2009). Easily simulated multivariate binary distributions with given positive and negative correlations. *Computational Statistics & Data Analysis* **53**, 999–1005.

Parke, W. R. (1986). Pseudo maximum likelihood estimation: The asymptotic distribution. *Annals of Statistics* **14**, 355–357.

Pearson, K. (1904). Mathematical contribution to the theory of evolution. XIII. On the theory of contingency and its relation to association and normal correlation. *Biometrics Series* I. Drapers Co. Research Memoirs.

Piccarreta, R. (2010). Binary trees for dissimilarity data. *Computational Statistics & Data Analysis* **54**, 1516–1524.

Pitt, M., D. Chan and R. Kohn (2006). Efficient Bayesian inference for Gaussian copula regression models. *Biometrika* **93**, 537–554.

Plackett, R. L. (1965). A class of bivariate distributions. *Journal of the American Statistical Association* **60**, 516–522.

Pohlmeier, W. and V. Ulrich (1995). An econometric model of the two-part decision-making process in the demand for health care. *Journal of Human Resources* **30**, 339–361.

Poon, W.−Y. and S.−Y. Lee (1987). Maximum likelihood estimation of multivariate polyserial and polychoric correlation coefficients. *Psychometrika* **52**, 409–430. (correction in **53**, p. 301).

Poon, W.−Y. and S.−Y. Lee (1992). Statistical analysis of continuous and polytomous variables in several populations. *British Journal of Mathematical and Statistical Psychology* **45**, 139–149.

Poon, W.−Y., S.−Y. Lee, A. A. Afifi and P. M. Bentler (1990). Analysis of multivariate polytomous variates in several groups via the partition maximum likelihood approach. *Computational Statistics & Data Analysis* **10**, 17–27.

Pourahmadi, M. (1999). Joint mean-covariance models with applications to longitudinal data: Unconstrained parameterisation. *Biometrika* **86**, 677–690.

Prentice, R. and R. Pyke (1979). Logistic disease incidence models and case-control studies. *Biometrika* **66**, 403–411.

Prentice, R. L. and L. P. Zhao (1991). Estimating equations for parameters in mean and covariances of multivariate discrete and continuous responses. *Biometrics* **47**, 825–839.

Price, C. J., C. A. Kimmel, R. W. Tyl and M. C. Marr (1985). The developmental toxicity of ethylene glycol in rats and mice. *Toxicology and Applied Pharmacology* **81**, 113–127.

Propert, K. J., A. J. Schaeffer, C. M. Brensinger, J. W. Kusek, L. M. Nyberg and J. R. Landis (2000). A prospective study of interstitial cystitis: results of longitudinal follow up of the interstitial cystitis data base cohort. The Interstitial Cystitis Data Base Study Group. *Journal of Urology* **163**, 1434–1439.

Quinn, K. (2004). Bayesian factor analysis for mixed ordinal and continuous responses. *Political Analysis* **12**, 338–353.

Rabe−Hesketh, S., A. Pickles and A. Skrondal (2001). GLLAMM: a class of models and a Stata program. *Multilevel Modelling Newsletter* **13**, 17–23.

Rabe−Hesketh, S., A. Skrondal and A. Pickles (2004). Generalized multilevel structural equation modeling. *Psychometrika* **69**, 167–190.

Rabe−Hesketh, S. and A. Skrondal (2005). *Multilevel and Longitudinal Modeling using Stata*. Stata Press.

Regan, M. M. and P. J. Catalano (1999a). Likelihood models for clustered binary and continuous outcomes: Application to developmental toxicology. *Biometrics* **55**, 760–768.

Regan, M. M. and P. J. Catalano (1999b). Bivariate dose-response modeling and risk estimation in developmental toxicology. *Journal of Agricultural, Biological and Environmental Statistics* **4**, 217–237.

Regan, M. M. and P. J. Catalano (2002). Combined continuous and discrete outcomes. In: *Topics in Modelling of Clustered Data* (M. Aerts, H. Geys, G. Molenberghs and L. M. Ryan, Eds.). pp. 233–261. Chapman & Hall/CRC Press.

Rizopoulos, D. (2012). *Joint Models for Longitudinal and Time-to-Event Data with Applications in*

R. Chapman & Hall/CRC Press.

Rizopoulos, D. and P. Ghosh (2011). A Bayesian semiparametric multivariate joint model for multiple longitudinal outcomes and a time-to-event. *Statistics in Medicine* **30**, 1366–1380.

Robert, C. P. (1995). Simulation of truncated normal variables. *Statistics and Computing* **5**, 121–125.

Robert, C. P. (2001). *The Bayesian Choice*. 2nd ed.. Springer–Verlag.

Robert, C. P. and G. Casella (2001). *Monte Carlo Statistical Methods*. 2nd ed.. Springer–Verlag.

Rochon, J. (1996). Analyzing bivariate repeated measures for discrete and continuous outcome variables. *Biometrics* **52**, 740–750.

Rokach, L. (2009). Taxonomy for characterizing ensemble methods in classification tasks: a review and annotated bibliography. *Computational Statistics & Data Analysis* **53**, 4046–4072.

Rossi, P. (1996). Existence of Bayes estimators for the binomial logit model. In: *Bayesian Analysis in Statistics and Econometrics,* (D. A. Berry, K. M. Chaloner and J. K. Geweke, Eds.). pp. 91–100. Wiley & Sons, Inc.

Rothman, A. J., E. Levina and J. Zhu (2010). Sparse multivariate regression with covariance estimation. *Journal of Computational and Graphical Statistics* **19**, 947–962.

Rotnitzky, A. and N. P. Jewell (1990). Hypothesis testing of regression parameters in semiparametric generalized linear models for cluster correlated data. *Biometrika* **77**, 485–497.

Roy, J. and X. Lin (2002). Analysis of multivariate longitudinal outcomes with nonignorable dropouts and missing covariates changes in methadone treatment practices. *Journal of the American Statistical Association* **97**, 40–52.

Ryan, L. M. (2002). Issues in modeling clustered data. In: *Topics in Modelling of Clustered Data* (M. Aerts, H. Geys, G. Molenberghs and L. M. Ryan, Eds.). pp. 37–45. Chapman & Hall/CRC Press.

Samejima, F. (1969). Estimation of latent ability using a response pattern of graded scores. *Psychometrika Monographs* 17.

Sammel, M. D. and L. M. Ryan (2002). Effects of covariance misspecification in a latent variable model for multiple outcomes. *Statistica Sinica* **12**, 1207–1222.

Sammel, M. D., L. M. Ryan and J. M. Legler (1997). Latent variable models for mixed discrete and continuous outcomes. *Journal of the Royal Statistical Society–B* **59**, 667–678.

Sammel, M., X. Lin and L. M. Ryan (1999). Multivariate linear mixed models for multiple outcomes. *Statistics in Medicine* **18**, 2479–2492.

Sanacora, G., R. M. Berman, A. Cappiella, D. Oren, A. Kugaya, N. Liu, R. Gueorguieva, D. Fasula and D. Charney (2004). Addition of the alpha2-antagonist yohimbine to fluoxetine: effects on rate of antidepressant response. *Neuropsychopharmacology* **29**, 1166–1171.

Sasieni, P. D. (1997). From genotypes to genes: doubling the sample size. *Biometrics* **53**, 1253–1261.

SAS/STAT User's Guide (1999). SAS Institute.

Savalei, V. (2006). Logistic approximation to the normal: The KL rationale. *Psychometrika* **71**, 763–767.

Savalei, V. and S. Kolenikov (2008). Constrained versus unconstrained estimation in structural equation modeling. *Psychological Methods* **13**, 150–170.

Schafer, J. L. (1997). *Analysis of Incomplete Multivariate Data*. Chapman & Hall/CRC Press.

Schwartz, P., C. Gennings and V. Chinchilli (1995). Threshold models for combination data from reproductive and developmental experiments. *Journal of the American Statistical Association* **90**, 862–970.

Seber, G. A. F. (1984). *Multivariate Observations*. Wiley & Sons, Inc.

Segal, M. and Y. Xiao (2011). Multivariate random forests. *Wiley Interdisciplinary Reviews: Data Mining and Knowledge Discovery* **1**, 80–87.

Segal, M. R. (1992). Tree-structured methods for longitudinal data. *Journal of the American Statistical Association* **87**, 407–418.

Shanno, D. F. (1970). Conditioning of quasi-Newton methods for function minimization. *Mathematics of Computation* **24**, 647–656.

Shapiro, A. (1984). A note on the consistency of estimators in the analysis of moment structures. *British Journal of Mathematical and Statistical Psychology* **37**, 84–88.

Shi, J.−Q. and S.−Y. Lee (2000). Latent variable models with mixed continuous and polytomous data. *Journal of the Royal Statistical Society−B* **62**, 77–87.

Shuo, C., M. W. Smith, T. H. Wagner and P. G. Barnett (2003). Spending for specialized mental health treatment in the VA: 1995–2001. *Health Affairs* **22**, 256–263.

Siciliano, R. and F. Mola (2000). Multivariate data analysis and modeling through classification and regression trees. *Computational Statistics & Data Analysis* **32**, 285–301.

Siroky, D. S. (2009). Navigating random forests and related advances in algorithmic modeling. *Statistics Surveys* **3**, 147–163.

Sklar, A. (1959). Fonctions de répartition à n dimensions et leurs marges. *Publications de l'Institut de Statistique de l'Université de Paris* **8**, 229–231.

Skrondal, A. and P. Laake (2001). Regression among factor scores. *Psychometrika* **66**, 563–576.

Skrondal, A. and S. Rabe−Hesketh (2004). *Generalized Latent Variable Modeling: Multilevel, Longitudinal, and Structural Equation Models*. Chapman & Hall/CRC Press.

Smith, M. and M. A. Khaled (2012). Estimation of copula models with discrete margins via Bayesian data augmentation. *Journal of the American Statistical Association* **107**, 290–303.

Smith, M. and R. Kohn (2002). Parsimonious covariance matrix estimation for longitudinal data. *Journal of the American Statistical Association* **97**, 1141–1153.

Song, P. X.−K. (2000). Multivariate dispersion models generated from gaussian copula. *Scandinavian Journal of Statistics* **27**, 305–320.

Song, P. X.−K. (2007). *Correlated Data Analysis: Modeling, Analytics, and Applications*. Springer.

Song, P. X.−K., M. Li and Y. Yuan (2009). Joint regression analysis of correlated data using Gaussian copulas. *Biometrics* **65**, 60–68.

Song, P. X.−K., P. Zhang and A. Qu (2007). Maximum likelihood inference in robust linear mixed-effects models using multivariate *t* distributions. *Statistica Sinica* **17**, 929–943.

Sozu, T., T. Sugimoto and T. Hamasaki (2012). Sample size determination in clinical trials with multiple co-primary endpoints including mixed continuous and binary variable. *Biometrical Journal* **54**, 716–729.

Spiegelhalter, D. J., N. G. Best, B. P. Carlin and A. van der Linde (2002). Bayesian measures of model complexity and fit. *Journal of the Royal Statistical Society−B* **64**, 583–639.

Spiess, M. and A. Hamerle (1996). On the properties of GEE estimators in the presence of invariant covariates. *Biometrical Journal* **38**, 931–940.

Stoel, R. D., F. G. Garre, C. Dolan and G. vandenWittenboer (2006). On the likelihood ratio test in structural equation modeling when parameters are subject to boundary constraints. *Psychological Methods* **11**, 439–455.

Su, X., M. Wang and J. Fan (2004). Maximum likelihood regression trees. *Journal of Computational and Graphical Statistics* **13**, 586–598.

Suzuki, A., R. Yamada, X. Chang, S. Tokuhiro, T. Sawada, M. Suzuki, M. Nagasaki,

M. Nakayama—Hamada, M. Kawaida, R.and Ono, M. Ohtsuki, H. Furukawa, S. Yoshino, M. Yukioka, S. Tohma, T. Matsubara, S. Wakitani, R. Teshima, Y. Nishioka, A. Sekine, A. Iida, A. Takahashi, T. Tsunoda, Y. Nakamura and K. Yamamoto (2003). Functional haplotypes of PADI4, encoding citrullinating enzyme peptidylarginine deiminase 4, are associated with rheumatoid arthritis. *Nature Genetics* **34**, 395–402.

Tan, M., Y. Qu and J. S. Rao (1999). Robustness of the latent variable model for correlated binary data. *Biometrics* **55**, 258–263.

Tanner, M. A. (1991). *Tools for Statistical Inference: Observed Data and Data Augmentation Methods*. Springer–Verlag.

Tate, R. F. (1954). Correlation between a discrete and a continuous variable. *Annals of Mathematical Statistics* **25**, 603–607.

Tate, R. F. (1955). Applications of correlation models for biserial data. *Journal of American Statistical Association* **50**, 1078–1095.

Teixeira—Pinto, A. and L. Mauri (2011). Statistical analysis of non-commensurate multiple outcomes. *Circulation: Cardiovascular Quality and Outcomes* **4**, 650–656.

Teixeira—Pinto, A. and S.—L. T. Normand (2009). Correlated bivariate continuous and binary outcomes: Issues and applications. *Statistics in Medicine* **28**, 1753–1773.

Teixeira—Pinto, A. and S.—L. T. Normand (2011). Missing data in regression models for non-commensurate multiple outcomes. *REVSTAT* **9**, 37–55.

Thompson, B. and J. G. Melancon (1996). Using item "testlets"/"parcels" in confirmatory factor analysis: An example using the PPSDQ-78. *ERIC Document* 404–349. Education Resources Information Center.

Train, K. (2003). *Discrete Choice Methods with Simulation*. Cambridge University Press.

Trégouët, D.—A., P. Ducimetière, V. Bocquet, S. Visvikis, F. Soubrier and L. Tiret (2004). A parametric copula model for analysis of familial binary data. *American Journal of Human Genetics* **64**, 886–893.

Treiblmaier, H., P. M. Bentler and P. Mair (2011). Formative constructs implemented via common factors. *Structural Equation Modeling* **18**, 1–17.

Trevisani, M. and A. E. Gelfand (2003). Inequalities between expected marginal log-likelihoods, with implications for likelihood-based model complexity and comparison measures. *Canadian Journal of Statistics* **31**, 239–250.

Trivedi, P. and D. Zimmer (2007). Copula modeling: An introduction for practitioners. *Foundations and Trends in Econometrics* **1**, 1–111.

Troxel, A. B., S. R. Lipsitz and D. P. Harrington (1998). Marginal models for the analysis of longitudinal measurements with nonignorable non-monotone missing data. *Biometrika* **85**, 661–672.

Tyl, R. W., C. Jones—Price, M. C. Marr and C. A. Kimmel (1983). Teratologic evaluation of Diethylhexyl phthalate (CAS no.111–81–7) in CD-1 mice. *Final Study Report for NCTR/NTP Contract No. 222–80–2031*. National Technical Information Service.

Van Dyk, D. A. and T. Park (2009). Partially collapsed Gibbs samplers: Theory and methods. *Journal of the American Statistical Association* **103**, 790–796.

Varin, C. (2008). On composite marginal likelihoods. *Advances in Statistical Analysis* **92**, 1–28.

Varin, C., N. Reid and D. Firth (2011). An overview of composite likelihood methods. *Statistica Sinica* **21**, 5–42.

Verbeke, G. and G. Molenberghs (2003). The use of score tests for inference on variance components. *Biometrics* **59**, 254–262.

Verikas, A., A. Gelzinis and M. Bacauskiene (2011). Mining data with random forests: A survey and results of new tests. *Pattern Recognition* **44**, 330–349.

Vuong, Q. (1989). Likelihood ratio tests for model selection and non-nested hypotheses. *Econometrica* **57**, 307–333.

Wagner, H. and C. Duller (2012). Bayesian model selection for logistic regression models with random intercept. *Computational Statistics & Data Analysis* **56**, 1256–1274.

Wagner, H. and R. Tüchler (2010). Bayesian estimation of random effects models for multivariate responses of mixed data. *Computational Statistics & Data Analysis* **54**, 1206–1218.

Wang, C. and M. J. Daniels (2011). A note on MAR, identifying restrictions, model comparison, and sensitivity analysis in pattern mixture models with and without covariates for incomplete data. *Biometrics* **67**, 810–818.

Wang, C. and M. J. Daniels (2012a). Correction to "A note on MAR, identifying restrictions, model comparison, and sensitivity analysis in pattern mixture models with and without covariates for incomplete data". *Biometrics* **68**, 994.

Wang, C., M. J. Daniels, D. O. Scharfstein and S. Land (2010). A Bayesian shrinkage model for incomplete longitudinal binary data with application to the breast cancer prevention trial. *Journal of the American Statistical Association* **105**, 1333–1346.

Wang, Y. and M. J. Daniels (2012b). Bayesian modeling of the dependence in longitudinal data via partial autocorrelations and marginal variances. *Biometrics*. In press.

Wei, G. C. G. and M. A. Tanner (1990). A Monte Carlo implementation of the EM algorithm and the Poor Man's Data Augmentation algorithms. *Journal of the American Statistical Association* **85**, 699–704.

Weir, B. (1996). *Genetic Data Analysis II: Methods for Discrete Population Genetic Data*. Sinauer Associates, Inc.

Whittaker, J. (1990). *Graphical Models in Applied Multivariate Statistics*. Wiley & Sons, Inc.

Wolfinger, R. and M. O'Connell (1993). Generalized linear mixed models: A pseudo-likelihood approach. *Journal of Statistical Computation and Simulations* **48**, 233–243.

Wong, F., C. K. Carter and R. Kohn (2003). Efficient estimation of covariance selection models. *Biometrika* **90**, 809–830.

Wooldridge, J. (2010). *Econometric Analysis of Cross Section and Panel Data*. 2nd ed.. MIT Press.

Wu, B. and A. R. de Leon (2012). Gaussian copula mixed models for clustered mixed outcomes, with application in developmental toxicology. Submitted.

Wu, J. and P. M. Bentler (2012). Application of H-likelihood to factor analysis models with binary response data. *Journal of Multivariate Analysis* **106**, 72–79.

Wu, L. (2010). *Mixed Effects Models for Complex Data*. Chapman & Hall/CRC Press.

Wu, M. C. and R. J. Carroll (1988). Estimation and comparison of changes in the presence of informative right censoring by modeling the censoring process. *Biometrics* **44**, 175–188.

Xing, C. and G. Xing (2009). Power of selective genotyping in genome-wide association studies of quantitative traits. *BMC Proceedings* **3**(Suppl 7), S23.

Yanai, H. and M. Ichikawa (2007). Factor analysis. In: *Handbook of Statistics–Psychometrics* (C. R. Rao and S. Sinharay, Eds.). Vol. 26. pp. 257–296. North–Holland.

Yang, Y. and J. Kang (2010). Joint analysis of mixed Poisson and continuous longitudinal data with nonignorable missing values. *Computational Statistics & Data Analysis* **54**, 193–207.

Yang, Y., J. Kang, K. Mao and J. Zhang (2007). Regression models for mixed Poisson and continuous longitudinal data. *Statistics in Medicine* **26**, 3782–3800.

Yuan, K.–H. and P. M. Bentler (2007). Structural equation modeling. In: *Handbook of Statistics–Psychometrics* (C. R. Rao and S. Sinharay, Eds.). Vol. 26. pp. 297–358. North–Holland.

Yuan, K.–H., P. M. Bentler and Y. Kano (1997). On averaging variables in a confirmatory factor analysis model. *Behaviormetrika* **24**, 71–83.

Yuan, K.–H., R. Wu and P. M. Bentler (2011). Ridge structural equation modeling with correlation matrices for ordinal and continuous data. *British Journal of Mathematical and Statistical Psychology* **64**, 107–133.

Zammuner, V. L. (1998). Concepts of emotion: "emotionness" and dimensional ratings of Italian emotion words. *Cognition and Emotion* **12**, 243–272.

Zeger, S. L. and K.–Y. Liang (1986). Longitudinal data analysis for discrete and continuous outcomes. *Biometrics* **42**, 121–130.

Zeger, S. L. and K.–Y. Liang (1992). An overview of methods for the analysis of longitudinal data. *Statistics in Medicine* **11**, 1825–1839.

Zeger, S. L., K.–Y. Liang and P. S. Albert (1988). Models for longitudinal data: A generalized estimating equation approach. *Biometrics* **44**, 1049–1060.

Zellner, A. (1962). An efficient method of estimating seemingly unrelated regressions and tests for aggregation bias. *Journal of the American Statistical Association* **57**, 348–368.

Zellner, A. (1963). Estimators for seemingly unrelated regression equations: Some exact finite sample results. *Journal of the American Statistical Association* **58**, 977–992.

Zhang, X., W. J. Boscardin and T. R. Belin (2006). Sampling correlation matrices in Bayesian models with correlated latent variables. *Journal of Computational and Graphical Statistics* **15**, 880–896.

Zhang, Y., J. Donohue, J. Newhouse and J. Lave (2009). The effects of the coverage gap on drug spending: A closer look at Medicare Part D. *Health Affairs* **28**, 317–325.

Zhang, Z. (1998). Classification trees for multiple binary responses. *Journal of the American Statistical Association* **93**, 180–193.

Zhao, L. P. and R. L. Prentice (1990). Correlated binary regression using a generalized quadratic model. *Biometrika* **77**, 642–648.

Zhao, L. P., R. L. Prentice and S. G. Self (1992). Multivariate mean parameter estimation by using a partly exponential model. *Journal of the Royal Statistical Society–B* **54**, 805–811.

Zheng, G., C. O. Wu, M. Kwak, W. Jiang, J. Joo and J. A. C. Lima (2012*a*). Joint analysis of binary and quantitative traits with data sharing and outcome-dependent sampling. *Genetic Epidemiology* **36**, 263–273.

Zheng, G., J. Joo, D. Zaykin, C. O. Wu and N. L. Geller (2009). Robust tests in genome-wide scans under incomplete linkage disequilibrium. *Statistical Science* **24**, 503–516.

Zheng, G., Y. Yang, X. Zhu and R. C. Elston (2012*b*). *Analysis of Genetic Association Studies*. Springer.

Zheng, G., Z. Li, J. L. Gastwirth and B. Freidlin (2006). Robust genomic control for association studies. *American Journal of Human Genetics* **78**, 350–356.

Zhu, H., M. Gu and B. Peterson (2007). Maximum likelihood from spatial random effects models via the stochastic approximation expectation maximization algorithm. *Statistics and Computing* **17**, 163–177.

Zimmer, D. (2012). The role of copulas in the housing crisis. *Review of Economics and Statistics* **94**, 607–620.

Zimmer, D. and P. Trivedi (2006). Using trivariate copulas to model sample selection and treatment effects: Application to family health care demand. *Journal of Business and Economic Statistics* **24**, 63–76.

Index

234

Milton Keynes UK
Ingram Content Group UK Ltd.
UKHW051950071024
449327UK00026B/2246